Monoelements

Scrivener Publishing
100 Cummings Center, Suite 541J
Beverly, MA 01915-6106

Publishers at Scrivener
Martin Scrivener (martin@scrivenerpublishing.com)
Phillip Carmical (pcarmical@scrivenerpublishing.com)

Monoelements

Properties and Applications

Edited by

Inamuddin, Rajender Boddula, Mohd Imran Ahamed and Abdullah M. Asiri

Scrivener Publishing

WILEY

This edition first published 2020 by John Wiley & Sons, Inc., 111 River Street, Hoboken, NJ 07030, USA and Scrivener Publishing LLC, 100 Cummings Center, Suite 541J, Beverly, MA 01915, USA

For more information about Scrivener publications please visit www.scrivenerpublishing.com.

Wiley Global Headquarters
111 River Street, Hoboken, NJ 07030, USA

For details of our global editorial offices, customer services, and more information about Wiley products visit us at www.wiley.com.

Library of Congress Cataloging-in-Publication Data

ISBN 978-1-119-65525-1

Cover image: Pixabay.Com
Cover design by Russell Richardson

Set in size of 11pt and Minion Pro by Manila Typesetting Company, Makati, Philippines

Contents

Preface

The development of new two-dimensional (2D) monoelements-based semiconductor materials, such as phosphorene, graphene, antimonene, etc., has attracted many researchers due to their wide range of applications in diverse sectors along with their promotion of novel innovations in the field of science. Due to their impressive physical, chemical, electronic, and optical properties these 2D monoelements have been identified as potential agents for a variety of applications such as electronics, theranostics, therapeutic delivery, bioimaging, sensors, field-effect transistors, the environment, energy conversion, storage, etc.

This edition of *Monoelements: Properties and Applications* explores the basic idea of 2D monoelements, classifications, and application in field-effect transistors for sensing and biosensing. Finally, various challenges, future developments, and research progress are also discussed. This book will be useful for beginners and experts—from undergraduate students to industrial engineers—working in the area of semiconductors, materials science, and engineering. Based on thematic topics, this edition contains the following eleven chapters:

Chapter 1 investigates recent advances in phosphorene. In particular, the effects of its strong anisotropy and its high reactivity on the physical properties of pure and functionalized phosphorene are discussed in detail. This largely distinguishes phosphorene from other 2D materials and makes it the ideal platform for emerging devices.

Chapter 2 provides insights into the recent exploration of the diverse properties of 2D antimonene, continuous updating of preparation methods, as well as further development of potential applications, and then looks ahead to the opportunities and challenges facing antimonene in the future.

Chapter 3 presents a brief review of graphene and its derivatives. Applications of graphene and graphene oxide in the electronic and biomedical fields are discussed, and their physical properties like thermal and electrical properties are outlined.

Chapter 4 discusses the molecular docking simulation study used to analyze the binding mechanisms of graphene oxide as a cancer drug carrier.

Chapter 5 aims to group the significant work reported in the last few years on metal-organic frameworks (MOFs)-derived carbon (graphene and carbon nanotubes) and MOF-carbon composite materials, with a special emphasis on the use of these nanostructures for energy storage devices (supercapacitors).

Chapter 6 summarizes the basic idea of 2D classification like graphene application in field-effect transistors for sensing and biosensing. These 2D monoelements have been identified as potential agents for a variety of applications such as theranostics, therapeutic delivery, bioimaging, gas sensors, biosensors, field-effect transistors, semiconductors, etc.

Chapter 7 describes various graphene-based ternary materials as a supercapacitor electrode. Different forms of functionalized graphene are discussed, including graphene oxide and reduced graphene oxide and its characteristic properties. A main focus of this chapter is the synthesis and electrochemical features of various graphene-based ternary composites with conducting polymer, metal oxide and other carbon-based materials.

Chapter 8 explains the physico-chemical properties such as electronic band structure, electrochemical influence of graphene doping, and chemistry behind graphite exfoliation. Furthermore, it extends to electrochemical sensing of analytes such as glucose, DNA and aptamer, pollutants, gas, pharmaceutics, and antioxidants.

Chapter 9 describes the structures, fundamental properties, and applications of germanene and aims to draw the attention of researchers towards the possibilities offered by the use of germanene-based materials for improving their working characteristics and for replacing rather expensive traditional materials used in energy storage devices.

Chapter 10 discusses the fundamental milestones of graphene origin, synthesis routes, advantages, and various literature cites toxicity studies. The role of 2D graphene in gene-delivery, tissue-engineering, prosthetic-implants, cancer-therapy, and biosensing fields are explored as well.

The major arena of focus is ascribed to *in vivo* and *in vitro* studies of graphene for biomedical use.

Chapter 11 describes graphene and graphene-based materials for energy devices. The chapter broadly discusses the role of graphene as electro-active anode and cathode materials with major focus on lithium ion batteries and supercapacitors. The chapter also includes basic synthesis and characterization techniques.

Editors
Inamuddin
Rajender Boddula
Mohd Imran Ahamed
Abdullah M. Asiri
September 2020

1

Phosphorene: A 2D New Derivative of Black Phosphorous

Lalla Btissam Drissi[1,2,3]*, Siham Sadki[1] and El Hassan Saidi[1,2,3]

[1]*Faculté des Sciences, Université Mohammed V de Rabat, Rabat, Morocco*
[2]*Academie Hassan II des Sciences et Techniques, Rabat, Morocco*
[3]*CPM, Centre of Physics and Mathematics, Faculty of Science, Mohammed V University, Rabat, Morocco*

Abstract

Phosphorene is a stable 2D elemental material obtained via the exfoliation of 3D layered phosphorus. Phosphorene exhibits several interesting features, including its unique highly buckled structural characteristics which lead to strong anisotropies in the transport, electronic, optical, mechanical, and thermal properties of this material along its two directions: zigzag and armchair. These excellent properties render phosphorene an ideal platform for various optoelectronic devices. However, under atmospheric conditions, for example, in the presence of oxygen, water, and light, phosphorene is very reactive due to the free non-bonding electrons existing on its surface. Consequently, the O-concentrations effect on the optoelectronic response, the elastic parameters, and thermal conductivity of phosphorene is significant and indicates interesting results.

Keywords: Phosphorene, synthesis, chemical functionalization, optoelectronic properties, mechanical response, thermal conductivity

1.1 Introduction

The 3D phosphorus is a very abundant element existing in several polymorphous forms. Among the allotropes of phosphorus, namely, white, red, violet, black, and blue; the black one (BP) constitutes the most

**Corresponding author*: b.drissi@academiesciences.ma; ldrissi@fsr.ac.ma

Inamuddin, Rajender Boddula, Mohd Imran Ahamed and Abdullah M. Asiri (eds.)
Monoelements: Properties and Applications, (1–26) © 2020 Scrivener Publishing LLC

thermodynamically stable phase under ambient conditions. This layered allotrope was discovered for the first time more than a century ago through the high-pressure [1]. Recently, 3D BP was synthesized from red phosphorus using the new sonochemical method [2]. In bulk BP, the layers are weakly stacked together via VDW interactions [3]. In each layer, the P atoms are connected to their three nearest neighbors by covalent bonds that form a rippled honeycomb structure [4]. BP is a semiconductor with a direct-gap, a strong in-plane anisotropy and a density greater than 2.5 g/cm^3 [5, 6].

Like its counterpart graphene, stable 2D phosphorene can be mechanically extracted from 3D BP. In 2014, phosphorene was synthesized, for the first time, using a scotch tape based microcleavage method [7–9]. The phosphorene's unit cell is composed of four P atoms and appears highly buckled in the armchair (AC) axis [10]. Because of its geometric characteristics, phosphorene exhibits highly anisotropic physical properties along its AC with respect to its zigzag one [11, 12]. Phosphorene is a p-type semiconductor [13–15] that shows a high flexibility, an important specific capacity and discharge potential that are very required for advanced battery applications [16–18]. In addition, it exhibits a strong excitonic effect [19], an optical gap located at 1.2 eV and its absorbs infrared to near ultraviolet radiation [20]. This new hexagonal material has great potential applications in optoelectronics and photovoltaic devices [21].

Furthermore, the puckered structure of phosphorene attributes its interesting elastic properties such as great structural flexibility and a resistance to 27% and 30% deformations along the zigzag and armchair directions, respectively [22, 23], which makes this material very suitable for wearable optoelectronic devices. Furthermore, the Young's modulus and Poisson ratio exhibit their maximum values along ZZ-axis indicating how it is difficult to strain it. Consequently, phosphorene is super flexible along the armchair axis [23]. It is also well to mention that phosphorene is an auxetic material [24, 25] and that its non-centrosymmetric point group leads to a large piezoelectric response [23] showing that phosphorene can convert mechanical energy into electrical one [26].

Despite all the exceptional properties of phosphorene, it is very reactive with oxygen due to the non-bonding pairs present at its surface [27]. This fact limits its applications in optoelectronics, sensors, energy conversion, photocatalytic, and so on. To overcome this obstacle, many different techniques have been used to fabricate air-stable phosphorene. The passivated phosphorene by graphene, h-BN, Al_2O_3, and the polymeric material is a promising technique to avoid chemical debasement and to modulate its features [28]. The measurements shown smaller degradation when phosphorene only exposes to O_2 or H_2O [29]. Phosphorene with different oxygen concentrations confers excellent new properties in these 2D materials [30, 31]. At high concentration, oxidation leads to a new family, namely, 2D planar and 1D tubular forms, with a transition in the band gap from semiconductors to insulators [32].

In this chapter, we first present pure phosphorene starting from its crystalline structures, its fabrication methods, its physical properties, and ending with certain applications. Secondly, we will investigate how the oxidation's arrangement and concentrations influence the electronic, elastic, and optical characteristics of phosphorene oxides.

1.2 Pristine 2D BP

Owing to its great buckle height, phosphorene has fascinating properties such as anisotropic optoelectronic and mechanical features which make it very attractive for nanodevices.

1.2.1 Synthesis and Characterization

Similar to graphene, 2D BP can be exfoliated from buckled material trough the top down method. The bottom-up method is still not promising for phosphorene CVD growth since most of the phosphorus precursors used in thermal depositions show a high amount of toxicity and cannot be adapted for CVD manufactures [33, 34]. It follows that the large-scale bottom-up method requires more effort from experimental scientists.

1.2.1.1 Top-Down Approaches

The mechanical exfoliation is an effective widely used method for cleaving 3D materials from mutilayers to some layers and then to isolate a single layer [34]. Graphene monolayer, for example, has been isolated from graphite simply by using adhesive tape [35, 36].

Monolayer, bi- and tri-BP sheets were successfully exfoliated using micromechanical cleavage of 3D BP with PDMS in 2014. This method was carried out using an adhesive tape in three steps.

First, the exfoliated phosphorene layers were transferred to PMMA/PVA (polymethyl methacrylate/Polyvinyl Alcohol) composites, and then, the resulting layers with the composites were moved to a SiN substrate with a thickness of 200 nm. Several chemicals are used to separate the obtained specimens from the PMMA/PVA composites and to ensure that no more scotch tapes was left [37]. Despite the success of the mechanical exfoliation process, it was found that it was not scalable and hence limited to be used in academic laboratories for fundamental studies. Thus, to obtain a phosphorene sheet, a more efficient manufacturing process has been introduced. In particular, an Ar+ plasma was used to produce monolayer phosphorene through thermal ablation. This process provides an improved means of controlling the phosphorene thickness, unlike it is still challenging for mass production [38, 39].

The interesting technique to fabricate large quantities of exfoliated phosphorene is the liquid phase preparation. The solution-based phosphorene synthesis is placed into the BP interlayers which enlarge the distance and allows the exfoliation. This approach is widely used to manufacture several 2D and 3D materials that have shown good performance in dispositive [40].

1.2.1.2 Bottom-Up Methods

Advanced chemical techniques were used intensively to fabricate large quantities of innovative devices based on new 2D sheets like graphene, germanene, borophene, silicene, and stanene [38]. For other synthesized 2D materials, this new processing route based on the deposition via thermal evaporation of their elemental forms is done on available suitable substrates/surfaces like Ag(111), Au(111), Pt(111), and Al(111) [34]. In parallel, other means, such as the successful epitaxial growth of graphene and TMDCs on insulating substrates made of sapphire or 300 nm of SiO_2 on Si (SiO2/Si) [41] open up also the way to a possible phosphorene. These bottom-up methods are very used for materials stable under moisturizing conditions and at high temperature. In contrast, large-scale phosphorene CVD and epitaxial growth are still incubating and breakthroughs due to various reasons, such as lack of suitable substrate, high toxicity of phosphorus, as well as instability of phosphorene in the presence of moisture under high pressure [38, 42].

1.2.1.3 Geometric Structure and Raman Spectroscopy

Crystallographic data and elemental details of phosphorene were gained both theoretically and also experimentally using different techniques such as X-ray cristallography, high performance spectrometers, SEM microscope, and EDX analysis. Phosphorene has been shown to be a nonplanar lattice along and seems to be a bilayer material in the zigzag direction as displayed in Figure 1.1a.

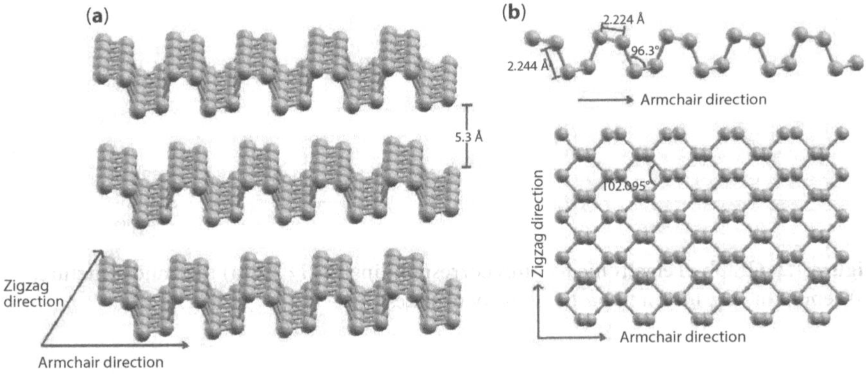

Figure 1.1 Optimized crystallographic structure of (a) 3D BP and (b) 2D BP.

Measurements made by means of preliminary X-ray investigations indicate lattice constants of 3.31 Å and 4.38 Å in ZZ- and AC-axes, respectively, with four atoms forming the unit cell of phosphorene [43, 44]. The experimental result concorde with the theoretical values obtained using *ab initio* DFT calculations [23, 45].

In phosphorene monolayer, each phosphorus atom is linked to first three nearest neighbor atoms to constitute an sp^3 hybridization in a covalent bound [10]. The non-planar geometry leads to two types of bonding, namely, the in-plane bond length d_1 is about 2.224 Å, and the out-of-plane bond length d_2 that is 2.244 Å, as illustrated in Figure 1.1b. The binding angles y and x are 96.3° and 102.095°, respectively. The height difference between the two half-layers is $d_Z = 2.10$ Å.

1.2.2 Physical Properties

1.2.2.1 Anisotropic Eectronic Behavior

Pristine phosphorene is a p-type semiconductor with a direct band gap [13, 46–48]. By using polarization-resolved photoluminescence excitation spectroscopy at room temperature, the quasi-particle band gap of phosphorene is measured to be 2.2 eV [49]. The same value is observed with the typical tunneling spectra of U shaped electronic spectra [48].

Pure phosphorene has no spin polarization, which is confirmed by symmetrical density of spin-up and -down states displayed in Figure 1.2b. Meanwhile, Figure 1.2a shows that the band dispersion is highly anisotropic around the electronic gap. Indeed, one can observe a much greater dispersion along the Γ-X direction for CBM and VBM with respect to the

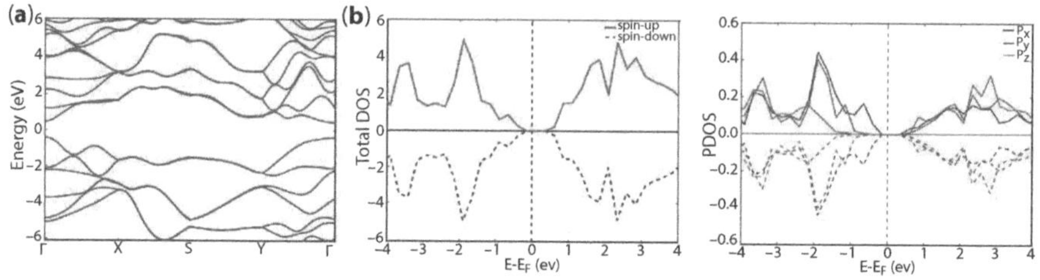

Figure 1.2 Graph of electronic features corresponding to 2D BP. (a) the band structure, (b) represents the total and partial density of states.

vertical bands in Γ-Y region. The partial density of states (p-DOS) plots clearly show that p_x orbital contribute mainly in the states of the unfilled C-band, while the p_z orbital of phosphorus dominates the valence band states [50]. The number of layers mainly affects the gap energy [24]. For instance, it takes the values of 1.51, 0.59, and 0.3 eV for the monolayer, the five layers, and the bulk black phosphorus [51]. Furthermore, the gap energy decreases with increasing the magnitude of external electrical field, which breaks the out-of-plane symmetry. Under biaxial strain and the two possible uniaxial strains, the deformed phosphorene shows a transition from the semiconducting to metallic phase [23].

1.2.2.2 Optical Properties

The puckered structure of phosphorene attributes it interesting optical properties. Phosphorene absorbs transverse radiation along its AC-axis, while it highly transmitted light along ZZ-axis [19, 52]. The photoluminescence excitation spectroscopy (PLE) measures an optical band gap of 1.31 eV owing to the exciton binding energy as discussed in [19, 49] and measured in [53]. Notice that the theoretical values are larger than the measured amounts because of the increased screening from the dielectric substrate that reduces the quasi-particle band gap and consequently the exciton binding energy [54]. Furthermore, phosphorene can absorbs the visible light since its optical absorption peak is located at 1.6 eV. All these features suggest phosphorene as a promising optoelectronic device for future applications.

The absorption peak can be tunable via strain as displayed in Figure 1.3. For deformed phosphorene, the absorption peak ranges from 0.38 to 2.07 eV under compressive and tensile strain revealing that the material absorbs both infrared and visible light. For the electric field vector $E_\perp$, the graph displaying the imaginary part of the dielectric function $E_2(\omega)$ shows

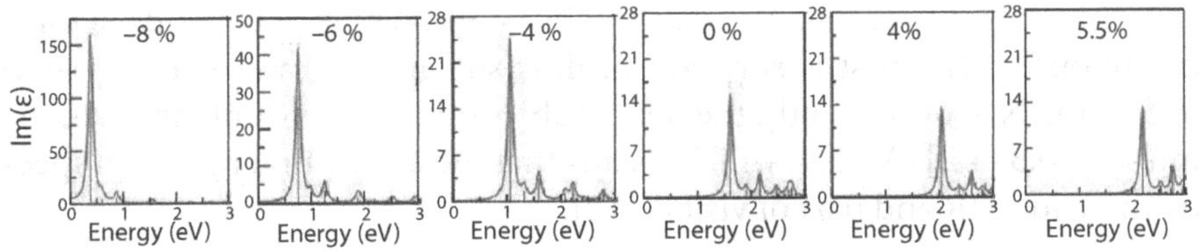

Figure 1.3 Absorption spectra for undeformed monolayer phosphorene with 0% strain and deformed mono-layer under compressive -8%, -6%, and -4% and tensile strain of 4% and 5.5%. The curves are obtained through the GW+BSE method.

a considerable shift of the first peak towards high energies when including quasi-particle corrections. However, the shape of $E_2(\omega)$ spectrum changes considerably when taking into account the electron-hole correlations (BSE) [19]. The exciton binding energy is 0.818 eV for the first active one and 0.66 eV for the first dark one. As displayed in Figures 1.4a and b, the dielectric screening enlarges both the gap and binding energy at $\omega = 0$. Furthermore, a large excitonic wave function distribution is observed for the first bright exciton in Figure 1.4c whose peak emerges in the IR part as shown in Figure 1.4d. The maximum reflectivity $R_{max}(\omega)$ of 38% occurs in the IR range while it did not exceed 22% for the visible light (see Figure 1.4e). The electron energy loss spectra in Figure 1.4f reveal that the first plasmon peaks in phosphorene sheet has a height of 11.003 dispersed in the IR range of the spectrum.

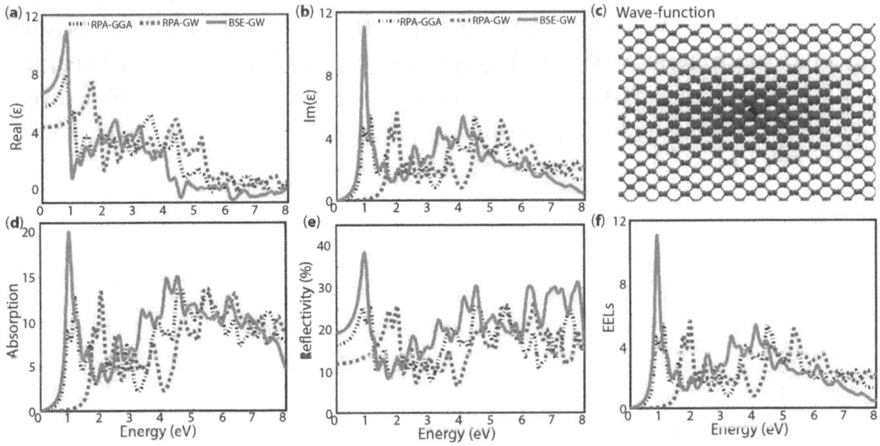

Figure 1.4 (a) and (b) dielectric function, (c) absorption coefficient, (d) reflectivity function, and (e) EELs function, obtained by using three approximations GW-BSE, GW-RPA, and GGA-RPA. (f) Represents the wave function of electron-hole.

In phosphorene multi-layers, the photoluminescence depends significantly on N. The first absorption peak is shifting to lower energies with increasing N. Consequently, the optical absorption coefficient ranges in the interval [0.3–1.2] eV [19], which means that the absorption radiation spectrum include IR and part of visible [52].

1.2.2.3 Elastic Parameters

The elastic properties of phosphorene are also very anisotropic since the Young values in AC-axis (x-direction) is four times lower than the one along the zigzag axis (y-axis), as indicated by the polar diagrams of $\Upsilon(\theta)$ illustrated in Figure 1.5a [23, 55]. Notice that the weakest P-P bond strength is the main cause of these small values of the Young parameter [22], compared to 1 TPa and 270 GPa reported for graphene and MoS_2, respectively. Furthermore, the Poisson's ratio, namely, 0.73 and 0.165 in the AC and ZZ directions, respectively, confirms also the high anisotropy in phosphorene [56]. However, the negative value observed in the small interval [5π, $10/3\pi$] as displayed in Figure 1.5b reveals that the material is auxetic. More precisely, for some particular stretch, phosphorene shows a lateral extension instead of longitudinal elongation as it is the case for conventional materials [57, 58]. Besides this, both the shear and compressional acoustic waves propagate more rapidly in the ZZ-axis as it is clearly deduced from the polar plot of the speed of sound in Figure 1.5c. Same anisotropic behavior is found for Debey temperature that is half times lower (see Figure 1.5d).

Besides the strong anisotropy of the elastic parameters, phosphorene exhibits a high elasticity compared with other monolayer materials such silicene, borophene MoS_2, and graphene [59]. Indeed, without breaking phosphorene can withstand large tensile strains along its two possible directions [23]. Whereas, MoS_2, for example, can only withstand deformation

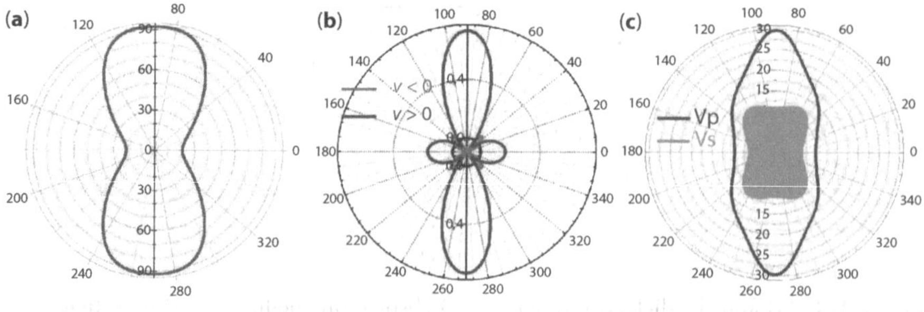

Figure 1.5 Polar plot of (a) Young modulus in J/m² and (b) positive and negative values of Poisson ratio, (c) speed of sound in km/s of pure phosphorene.

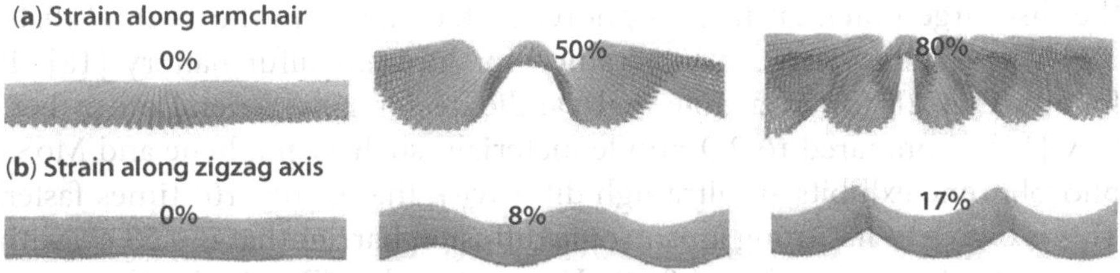

Figure 1.6 Monolayer phosphorene under different values of in-plane compressive strain at 300 K in the two directions.

up to 13%. At 300 K and under a small magnitude of strain (E), Figure 1.6 depicts small ripples on the flat surface of phosphorene. When the compressive E grows, the buckling parameter increases. Interestingly, phosphorene maintains its structural stability in AC-axis at large compressive force up to 80%, but it breaks along the ZZ direction for a 17% deformation, which reveals the super flexible character of this material [60].

1.2.3 Applications

The remarkable properties and the strong anisotropy observed in phosphorene make it an ideal candidate for photodetectors, modulators, and sensors. Below, we will report some applications.

1.2.3.1 Gas Sensors

In 2D materials, the large ratio of surface area to volume renders them promising for gas sensors. In addition, the unique supplementary advantages of phosphorene, like its in-plane anisotropy, structural stability, and high chemical reactivity with molecules, make it highly desirable as a superior gas sensor [61]. Indeed, under gases exposition, phosphorene undergoes multiple modifications [25]. For example, the gas molecules adsorption on phosphorene leads to a reduction or an increase of the resistance that is very required for the markers in sensing applications. Furthermore, phosphorene depends mainly on certain toxic gases because of the high binding strength gas molecules. As a result, the selective behavior of the phosphorene in adsorbing gases influences significantly its transport properties along the two axis directions [61].

1.2.3.2 Battery Applications

According to [16, 17], phosphorene exhibits a specific capacity of 2,596 mAh/g, which is larger than the ones of sulfur and graphene. In addition, the discharge potential in phosphorene, that ranges in the interval [0.4–1.2] V, is smaller than 2.1 V obtained for lithium/sulfur battery [18]. It follows that 2D BP is a potential candidate for new generation of battery [10]. Compared to 2D anode materials, such as graphene and MoS_2, phosphorene exhibits an ultrahigh diffusivity, that is 10^2–10^4 times faster. This is owing to its strong anisotropic diffusion barrier that is 0.68 eV, with respect to the small value of 0.08 eV found in the ZZ-axis. Furthermore, phosphorene-based Li battery shows a voltage of 2.9 V which is larger than the one of other 2D layered materials, making this kind of battery of potential use as a rechargeable battery for different electronic and energetic devices.

1.2.3.3 FETs

Another significant application of phosphorene is the fabrication of field-effect transistors (FETs) [38]. Phosphorene devices can offer many advantages over graphene transistors due to the good saturation of the current and their band gap [62]. Phosphorenes have attractive characteristics that are critical for advanced circuits and sophisticated amplifiers [10]. In particular, phosphorene exhibits drain current modulation of 10^5, high flexibility, and high carrier mobilities of about 1,000 $cm^2V^{-1}s^{-1}$ which is larger than the other flexible transistors based on 2D monolayers like WSe_2 and MoS_2. Furthermore, when the length of channel is 300 nm, the measurements show that phosphorene exhibits a cutoff frequency of 12 GHz for the short-circuit current while frequency oscillation reaches the value of 30 Hz.

Beyond the multi-GHz frequency, phosphorene constitutes one of the best candidates for future generations of ultrathin layer transistors [10]. Moreover, phosphorene is not only used in field-effect transistor applications, but also in other electronic devices based on semiconductor materials due to its electronic properties and its charge mobility.

1.3 Phosphorene Oxides

The non-bonding pairs of electrons present on the surface of phosphorene leads to degradation of this 2D monolayer under ambient conditions, namely, oxygen, water, and light. This impedes phosphorene from some of its potential applications. To overcome this obstacle, phosphorene oxides with different O-concentration were investigated. It follows that phosphorene is stable at low O-concentrations. More precisely, half-oxidation is the best concentration to construct a stable material.

1.3.1 Challenges: Degradation of Phosphorene

In contrast to the unique properties and great potential of phosphorene, which distinguish it from other 2D materials, phosphorene remains unstable under atmospheric conditions, for example, in the presence of oxygen, water, and light, due to the non-bonding pairs of electrons present at its surface [27, 63]. The unprotected surface of phosphorene develops significant roughness, causing important changes and consequent degradation in the compositional and physical features of the material. In some cases, the degradation poses a serious performance problem of phosphorene-based devices [64].

1.3.1.1 Light Exposure

In phosphorene, the ambient degradation in the atmosphere is divided into three stages. In the first stage, the reaction induced by the ambient light O_2 leads to the formation of oxygen. In that case, the reaction expressing the transfer if charge is given by: $\left(O_2 + h\upsilon \rightarrow O_2^- + h^+\right)$ where P corresponds to phosphorene and h^+ is a hole with positive charge. In the second stage, the oxygen molecule is separated at the surface leading to the following: $O_2^- + P + h^+ \rightarrow P_xO_y$. Finally, in the last step that is a hydrogen-bond interaction, the P atom is removed from the surface and the bonded O is absorbed by water molecules. It follows that the top layer of phosphorene is broken and excitons can be produced under ambient light.

To evaluate the evolution of BP degradation for various light's wavelengths and at different time scales, six representative BP flakes were studied individually. Using atomic force microscopy (AFM), the exposures were evaluated in a dark room at six values of wavelengths ranging from 280 to 1,050 nm and imaged before and after identical exposure durations varying from 30 to 120 min with a step of 30 min [65]. The maximum degradation was observed for the UV light (280 nm), then for the blue one (455 nm). In contrast, phosphorene does not show any degradation for green, red, and infrared light, namely, 565, 660, 850, and 1,050 nm. Consequently, the UV light is the predominant contributor to the degradation of BP.

In addition, the engineering of the phosphorene's band gap renders this material a good candidate for a photodetector, with a large spectral response ranging from the UV towards IR region. For instance, phosphorene photon detector shows a very fast response of 1.82 A/W in the presence of visible light irradiation of 550 nm. With photon energy and a bias of 0.1 V, the photoresponsivity attains the value of 175 A/W in the NIR regime, and at a higher bias of 3 V, it reaches 9×10^4 A/W offering phosphorene potential as a UV detector [66].

1.3.1.2 Phosphorene vs Air

When phosphorene is exposed directly to air, its reaction causes rapid degradation of phosphorene-based devices. Moreover, the exothermic process reveals that H_2O will react with oxidized phosphorene. Both theoretical calculations and experiments have shown that at room temperature, phosphorene undergoes a spontaneous oxidation when it is exposed to O_2. The oxidation pathway leads to the formation of phosphoric acid and defective phosphorene [29]. Besides, humidity (the presence of H_2O) is very important to determine the stability of phosphorene in air. To avoid surface degradation of phosphorene and overcoming the oxidation barrier, some passivation techniques must be introduced. For example, graphene, h-BN, AlO_x, Al_2O_3, P_xO_y, and polymeric materials are used to protect it from mechanical and chemical degradation [28]. Under low oxidation, phosphorene is stable and tend to be less stable when increasing the oxygen concentrations [25]. Consequently, 50% oxidation of phosphorene is the best amount to stabilize phosphorene after a two-day exposure to the atmosphere [67].

1.3.1.3 Functionalized Phosphorene

Non-metallic adatoms can also be strongly bound to phosphorene due to the lone electrons pair. Functionalization of phosphorene by adsorption of non-metallic atoms with a [He] core electronic, namely, B, C, N, F, and Al showed different site preferences as schematically illustrated in Figure 1.7. Indeed, while the adatoms B, C, and Al prefer to adsorb to the hollow (H) site, F adatom is adsorbed at the top site and N adatom prefers the bridge one [68]. Furthermore, the chemical functionalization of these non-metallic adatoms exists in three classes [69]. In the first one, the C and B adatoms get located at the interstitial site after breaking the P-P bonds. However in the second group, the N and F atoms remain on the surface of the P atoms and preserve the lattice structure of phosphorene.

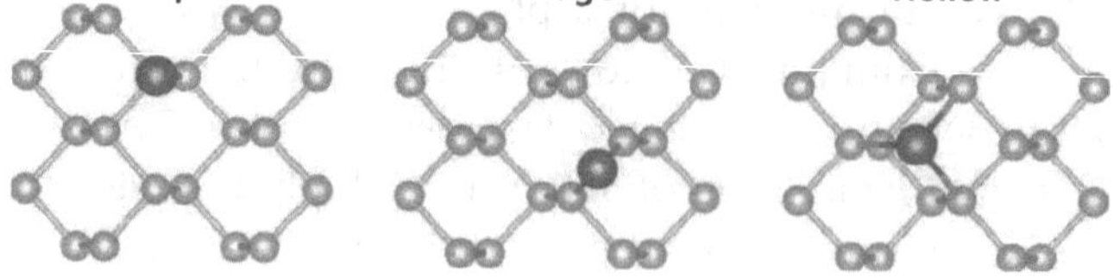

Figure 1.7 The three possible adsorption sites.

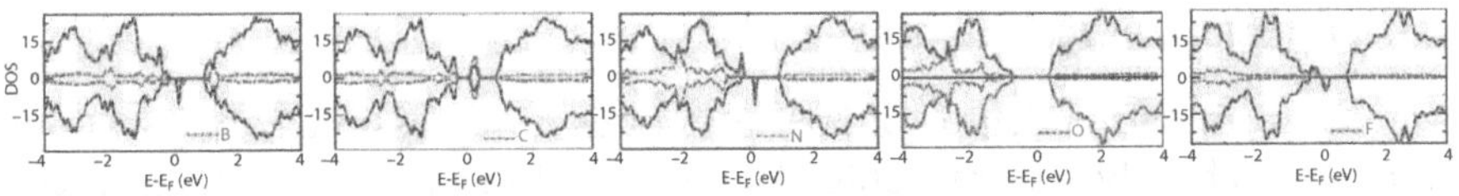

Figure 1.8 DOS of spin up and down of adatoms.

The last group is formed by the Al impurity, located at the top of the centre of the hexagon. The interatomic distances show that the smaller the adatom, the closer it is to the P monolayer, which implies a higher binding energy compared to the larger ones. This result is confirmed by the calculations of the binding energy (E_b).

For B, C, N, F, and Al adatoms on phosphorene, E_b is −5.08, −5.16, −2.98, −2.30, and −3.18 eV, respectively [68, 70]. The adsorption process is more stable in phosphorene since the values of E_b are much greater than the case of adsorbed graphene [71–73]. The higher values of E_b are mainly deserved to the buckled sp^3 configuration of the reactive material as reported in [68]. Mid-gap states are observed in the spin-polarized density of states plotted in Figure 1.8 with 1 μ_B for B, N, and F systems. However, the curves for the C and Al impurities reveal the same number of electrons having up-spin and down-spin, indicating the absence of magnetic order in these configurations [70].

Moving to 3D transition metal (TM), such as Cu, Ti, V, Ni, Cr, and Fe adsorbed at the H site in phosphorene. According to [69], TM adatoms induce a magnetic moment ranging from 1.00 to 4.93 μ_B. In particular, the Ti adatom states contribute in the midgap and the conduction band (CB), which reduces the band gap to 0.41 eV in the presence of a magnetic order of 1.87 μ_B [70]. A magnetic of 2.00 μ_B is observed for Fe adatom systems. In the case of Cr and V adatoms, the spin-down is observed in CB. However, the spin-up of V states dominates the Fermi level and splits into two peaks for Cr adatom. The situation is different for the Ni and Cu adatoms which exhibit no spin-polarization.

1.3.2 Half-Oxided Phosphorene

Similar to graphene oxide, oxygen adsorption on phosphorene can be used efficiently to tune the optoelectronic properties as well as the protective layer of phosphorene. The absorption of a single oxygen atom on phosphorene can occupy numerous positions like interstitial, horizontal, diagonal ones [74]. As mentioned previously, phosphorene is stable at low oxygen concentrations.

1.3.2.1 Electronic Structure

An oxidation with a degree of 50% generates nine possible configurations among which only six are stable. As illustrated in Figure 1.9, the unit cell comprises four P-atoms assigned as P_1, P_2, P_3, and P_4 and two O-atoms, namely, O^A and O^B. Oxygen atoms bind to two P-atoms on the same side (more precisely, they are attached either to the up-side $\left(O_U^A \text{ and } O_U^B\right)$ or to the down-side of the surface and $\left(O_D^A \text{ and } O_D^B\right)$ or they are in opposite sides, namely $\left(O_U^A \text{ and } O_D^B\right)$. The index U and D referred to the up and the down side of the P-atoms. For example, to get $P_2O_d^B$, one should place the O^A-atom up on P_1 and the O^B down on P_2 (O^B is in the opposite side of O^A) forming a O_u^A/O_d^B fashion on either side of the plane. The resulting new derivatives can be divided onto three main groups. The first group contains the structures $P_2O_U^B$, $P_3O_D^B$ and $P_4O_D^B$ that have only dangling bonds $P = O$. In three conformers $P_2O_D^B$, $P_3O_U^B$, $P_4O_U^B$ constituting the second group, dangling oxygen motif and bridging bond alternate, respectively. The third class concerns the configurations exhibiting only bridging bonds which are energetically less favorable (see [56] for more details). All the half-oxidized conformers exhibit a high buckling parameter confirming the anisotropic behavior of their properties. It was found, using different methods [24, 56], that half-oxidation of phosphorene (P_4O_2) allows to build a stable material.

Half O-functionalization influences significantly the electronic structure of phosphorene mono-layer as depicted in Figure 1.10. The oxidation induces a band gap modulation with the highest value observed in the

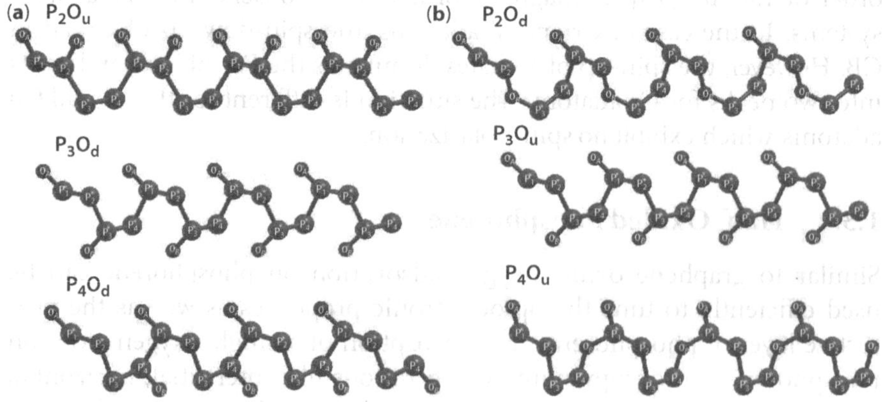

Figure 1.9 Configurations of 50% oxidized phosphorene: (a) dangling structures and (b) bridge structures.

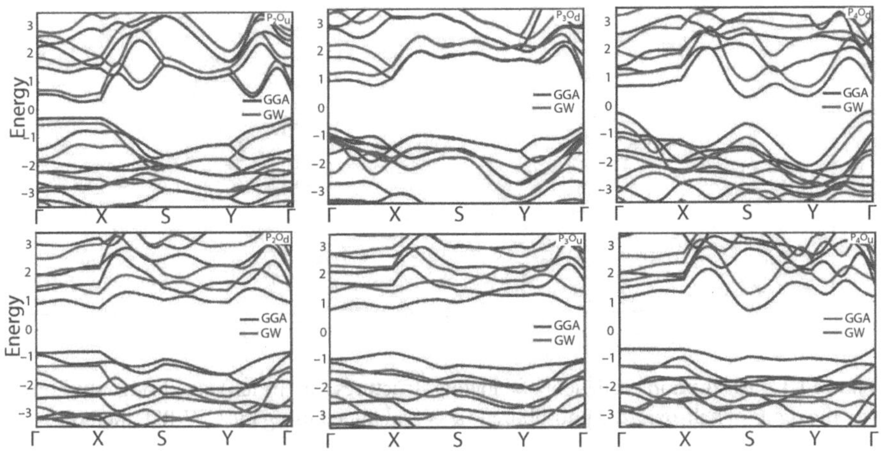

Figure 1.10 GGA and GW band structures of half-oxidized structures.

bridge structures [56]. In all structures the band gap ranges from 0.54 (1.19) to 1.57 (2.88), calculated by GGA (GW) approximation. Moreover, a half O-functionalization tunes the band gap from direct to indirect in all the conformers, except P_2O_D which presents a direct band gap.

POs can provide oxygen in its solid-phase to valve regulated Li O_2 batteries. In addition, when the Li atom is absorbed at the surface of the POs, it binds strongly to the O atoms indicating a strong ionic characteristic of the bond between oxygen and lithium [75]. The absolute values of binding energies of the Li atom adsorbed on the PO surface are greater than those of the Li atom on pure phosphorene, MoS_2 and graphene [76–78]. POs promise high diffusivity owing to the anisotropy of POs cathode barrier that is reduced by half with respect with the armchair axis for Li diffusion on POs. Besides, Li-PO structures with a number of Li atoms lower than O atoms show stable discharge products for PO cathodes [75].

1.3.2.2 Optical Response

The optical absorption spectra of half oxidized phosphorene in Figure 1.11 show maximum values for dangling structures observed at 1.46, 1.71, and 2.62 eV in $P_2O_u^B$, $P_4O_d^B$, and $P_3O_d^B$, respectively.

This absorption behavior is required for photodetector with high efficiency. In contrast, pics of spectra describing the bridge structures coincide with the ultraviolet part and the visible light, since the they are located at 1.81, 2.03, and 3.18 eV for $P_3O_u^B$, $P_2O_d^B$, and $P_4O_u^B$, respectively. One deduces

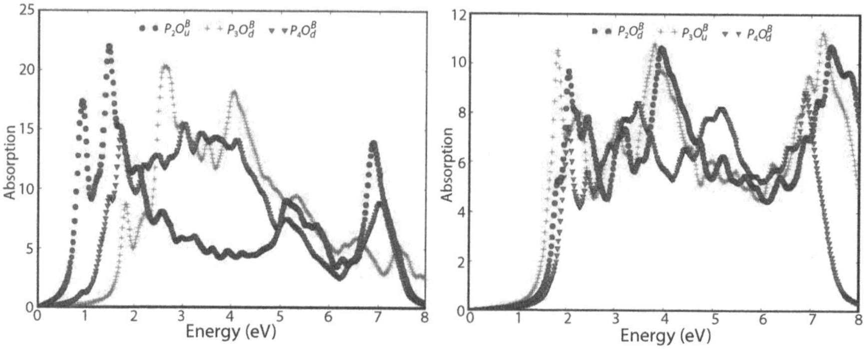

Figure 1.11 Absorption coefficient of dangling structures (on the left) and bridge structures (on the right) of half-oxidized phosphorene sheets obtained using the GW-BSE methods.

that 50% oxidation is an effective manner to enlarge the absorption range of phosphorene along the light spectrum [20].

Half oxidation is also used to modify the reflectivity of phosphorene as illustrated in Figure 1.11. Indeed, its maximum value in the UV region is around 38%, 50%, and 34% located at 8.21, 8.04, and 7.06 eV in $P_2O^B_d$, $P_3O^B_u$, and $P_4O^B_u$, respectively. In the visible part, these structures show a reflectivity lower than 15% indicating their potential use for transparent electronics. The situation is different for the dangling configurations P_2O_U, P_3O_D, and P_4O_D structures that reflect the visible light with a maximum value of 42%, 38%, and 39% found for the energy 1.48, 2.58, and 1.68 eV, respectively [20].

In contrast to pure phosphorene, the first peak of the optical absorption in all P_4O_2 structures is characterized by a dark exciton with long-lifetime. This result makes these new systems promising candidates as molecular sensors or applications in on-chip communication. Studying the excitonic effects of half-oxidized phosphorene conformers reveals that the wave function in the dangling phosphorene extends along the armchair direction, which is similar to pure phosphorene (see Figure 1.12). These results indicate that six half-oxidized phosphorene conformers are potential candidates for electronic devices and photovoltaic applications [20].

1.3.2.3 Strain Effect

Besides the high flexibility and strong anisotropic elastic properties of phosphorene, oxidation is so important to tune its elastic properties and extending the corresponding applications. Phosphorene half-oxides can be stretched becoming ideal for devices requiring flexibility [79]. Moreover,

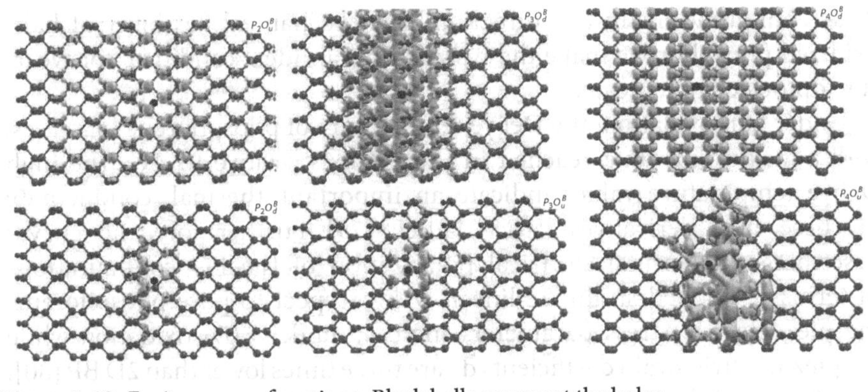

Figure 1.12 Excitons wave functions. Black balls represent the holes.

the degree of oxidation influences significantly the elastic parameters [30–80]. The polar plot of Young modulus and Poisson ratios reveals that the maximal values are attempted for armchair-strain resulting super flexible structures. However, it is hard to implement zigzag deformation direction that shows minimal values of elastic parameters (see Figures 1.13a, b). Importantly, the Poisson ratios of $P_3O_d^B$ and $P_4O_d^B$ conformers take negative values for some ranges of angles, which lead to an auxetic behavior (see [79]). Moreover, in all the conformers, the Poisson ratio is lower than 0.5, which correspond to incompressible materials. Stress-strain responses, under the armchair and zigzag tensile strain, show that half oxidation leads to much higher critical points, compared to pure phosphorene. It is found that both the dangling and bridge structures resist to a large axial deformation up to 27%–39% along the AC-axis with respect to the ZZ one. More precisely, the maximum values coincide with the dangling structures P_4O_U, P_2O_D, and P_3O_U which can resist a tensile strain up to 30%, 33%, and 39%, respectively, showing a high flexibility in the armchair direction. This result is owing to the high buckled honeycomb structures that exhibit

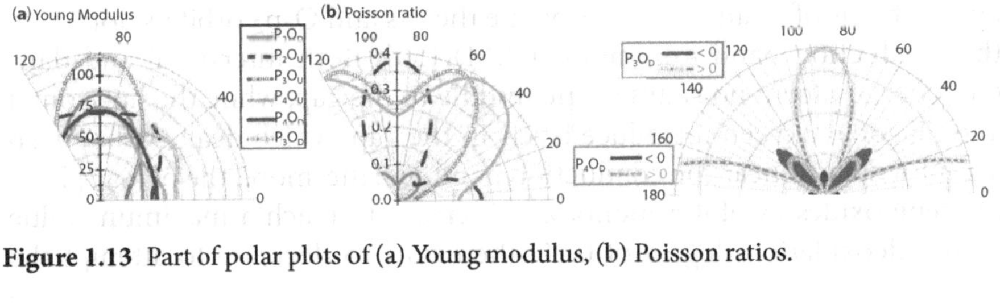

Figure 1.13 Part of polar plots of (a) Young modulus, (b) Poisson ratios.

these configurations in this direction. The mechanical flexibility of half-oxidized phosphorene make these structures an ideal candidate for wearable optoelectronic devices.

Under half oxidation, the Debye temperature of phosphorene increases, with a maximum value reached in the ZZ-axis relative to AC. The high Debye temperature values indicate an important thermal conductivity in these new derivatives lattice lattice [79]. Furthermore, the curves describing the normal electrical polarization of the PO configurations in terms of applied strain are linear. With respect to pure phosphorene, the piezoelectric stress parameters increase under 50% oxidation while the piezoelectric strain coefficients d_{11} are three times lower than 2D BP [80].

When axial deformation is implemented, the electronic features of the POs become modulated. The band gap of dangling and bridge structures increases with low tensile strain, then it reduces to achieve a metallic state for large deformations. Besides, both groups of POs can maintain the semiconductor behavior along the armchair direction for a strain ranging from 20% to 40% [81]. One can deduce that the adjustment of the phosphorene oxides features makes this class of materials potential candidates for advanced devices.

1.3.3 Surface Oxidation on Phosphorene

In contrast to the functionals H, F, and -OH which work like scissors by breaking down phosphorene into nanoribbons [82], a complete oxidation maintains the initial phosphorene configuration and shifts the lattice constants without breaking the Phosphorous bonds connecting the two P-half-layers, namely, the upper and lower ones [30].

1.3.3.1 Optoelectronic Features

At high concentration, oxidation leads to new derivatives of oxided phosphorene. As shown in Figure 1.14, up to fully oxidation, the interatomic P-P lengths increase to 2.32 and 2.37 Å while the direct gap located at point Γ reduces with respect to pure phosphorene. VBM is characterized by p_y orbitals of P- and O-atoms, while the P-s and O-p_z orbitals dominate the conduction band minimum (CBM) [30]. Both interstitial and dangling oxygen form no states in the middle of the gap, while the horizontal and diagonal oxygen introduce levels in the gap, which deals with a deep acceptor state at near the conduction band. Furthermore, the planar phosphorene oxides exhibit a monotonic increase to reach a maximum value with a deoxidation degree of 0.25, then start to decrease to attempt the

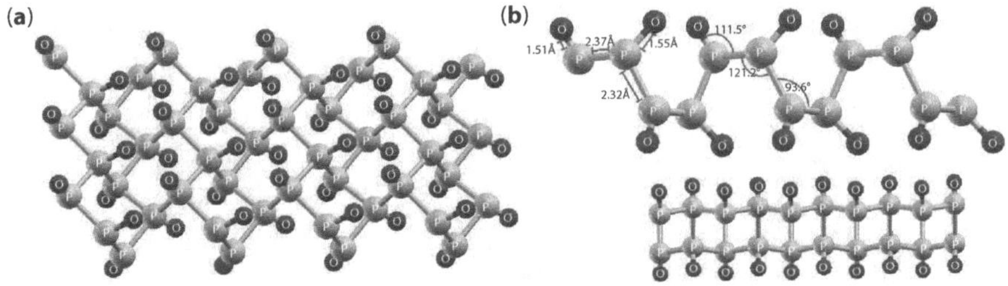

Figure 1.14 (a) Top and (b) side views of phosphorene oxides PO.

value of 0.6 eV in a fully oxidized PO structure. For the tubular structure, the band gaps take the values from 0.4 (1.62) to 5.56 (7.78) eV at PBE (HSE level) [32]. Interestingly, the GW corrected band gap shows that the increasing oxygen coverage leads to an increase in the band energy from 4 eV to 10 eV, indicating that the VBM and CBM part become more localized [83].

The application of electric field reduces the gap energy of PO to a minimum of about 0.4 eV for a field E = 1.5 V /Å. The band gap fluctuates also from direct found for 100% to indirect for O-concentrations of 12.5%, 25%, and 50 %. Also, the work function in phosphorene increases linearly with the increased of the oxidation degree. The calculated values for $PO_{0.125}$, $PO_{0.25}$, and $PO_{0.5}$, are 4.9, 5.2, and 5.8, respectively, compared to PO that has 7.2 eV [30].

Under ambient conditions, phosphorene oxide is a stable material that did not exhibit any negative frequencies in its phonon dispersion curve [30]. Moreover, the simulation indicates that oxided structure is still robust and intact at low temperature, confirming its stability, while the material cut for large temperature values [84]. Unlike pure phosphorene, the phonon dispersion of PO exhibits three main regions as displayed in Figure 1.15b, namely, (i) the acoustic region, (ii) the middle region, and

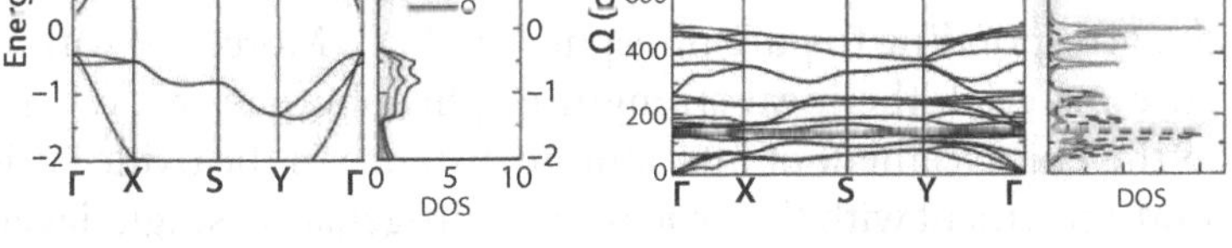

Figure 1.15 Phosphorene oxide (a) band structure and density of states, (b) phonon dispersion curves and density of states.

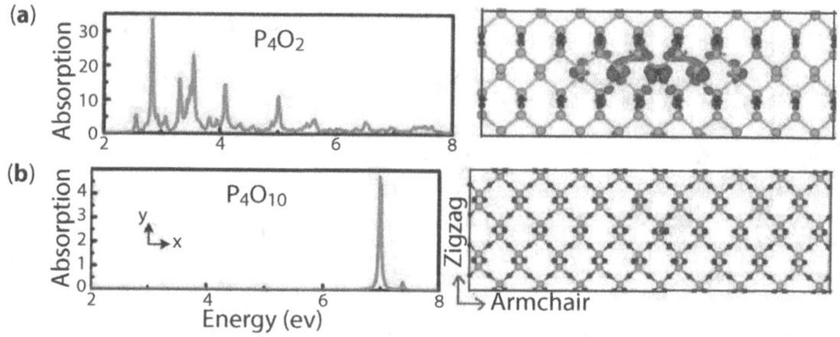

Figure 1.16 Absorption spectrum and exciton wavefunction for the first transition peak for (a) P_4O_2 and (b) P_4O_{10} structures.

(iii) the high frequency range. Moreover, in contrast to the electron effective mass, the effective masses of holes are anisotrope [30].

Besides the band structure modification, the oxidation tunes also the optical features of phosphorene. In PO systems, the absorption spectrum reveals that the 1st absorption peak is located at 2.7 eV, in P_4O_2, it is also found that both phosphorus and oxygen atoms contribute in the transition and extend the wavefunction along the AC-axis (see Figure 1.16a). Further, in the P_4O_{10} system, the absorption peak moves to high energy with a peak located at 7.0 eV. The wavefunction only localized on oxygen atoms adsorbed at the surface (see Figure 1.16b). This changes the binding energy E_b from −1.4 eV to reach −3.0 eV for P_4O_2 and P_4O_{10}. The electronic and optical band gap as well as the binding energy of P_4O_2 and P_4O_{10} are very close to those of benzene [83].

1.3.3.2 Stress vs Strain

When the surface is oxidized, the electrons get transferred forming ions in phosphorene which influences mainly the mechanical response of the material describe by its stiffness against externally applied strains. It results that the oxidation changes the elastic moduli leading to a higher flexible structure [31]. This is also the case for reduced concentrations. Indeed, phosphorene with an oxidation degree of 12.5% can resist to a deformation up to 32% and 35% in AC- and ZZ-axes, respectively, which are higher than that corresponding to pure phosphorene [31]. Moreover, with respect to the pure material, the ideal strength in phosphorene oxide is reduced owing to the enhancement of interatomic distance in the oxide lattice [30, 31] in good agreement with the process of hydrogenating single-layer h-BN [85]. Therefore, the oxidation causes a two times reductions in the value of ideal strength [31].

For a tensile strain varying from −8% to 8%, the band gap in PO increases under compressive strain and decreases with tensile strain, ranging from 0.85 to 0.1 eV for the strain values of [30]. The variation of the gap width results only from the CBM since the VBM is not influenced by the elastic strength. For small tensile strain interval of [−0.006, 0.006], the linear change of polarizations of PO reveals two values of stress piezoelectric responses, namely, e_{11} = 20.13 10^{-10}C/m and e_{31} = 4.06 10^{-10}C/m that correspond to the piezoelectric coefficients given in [86]. All these results indicate that phosphorene oxide are excellent candidates for potential applications requiring the conversion of energy [86].

1.3.3.3 *Thermal Conductivity*

Compared with pure phosphorene (P), phosphorene oxide (PO) exhibits a much lower thermal conductivity over the whole temperature range [87]. Indeed, the values of the thermal conductivityfor both P and PO along the armchair axis, namely κ_{PO}^{A} and κ_{P}^{A} are 2.5 times smaller than κ_{PO}^{Z} and κ_{P}^{Z} along the zigzag one. At room temperature T = 300 K, the κ_{PO}^{A} takes the value of 2.42 W/mK, which is very small compared to 65 W/mK reported for pure phosphorene as well as that of other 2D materials such as silicene (26 W/mK) [88] and MoS_2 (34.5 W/mK) [89]. Low thermal conductivity renders PO an advantageous novel low dimensional candidate for high-performance thermoelectric materials [87].

To highlight the main responsible of such a low lattice thermal conductivity in PO, one should examine the various phonon modes. The puckered structure of PO allows more phonon-phonon scattering of the ZA mode with a contribution of 15% to 17%, while the longitudinal and transverse acoustic modes are the most dominant ones. Furthermore, the lattice thermal conductivity in a material results on the use of different phonon scattering sources. For the case of PO, only the phonon-phonon scattering is considered, since the other sources, such as Umklapp scattering, phonon-electron scattering, impurity effect and boundary effect are so negligible. As shown in Figure 1.17, the anharmonic relaxation times of as a function of frequency indicates that the phonon lifetime corresponding to three acoustic modes (ZA, TA, and LA) and the other higher modes of PO is more lower than that of pure phosphorene. This reduction is mainly owing to dangling bond connecting oxygen atom to its phosphorous neighbor which allow to O not only in the optical P-O vibration, but also vibrate along the in-plane directions together with phosphorous atoms contributing to the acoustic modes. It follows that this contribution is responsible for the acoustic phonon softening which decreases the thermal conductivity of PO [87].

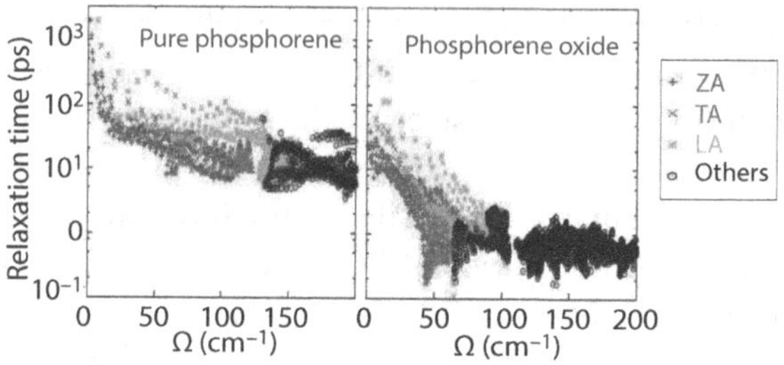

Figure 1.17 Mode-dependent anharmonic phonon relaxation time for acoustic modes.

1.4 Conclusion

In this chapter, we have presented an overview of pure phosphorene, its geometric structure, its physical properties, its fabrication methods, and several applications. We have also shown thatowing to its puckered structure and its strong anisotropic electronic, mechanical, magnetic, and optical properties, phosphorene constitutes an ideal candidate for potential applications, including gas sensor, field-effect transistor, and solar cell application. Unstable under atmospheric conditions, we have reported phosphorene oxides and demonstrated how O-functionalization is a promising technique to enhance the features of this novel material.

Acknowledgment

Lalla Btissam *Drissi et al.* thank "Académie Hassan II des Sciences et Techniques-Morocco" for financial support.

References

1. Bridgman, P.W., Two new modifications of phosphorus. *Chem. Soc.*, 36, 1344–1363, 1914.
2. Aldave, S.H., Yogeesh, M.N., Zhu, W., Kim, J., Sonde, S.S., Nayak, A.P., Akinwande, D., Characterization and sonochemical synthesis of black phosphorus from red phosphorus. *2D Mater.*, 3, 1, 014007, 2016.
3. Morita, A., Semiconducting black phosphorus. *Appl. Phys. A Mater. Sci. Process*, 39, 4, 227, 1986.
4. Liu, Y., Qiu, Z., Carvalho, A., Bao, Y., Xu, H., Tan, S.J., Lu, J., Gate-tunable giant stark effect in few-layer black phosphorus. *Nano Lett.*, 17, 3, 1970, 2017.

5. Castellanos-Gomez, A., Vicarelli, L., Prada, E., Island, J.O., Narasimha-Acharya, K.L., Blanter, S.I., Zandbergen, H.W., Isolation and characterization of few-layer black phosphorus. *2D Mater.*, 1, 2, 025001, 2014.
6. Liu, H., Du, Y., Deng, Y., Peide, D.Y., Semiconducting black phosphorus: Synthesis, transport properties and electronic applications, Chemical Society Reviews. *Chem. Soc. Rev.*, 44, 9, 2732, 2015.
7. Hu, W. and Yang, J., Defects in phosphorene. *J. Phys. Chem. C*, 119, 35, 20474–20480, 2015.
8. Elahi, M., Khaliji, K., Tabatabaei, S.M., Pourfath, M., Asgari, R., Modulation of electronic and mechanical properties of phosphorene through strain. *Phys. Rev. B*, 91, 11, 115412, 2015.
9. Reich, E.S., Phosphorene excites materials scientists. *Nat. News*, 506, 7486, 19, 2014.
10. Kou, L., Chen, C., Smith, S.C., Phosphorene: Fabrication, properties, and applications. *J. Phys. Chem. Lett.*, 6, 2794, 2015.
11. Fei, R.X., Faghaninia, A., Soklaski, R., Yan, J.A., Lo, C., Yang, L., Enhanced thermoelectric efficiency via orthogonal electrical and thermal conductances in phosphorene. *Nano Lett.*, 14, 6393, 2014.
12. Srivastava, A., Khan, M.S., Gupta, S.K., Pandey, R., Unique electron transport in ultrathin black phosphorene: Ab-initio study. *Appl. Surf. Sci.*, 356, 881, 2015.
13. Pan, Y., Wang, Y., Ye, M., Quhe, R., Zhong, H., Song, Z., Lu, J., Monolayer phosphorene–metal contacts. *Chem. Mater.*, 28, 2100, 2016.
14. Lv, H.Y., Lu, W.J., Shao, D. F., Sun, Y. P., Enhanced thermoelectric performance of phosphorene by strain-induced band convergence. *Phys. Rev. B*, 90, 8, 085433, 2014.
15. Qiao, J., Kong, X., Hu, Z.X., Yang, F., Ji, W., High-mobility transport anisotropy and linear dichroism in few-layer black phosphorus. *Nat. Commun.*, 5, 4475, 2014.
16. Yoo, E., Kim, J., Hosono, E., Zhou, H.S., Kudo, T., Honma, I., Large reversible Li storage of graphene nanosheet families for use in rechargeable lithium ion batteries. *Nano Lett.*, 8, 8, 2277, 2008.
17. Wang, H., Yang, Y., Liang, Y., Robinson, J.T., Li, Y., Jackson, A., Dai, H., Graphene-wrapped sulfur particles as a rechargeable lithium–sulfur battery cathode material with high capacity and cycling stability. *Nano Lett.*, 11, 7, 2644, 2011.
18. Yin, Y.X., Xin, S., Guo, Y.G., Wan, L.J., Lithium–sulfur batteries: Electrochemistry, materials, and prospects. *Angew. Chem. Int. Ed.*, 52, 50, 13186, 2013.
19. Tran, V., Soklaski, R., Liang, Y., Yang, L., Layer-controlled band gap and anisotropic excitons in few-layer black phosphorus. *Phys. Rev. B*, 89, 235319, 2014.
20. Sadki, S. and Drissi, L.B., Tunable optical and excitonic properties of phosphorene via oxidation. *J. Phys.: Condens. Matter*, 30, 25, 255703, 2018.

21. Zhang, W., Yin, J., Zhang, P., Ding, Y., Strain/stress engineering on the mechanical and electronic properties of phosphorene nanosheets and nanotubes. *RSC Adv.*, 7, 51466–51474, 2017.
22. Peng, X., Wei, Q., Copple, A., *Phys. Rev. B*, 90, 085402, 2014.
23. Drissi, L.B., Sadki, S., Sadki, K., Strain-engineered direct-indirect band gap transition and its mechanism in two-dimensional phosphorene. *J. Phys. Chem. Solids*, 112, 137–142, 2018.
24. Carvalho, A., Wang, M., Zhu, X., Rodin, A.S., Su, H., Neto, A.H.C., Phosphorene: From theory to applications. *Nat. Rev. Mater.*, 1, 16061, 2016.
25. Sorkin, V., Cai, Y., Ong, Z., Zhang, G., Zhang, Y.W., Recent advances in the study of phosphorene and its nanostructures. *Crit. Rev. Solid State Mater. Sci.*, 42, 1, 1, 2017.
26. Hu, T. and Dong, J., Two new phases of monolayer group-IV monochalcogenides and their piezoelectric properties. *Phys. Chem. Chem. Phys.*, 18, 47, 32514, 2016.
27. Koenig, S.P., Doganov, R.A., Schmidt, H., Castro Neto, A.H., Oezyilmaz, B., Electric field effect in ultrathin black phosphorus. *Appl. Physics Lett.*, 104, 10, 103106, 2014.
28. Cho, K., Yang, J., Lu, Y., Phosphorene: An emerging 2D material. *J. Mater. Res.*, 32, 15, 2839–2847, 2017.
29. Wang, G., Slough, W.J., Pandey, R., Karna, S.P., Degradation of phosphorene in air: Understanding at atomic level. *2D Mater.*, 3, 2, 025011, 2016.
30. Wang, G., Pandey R., Karna S.P., *J. Appl. Phys.*, 118, 234304, 2015.
31. Hao, F. and Chen, X., *J. Appl. Phys.*, 118, 234304, 2015.
32. Ziletti, A., Carvalho, A., Trevisanutto, P.E., Campbell, D.K., Coker, D.F., Neto, A.C., *Phys. Rev. B*, 91, 8, 085407, 2015.
33. Pang, J., Bachmatiuk, A., Yin, Y., Trzebicka, B., Zhao, L., Fu, L., Rummeli, M.H., *Adv. Energy Mater.*, 8, 8, 1702093, 2018.
34. Khandelwal, A., Mani, K., Karigerasi, M.H., Lahiri, I., *Mater. Sci. Eng.: B*, 221, 17–34, 2017.
35. Geim, A.K., *Rev. Mod. Phys.*, 83, 3, 851, 2011.
36. Novoselov, K.S., *Rev. Mod. Phys.*, 83, 3, 837, 2011.
37. Batmunkh, M., Bat-Erdene, M., Shapter, J.G., *Adv. Energy Mater.*, 8, 5, 1701832, 2018.
38. Bagheri, S., Mansouri, N., Aghaie, E., *Int. J. Hydrogen Energy*, 41, 7, 4085–4095, 2016.
39. Lu, W., Nan, H., Hong, J., Chen, Y., Zhu, C., Liang, Z., Zhang, Z., *Nano Res.*, 7, 6, 853–859, 2014.
40. Coleman, J.N., Lotya, M., O'Neill, A., Bergin, S.D., King, P.J., Khan, U., Shvets, I.V., *Science*, 331, 6017, 568–571, 2011.
41. Liu, K.K., Zhang, W., Lee, Y.H., Lin, Y.C., Chang, M.T., Su, C.Y., Lai, C.S., *Nano Lett.*, 12, 3, 1538–1544, 2012.
42. Cudazzo, P., Attaccalite, C., Tokatly, I.V., Rubio, A., *Phys. Rev. Lett.*, 104, 226804, 2010.

43. Wang, V., Kawazoe, Y., Geng, W.T., *Phys. Rev. B*, 91, 4, 045433, 2015.
44. Kopf, M., Eckstein, N., Pfister, D., Grotz, C., Kruger, I., Greiwe, M., Nilges, T., *J. Cryst. Growth*, 405, 6–10, 2014.
45. Wei, Q. and Peng, X., *Appl. Phys. Lett*, 104, 251915, 2014.
46. Sisakht, E.T., Zare, M.H., Fazileh, F., *Phys. Rev. B*, 91, 8, 085409, 2015.
47. Rodin, A.S., Carvalho, A., Castro, A.H., *Neto Phys. Rev. Lett.*, 112, 176801, 2014.
48. Liang, L., Wang, J., Lin, W., Sumpter, B.G., Meunier, V., Pan, M., *Nano Lett.*, 14, 11, 6400, 2014.
49. Wang, X., Jones, A.M., Seyler, K.L., Tran, V., Jia, Y., Zhao, H., Xia, F., *Nat. Nanotechnol.*, 10, 6, 517, 2015.
50. Khan, I. and Hong, J., *New Journal of Physics*, 17, 2, 023056, 2015.
51. Dhanabalan, S.C., Ponraj, J.S., Guo, Z., Li, S., Bao, Q., Zhang, H., *Adv. Sci.*, 4, 1600305, 2017.
52. Cakir, D., Sahin, H., Peeters, F.M., *Phys. Rev. B*, 90, 20, 205421, 2014.
53. Castellanos-Gomez, A., Vicarelli, L., Prada, E., Island, J.O., Narasimha-Acharya, K.L., Blanter, S.I., Zandbergen, H.W., *2D Mater.*, 1, 2, 025001, 2014.
54. Rodin, A.S., Carvalho, A., Neto, A.C., *Phys. Rev. B*, 90, 7, 075429, 2014.
55. Wang, L., Kutana, A., Zou, X., Yakobson, B.I., *Nanoscale*, 7, 21, 9746–9751, 2015.
56. Drissi, L.B., Sadki, S., Sadki, K., *J. Phys.: Condens. Matter*, 28, 14, 145501, 2016.
57. Cadelano, E., Palla, P.L., Giordano, S., Colombo, L., *Phys. Rev. B*, 82, 235414, 2010.
58. Evans, K.E. and Alderson, A., *Adv. Mater.*, 12, 617, 2000.
59. Yuan, J., Yu, N., Xue, K., Miao, X., *RSC Adv.*, 7, 8654, 2017.
60. Wang, G., Loh, G.C., Pandey, R., Karna, S.P., *Nanotechnology*, 27, 5, 055701, 2015.
61. Kou, L., Frauenheim, T., Chen, C., *J. Phys. Chem. Lett.*, 5, 15, 2675, 2014.
62. Wang, H., Wang, X., Xia, F., Wang, L., Jiang, H., Xia, Q., Han, S.J., *Nano Lett.*, 14, 11, 6424, 2014.
63. Favron, A., Gaufres, E., Fossard, F., Phaneuf-L'Heureux, A.L., Tang, N.Y., Levesque, P.L., Martel, R., *Nat. Mater.*, 14, 8, 826, 2015.
64. Kim, J.S., Liu, Y., Zhu, W., Kim, S., Wu, D., Tao, L., Akinwande, D., *Sci. Rep.*, 5, 8989, 2015.
65. Ahmed, T., Balendhran, S., Karim, M.N., Mayes, E.L., Field, M.R., Ramanathan, R., Walia, S., *npj 2D Mater. Appl.*, 18, 13–16, 2017.
66. Lu, J., Yang, J., Carvalho, A., Liu, H., Lu, Y., Sow, C.H., *Acc. Chem. Res.*, 49, 9, 1806, 2016.
67. Edmonds, M.T., Tadich, A., Carvalho, A., Ziletti, A., O'Donnell, K.M., Koenig, S.P., Fuhrer, M.S., *ACS Appl. Mater. Interfaces*, 7, 27, 14557–14562, 2015.
68. Wang, G., Pandey, R., Karna, S.P., *Appl. Physics Lett.*, 106, 17, 173104, 2015.
69. Kulish, V.V., Malyi, O.I., Persson, C., Wu, P., *Phys. Chem. Chem. Phys.*, 17, 2, 992–1000, 2015.

70. Rastogi, P., Kumar, S., Bhowmick, S., Agarwal, A., Chauhan, Y.S., *IETE J. Res.*, 63, 2, 205–215, 2017.
71. Nakada, K. and Ishii, A., *Solid State Commun.*, 151, 13–16, 2011.
72. Lehtinen, P.O., Foster, A.S., Ayuela, A., Krasheninnikov, A., Nordlund, K., Nieminen, R.M., *Phys. Rev. Lett.*, 91, 017202, 2003.
73. Wu, M., Liu, E.-Z., Jiang, J.Z., *Appl. Phys. Lett.*, 93, 082504, 2008.
74. Ziletti, A., Carvalho, A., Campbell, D.K., Coker, D.F., Neto, A.C., *Phys. Rev. Lett.*, 114, 4, 046801, 2015.
75. Li, Y., Ma, F., Wang, L.W., *J. Mater. Chem. A*, 6, 17, 7815, 2018.
76. Li W, W., Yang, Y., Zhang, G. *et al.*, *Nano Lett.*, 15, 3, 1691, 2015.
77. Yang, G., Fan, X., Liang, Z. *et al.*, *RSC Adv.*, 6, 32, 26540, 2016.
78. Li, Y., Wu, D., Zhou, Z. *et al.*, *J. Phys. Chem. Lett.*, 3, 16, 2221, 2012.
79. Sadki, K., Sadki, S., Drissi, L.B., *J. Phys. Chem. Solids*, 130, 13, 2019.
80. Yin, H., Zheng, G.P., Gao, J., Wang, Y., Ma, Y., *Phys. Chem. Chem. Phys.*, 19, 40, 27508, 2017.
81. Sadki, S., Sadki, K., Drissi, L.B., Superlattices Microstruct., 126, 186, 2019.
82. Peng, X. and Wei, Q., *Mater. Res. Express*, 1, 4, 045041, 2014.
83. Lu, Y. and Zhu, X., *J. Phys. Chem. A*, 123, 1, 21–25, 2018.
84. Dai, J. and Zeng, X.C., *RSC Adv.*, 4, 89, 48017–48021, 2014.
85. Noor-A-Alam, M., Kim, H.J., Shin, Y.H., *Phys. Chem. Chem. Phys.*, 16, 6575, 2014.
86. Li, J., Zhao, T., He, C., Zhang, K., *J. Phys. D: Appl. Phys.*, 51, 12, 12LT01, 2018.
87. Lee, S., Kang, S.H., Kwon, Y.K., *Sci. Rep.*, 9, 1, 5149, 2019.
88. Li, X., Mullen, J.T., Jin, Z., Borysenko, K.M., Nardelli, M.B., Kim, K.W., *Phys. Rev. B*, 87, 11, 115418, 2013.
89. Liu, T.-H., Chen, Y.-C., Pao, C.-W., Chang, C.-C., *Appl. Phys. Lett.*, 104, 201909, 2014.

2

Antimonene: A Potential 2D Material

Shuai Liu, Tianle Zhang and Shengxue Yang*

School of Materials Science and Engineering, Beihang University, Beijing, China

Abstract

Since the two-dimensional antimonene was reported, it has become the focus of theoretical research. The predicted various interesting properties have inspired many experimental scientists to confirm and better understand this material. Recent studies on the preparation and application of such material have yielded many important results. This chapter is mainly on how to produce and apply antimonene experimentally. First, the theoretically calculated fundamental characteristics of antimonene are given. Second, different preparation methods of antimonene from traditional mechanical exfoliation to other novel approaches are listed. Finally, it summarizes the potential applications in several technological fields such as optical, optoelectronic, electronic, and biomedical fields. In addition, it provides insights into further exploration of the diverse properties of antimonene, continuous updating of preparation methods, as well as further development of potential applications, and then looks ahead to the opportunities and challenges of antimonene facing in the future.

Keywords: Antimonene, structural characteristics, band structure, preparation methods, potential applications

2.1 Introduction

Two-dimensional (2D) materials are different from traditional three-dimensional (3D) bulk materials because the movement of charges is confined to the atomic layer, resulting in many novel physical phenomena. As the most typical 2D material, monolayer graphene was successfully isolated

**Corresponding author*: sxyang@buaa.edu.cn

Inamuddin, Rajender Boddula, Mohd Imran Ahamed and Abdullah M. Asiri (eds.)
Monoelements: Properties and Applications, (27–56) © 2020 Scrivener Publishing LLC

in 2004 [1], which paves the way to discover and study more families of 2D materials. However, the valence band intersects the conduction band at the Dirac point [2], making graphene a semi-metallic material with no band gap, which limits its applications in electronic and optoelectronic devices.

Afterwards, many researchers hope to find novel 2D materials with a certain band gap. Transition metal dichalcogenides (TMDs) are also a typical class of 2D materials, these materials have tunable band gap ranging from 1.5 to 2.5 eV and some of which are semiconductors with direct band gap [3]. However, TMDs are still not very suitable for optoelectronic applications. Therefore, other groups of 2D materials have aroused interest in research. Among them, the earliest concerned and widely explored 2D material is black phosphorus (BP), which is the most stable allotrope of phosphorus. The band gap of BP can be adjusted in a large range by the number of layers to achieve light absorption from the near infrared to the visible region. BP also has a high carrier mobility of up to 10^3 cm^2 $V^{-1}s^{-1}$, and a large on-off current ratio of 10^5 [4]. Therefore, it is a very promising electronic and optoelectronic materials. In addition, unlike other 2D materials, BP possesses crystal orientation-dependent carrier mobility, light absorption, and other properties as well, due to its anisotropic nature [5, 6]. Unfortunately, a fatal disadvantage of BP is its easy degradation, due to the joint influence of oxygen, water, and light [7].

In view of this, the researchers turned their attention to the other group 15 elements. One of the most representatives is antimonene with strong spin-orbit coupling. In 2015, the novel 2D antimonene was first proposed theoretically [8]. The theoretical studies show that when the antimonene is thinned from bulk to monolayer, the band structure will be abrupt, resulting in its conversion from semi-metal to semiconductor, and the 2D form is still extremely stable. Monolayer antimonene is predicted that having a band gap of 2.28 eV [8]. This makes antimonene very interesting for fabrication in field-effect-transistors (FETs) and optoelectronic devices [9, 10]. At the same time, theoretical predictions have found that Group 15 monoelements have higher carrier mobility [11]. Apart from this, 2D antimonene has also achieved many significant results in experiments. Here, we discuss the research on 2D antimonene in recent years, including four parts as below. Section 2.1 gives the predicted structure and electronic band of antimonene based on the theoretically calculations. Section 2.2 summarizes the experimental synthesis of antimonene with different thicknesses and shapes, including mechanical exfoliation, liquid phase exfoliation, epitaxial growth, and some novel technologies [12]. In Section 2.3, we discuss various potential applications of antimonene, such as nonlinear optics, electrocatalysis, energy storage, biomedicine, and magneto-optic storage.

Moreover, a perspective of the future 2D antimonene research is provided in Section 2.4.

2.2 Fundamental Characteristics

2.2.1 Structure

By means of density functional theory (DFT) calculations, one can see that honeycomb-structure antimonene monolayers are composed of different allotropic forms, which are shown in Figure 2.1a [11]. In these allotropes, the theoretically and experimentally studied antimonene structures are α and β phases. β-antimonene with a buckled structure is the lowest-energy configuration, while α-antimonene has a puckered structure (Figure 2.1b). From the calculated phonon spectra of these two phases, it is observed that there are no appreciable imaginary phonon modes, indicating that both α- and β-phase antimonene structures are thermodynamically stable (Figure 2.1c). Furthermore, each Sb atom is bonded with three adjacent Sb atoms in one atomic layer, endowing them with octet stability [8]. The buckled honeycomb structure of β-antimonene stabilizes further the layered structure, which turns β phase to be the most stable structure. Due to the weak interlayer interactions, α- and β-antimonene monolayers can be easily exfoliated from their bulk crystals.

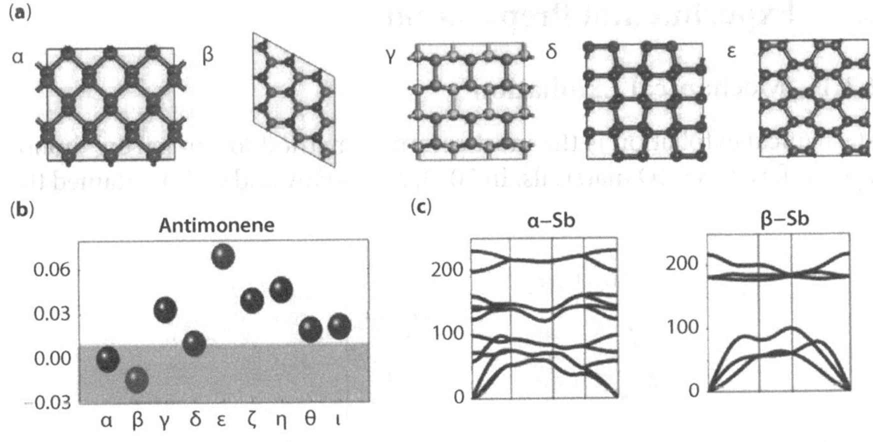

Figure 2.1 (a) Top views of the relaxed antimonene monolayer allotropic forms with five typical honeycomb structures (α, β, γ, δ, ε). (b) Calculated average binding energies of antimonene allotropes with different phases (α, β, γ, δ, ε, ζ, η, θ, ι). (c) Phonon band dispersions of α and β phases of antimonene monolayer. Reproduced with permission [11]. Copyright 2016, Wiley-VCH.

2.2.2 Electronic Band Structure

Based on first-principles calculations, bulk antimony is a typical semimetal material, while it is interesting that it will be transformed into a semiconductor when thinned to be a monolayer antimonene. Zhang *et al.* reported the electronic band structure of β-antimonene, which was calculated by the hybrid functional theory (HSE06) method [8]. As illustrated in Figure 2.2, trilayer and bilayer antimonene are still semi-metallic, where both valence-band tops and conduction-band bottoms cross the Fermi level at several points, causing a band gap of 0 eV in the Brillouin zone. However, for monolayer antimonene, the valence band and conduction band shift respectively down and up, resulting in the formation of a wide band gap of 2.28 eV. The valence band maximum (VBM) locates at K point, while the conduction maximum (CBM) is at Γ point, showing that monolayer antimonene is an indirect semiconductor. Due to the wide and indirect band gap, monolayer antimonene-based optoelectronic devices prefer to respond to the blue and ultraviolet light. After applying a small biaxial tensile strain, antimonene will experience an indirect-to-direct band-gap transition, making it more suitable for the applications of optoelectronic devices. The calculated electron effective masses of monolayer antimonene are $m_e^{\Gamma K} = 0.20\,m_o$ and $m_e^{\Gamma M} = 0.24\,m_o$, indicating that it owns a high carrier mobility.

2.3 Experimental Preparation

2.3.1 Mechanical Exfoliation

Mechanical exfoliation is the most common method for preparing monolayer or few-layer 2D materials. In 2004, Novoselov and Geim obtained the

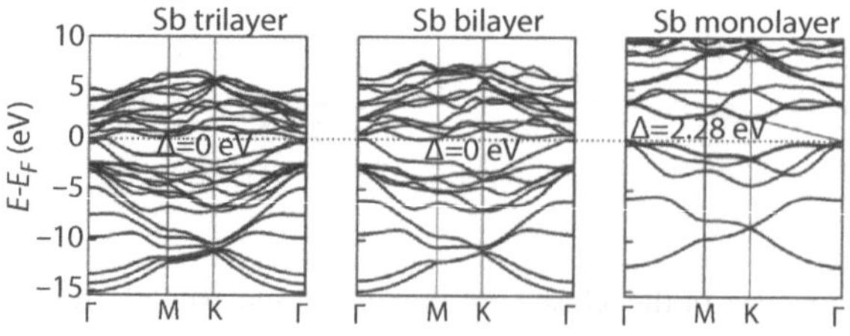

Figure 2.2 HSE06 calculated electronic band structures of trilayer, bilayer, and monolayer antimonene. Dots: Fermi levels. Reproduced with permission [8]. Copyright 2015, Wiley-VCH.

first monolayer graphene *via* this method in human history [1]. Thereafter, this method is widely applied to the preparation of other layered materials. The van der Waals bond between the layers of the layered material is weak, and the binding energy is only 40–70 meV, so the layers and layers are more easily isolated by the external force [13]. The mechanical exfoliation is very suitable for scientific research because of its simple method, and the obtained sample is free from contamination and has a higher crystal quality.

Since the calculated binding energy of β-antimony is smaller than 30 meV, this method can be used to peel off its monolayer form [11]. Ares *et al.* first obtained monolayer antimonene by a modified mechanical exfoliation with a double-step transfer procedure [14]. The specific preparation process is shown in Figure 2.3a, it started by repetitive peeling a freshly cleaved antimony crystal using adhesive tape, where sub-millimeter flakes were easily obtained. Instead of direct transfer of these flakes onto a SiO_2/Si substrate, an initial transfer from the adhesive tape to a viscoelastic stamp (Gel-park Gel-film) was needed. The stamp was then pressed against the surface of the SiO_2/Si substrate and slowly peeled off from it, consequently, large-area antimony thin flakes were prepared in a more controlled way [15]. The isolated antimony flakes can be identified by optical microscopy,

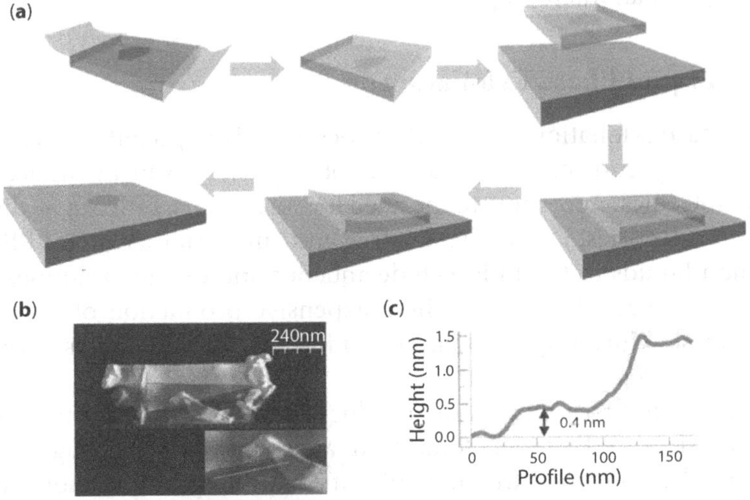

Figure 2.3 (a) Diagram of the steps involved in the sophisticated version of mechanical exfoliation. (b) AFM image of folded antimonene flake. (c) Profile along the green line in the inset of (b). (a) Reproduced with permission [15]. Copyright 2014, IOP Publishing Ltd. (b–c) Reproduced with permission [14]. Copyright 2016, Wiley-VCH.

where different colors represent different thicknesses. The thickness was determined by atomic force microscopy (AFM), and it was found that the height of a monolayer terrace was 0.9 nm due to the presence of water layers. By measuring the step height of single folds, the thickness of a monolayer antimonene was considered to be 0.4 nm (Figure 2.3b, c). The isolated flakes showed good stability in ambient conditions, even if they were immersed in water.

Mechanical exfoliation process usually causes antimony flakes with different thicknesses, but Raman signals of these flakes are too weak to be detected which cannot provide their thickness information. Another work by Ares *et al.* developed a simple and quite accurate method to identify the thicknesses of isolated antimony flakes using optical microscopy [16]. Comparing the optical contrast *versus* thickness measurements with a Fresnel Law model, the refractive index and absorption coefficient of these flakes in the visible spectrum can be yielded, which are obviously different in thin and thick flakes, then being used to distinguish various thicknesses. After that, Abellán *et al.* prepared few-layer antimonene flakes on the SiO_2/Si and gold substrates by mechanical exfoliation and then functionalized their surface with a perylene bisimide (PDI) [17]. This noncovalent functionalization process increases the optical contrast of antimonene under white-light illumination and leads to an obvious quenching of the perylene fluorescence, allowing easy characterization of the flakes in seconds by scanning Raman microscopy.

2.3.2 Liquid Phase Exfoliation

Liquid phase exfoliation (LPE) is the process of placing a bulk material into a liquid and peeling off large quantities of dispersed layers by the action of liquid molecules. According to the need for surfactants, LPE can be divided into two categories, i.e., surfactant-free and surfactant-assisted LPE [18]. Common liquids in the LPE include aqueous and organic solutions. This method is expected to realize the inexpensive production of large-scale 2D materials. Currently, monolayer and few-layer 2D materials have been successfully prepared by the LPE.

Gibaja *et al.* first reported the production of few-layer antimonene by surfactant-free LPE under sonication assistance [19]. Through several attempts, they obtained the best solvent for peeling off antimonene that is the 2-propanol/water mixture with volume ratio of 4:1. In addition to solvent selection, other factors such as sonication time, initial quantity of antimony crystals, and centrifugation conditions were also considered to optimize the LPE process. Ground antimony crystals in selected solvent were sonicated (400 W, 24 kHz) for 40 min, yielding a colorless dispersion with a Faraday-Tyndall effect (Figure 2.4a). After removing the unpeeled bulk materials by centrifugation (3,000 rpm) for 3 min, a stable-dispersion suspensions of micro-scale few-layer antimonene can be obtained with a concentration of ~1.74×10^{-3} gL^{-1}. The surface topography of few-layer antimonene flakes was observed by AFM (Figure 2.4b), confirming the successful exfoliation of antimony crystals using the LPE method. Analyzing the step heights of these flakes in Figure 2.4c, multiples of which were about 4 nm without showing typical terrace characteristics. These antimonene flakes possessed well-defined structures and their lateral dimensions were greater than 1–3 μm^2. From the high-angle annular

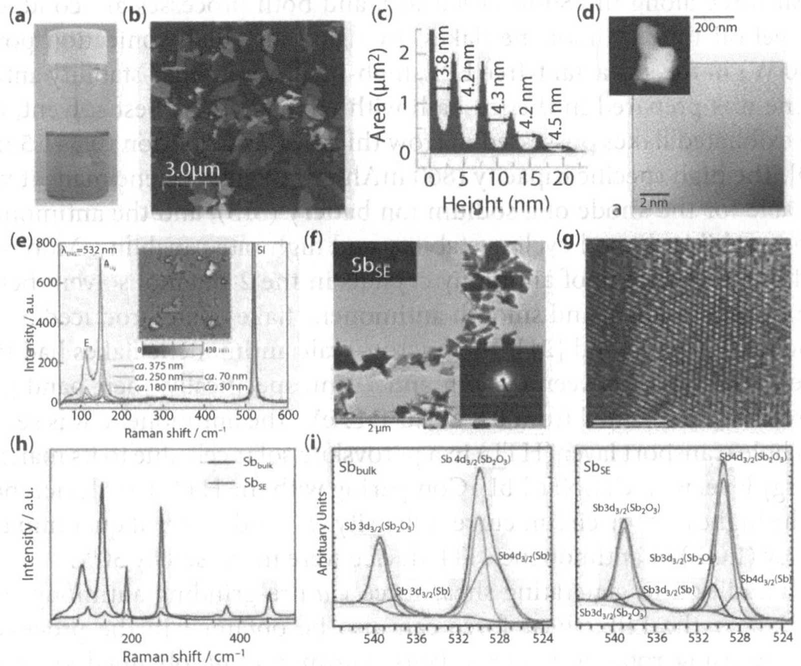

Figure 2.4 (a) Optical image of a dispersion of exfoliated few-layer antimonene. (b) AFM image of few-layer antimonene flakes drop-casted onto a SiO_2/Si substrate. (c) Height histogram of the image in panel (b). (d) Low-magnification (top left) and an atomic-resolution (down right) HAADF images of a typical antimonene flake taken along the [0–12] direction. (e) Single-point spectra of different thicknesses were measured by studying AFM images (inset). The TEM (f) and HRTEM (g) images of exfoliated antimonene nanosheets. Raman spectra (h) and high-resolution XPS (i) of Sb_{bulk} and Sb_{SE}. (a–e) Reproduced with permission [19]. Copyright 2016, Wiley-VCH. (f–i) Reproduced with permission [21]. Copyright 2017, Wiley-VCH.

dark field (HAADF) images of a typical antimonene flake in Figure 2.4d, it can be seen that the crystal structure conforms to β-antimonene and the flake is highly crystalline with no major defects. Figure 2.4e shows the Raman spectra of exfoliated antimonene flakes with different thicknesses, it is observed that the Raman intensity has a certain dependence on the thickness, where the peak intensity increases with the increase of the thickness, only for the flakes with thicknesses greater than 70 nm.

It usually takes a long time for the sonication process in the LPE. If the sonication power is increased during the LPE or the antimony crystals are pretreated before the LPE, the exfoliation time will be greatly shortened and the yield can be remarkably improved [20, 21]. High sonication power affords sufficient energy to break the van der Waals interactions between the Sb-Sb layers, while the pre-grinding of antimony crystals provides a shear force along the Sb-layer surface, and both processes are conducive to peel off thin antimonene flakes. By using ultrahigh sonication power (850 W) in the surfactant-free LPE, high-quality and high-stability antimonene was prepared in the ice-bath with ethanol as the best solvent, and the exfoliated flakes possessed narrow thickness distribution (0.5–1.5 nm) [20]. The high specific capacity (860 mAh g^{-1}) of antimonene made it very suitable for the anode of a sodium ion battery (SIB), and the antimonene anode exhibited good cycling stability and high rate capability. Moreover, adding pre-grinding of antimony crystals in the 2-butanol solvent before sonication, uniform and smooth antimonene flakes were produced by the modified LPE method [21]. The micron-scale antimonene flakes had tunable thicknesses between 0.5 nm and 7 nm, specifically, their band gaps were also finely tuned from 0.8 eV to 1.44 eV. The antimonene was served as a hole transport layer (HTL) in a perovskite solar cell, due to its matched energy level with $CH_3NH_3PbI_3$. Comparing with the HTL-free device, both of the highest short-circuit current density (J_{sc}) and external quantum efficiency (EQE) of antimonene-HTL device were increased by 30%.

In addition to generating shear force *via* pre-grinding antimony crystals before the LPE, shear force can also be obtained in the process of LPE by using rotating blades mixers. Gusmão *et al.* obtained arsenene, antimonene, and bismuthene exfoliated nanosheets by the surfactant-assisted LPE method under the action of shear force generated by rotating household kitchen blenders [22]. As the transmission electron microscopy (TEM) shown in Figure 2.4f, the morphology of exfoliated antimonene (Sb_{SE}) nanosheets consisted of shapes with defined angles and nanostripes because the preferential cleavage of antimony crystals was easy to occur in different crystallographic directions. The size distribution of antimonene nanosheets exhibited a broad range from 100 nm to 900 nm and the maxima value appeared at around 200 nm. Then, the high-resolution TEM (HRTEM) was employed to determine the structure of antimonene (Figure 2.4g). The Raman intensity of Sb_{SE} was lower than that of bulk antimony (Sb_{bulk}), and the Raman peak shifted to higher frequency owing to partial oxidation (Figure 2.4h). The bonding states of Sb_{bulk} and Sb_{SE} were also investigated by the high-resolution X-ray photoelectron spectra (XPS), it can be seen that both the element phase (Sb $3d_{5/2}$, $3d_{3/2}$) and partial oxidation phase (Sb_2O_3) existed in the Sb_{SE} (Figure 2.4i). As a pnictogen, Sb_{SE} can be used for the electrochemical detection of ascorbic acid due to its high electron transfer properties and low onset oxidation potential. Also, Sb_{SE} can be considered as a wide-pH-range catalyst for the hydrogen evolution reaction (HER).

2.3.3 Epitaxial Growth

Epitaxial growth is a method that a crystalline layer can be directly deposited onto a crystalline substrate [23]. This method is widely used in the preparation of high-quality crystalline 2D materials with large capacity. At present, van der Waals epitaxy (vdWE) and molecular beam epitaxy (MBE) are explored to grow few- or even single-layer antimonene on different crystalline substrates.

Traditional epitaxial growth requires very similar lattices to realize the matching between the substrate and the epitaxial layer. Koma *et al.* found that the vdWE proceeded with the van der Waals force can almost avoid the limitation of the substrate and can easily grow various layered materials on the substrates without dangling bonds [24]. Ji *et al.* first used the vdWE method to synthesize few-layer β-antimonene monocrystalline polygons on different substrates and highlighted their atomic structure and ambient stability [25]. The synthesis process was carried out in a two-zone tube furnace for 60 min, where antimony powders in the T_1 zone was heated up to 660°C to generate antimony vapor and the substrates (fluorophlogopite mica, silicon, sapphire) were placed in the downstream T_2 zone (380°C) (Figure 2.5a). Antimony vapor was carried from T_1 zone to T_2 zone by Ar/H_2 (70%/30%) mixed gas and then deposited on the substrates to grow various antimonene polygons. The absence of dangling bonds and low migration energy barrier of antimony atoms on mica substrate are beneficial to a fast growth of antimonene on mica. From the optical microscope images shown in Figure 2.5b, it can be seen that few-layer antimonene polygons exhibit various shapes, including triangles, hexagons, rhombus, and trapezoids with lateral sizes about 5–10 μm. The thicknesses of these polygons are down to 4 nm, while the thinnest one is found to be 1 nm,

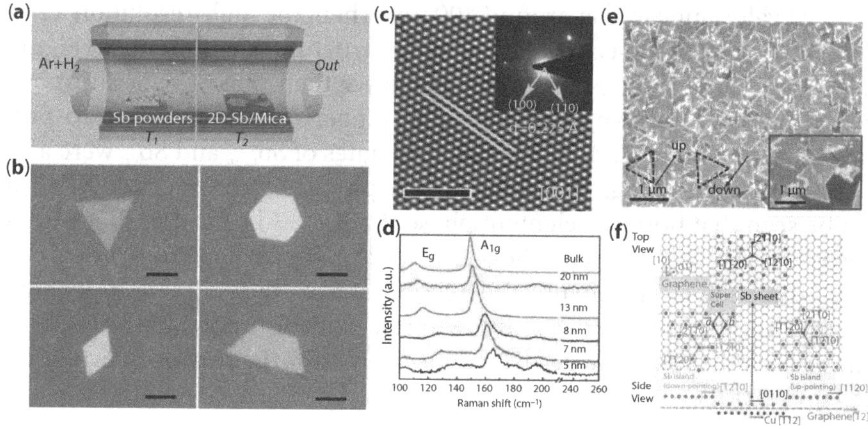

Figure 2.5 (a) Schematic diagram of the synthesis process of antimonene by the vdWE method. (b) Optical microscope images of few-layer antimonene polygons with various shapes. The scale bar is 5 μm. (c) HRTEM image of a typical antimonene polygon. Inset: selected area electron diffraction (SAED) pattern along the [001] axis. The scale bar is 2 nm. (d) Raman spectra of antimonene polygons with different thicknesses. (e) Scanning electron microscopy (SEM) image of 10-nm thick antimonene on an as-grown graphene substrate. (f) Atomic models of the epitaxial alignments between Sb (0001) and graphene in three morphs of Sb. (a–d) Reproduced with permission [25]. Copyright 2016, Nature Publishing Group. (e, f) Reproduced with permission [26]. Copyright 2018, American Chemical Society.

corresponding to a monolayer antimonene. The HRTEM image of a typical antimonene polygon is shown in Figure 2.5c, it is extracted that the synthesized antimonene belongs to the rhombohedral structure, namely, β-antimonene. The Raman spectra of antimonene polygons also show a thickness dependence, in which the Raman peaks move to the higher frequency region with the decrease of thickness (Figure 2.5d). By comparing optical microscopy, AFM, Raman spectroscopy, and XPS results of antimonene before and after one-month aging, it is observed that few-layer antimonene show outstanding stability in ambient condition. Moreover, the synthesized antimonene polygons exhibited high electrical conductivity up to 1.6×10^4 S m^{-1} and good transparency in the visible range.

In particular, there are no dangling bonds in 2D materials, thus the 2D materials can also be employed as the substrates to grow antimonene. Sun *et al.* grew a series of antimonene (Sb) with various morphs on two types of single-crystalline graphene substrates by the vdWE method [26]. One of the substrates was the as-grown graphene on Cu (111)/*c*-sapphire *via* chemical vapor deposition (CVD), and another was the transferred graphene on SiO_2/Si. On the as-grown graphene, triangle Sb islands (up-pointing and down-pointing triangles) were grown in Volmer-Weber (VW) modes with height of ~17.5 ± 0.7 nm, while Sb sheets were grown in Frank-van der Merve (FM) modes with height of ~4.5 ± 0.7 nm (Figure 2.5e). The growth of Sb sheets was probably caused by the remote epitaxy between Sb and underlying Cu. The epitaxial alignments between Sb (0001) and graphene were different in various morphs of Sb, i.e., Sb [1ii00]‖graphene [10] and [10i0]‖graphene [10] for up-pointing and down-pointing Sb islands, [2ii0]‖graphene [10] for Sb sheets (Figure 2.5f). In contrast, only VW growth of Sb islands was found on the transferred graphene due to the absence of Cu. In addition, high-quality Sb thin films were also grown on both substrates, and these two films showed no significant difference. The epitaxial alignment in Sb thin films was the same as that in Sb islands. In another work, β-antimonene was also grown on WSe_2 substrate to form the Sb/WSe_2 heterostructure through van der Waals interactions [27].

The earliest MBE growth of antimonene was reported by T. Lei *et al.* in 2016 [28]. In this work, bilayer antimonene was grown smoothly on 3D topologically insulated Bi_2Te_3 (111) and Sb_2Te_3 (111) substrates with small lattice mismatch. The clear low-energy electron diffraction (LEED) pattern indicated the 1 × 1 periodicity of the antimonene/Sb_2Te_3 (111) surface, showing that high-quality epitaxial antimonene was formed. The surface states of MBE grown antimonene were probed by angle-resolved photoemission spectroscopy (ARPES), and it showed similar electronic structures on the surfaces of two substrates, but the bands of antimonene on Bi_2Te_3 (111) surface were flatter than on Sb_2Te_3 (111) surface due to the larger lattice constants of Bi_2Te_3 (111) substrate. The interfacial strain and charge transfer induced a change in band dispersions. This work is a milestone in this kind of study, which is critical for the understanding of the electrical structure of antimonene.

Subsequently, monolayer-to-multilayer antimonene were successfully grown on 2D substrates by the MBE method, such as $PdTe_2$, MoS_2, WTe_2, and graphene [29–32]. Wu *et al.* grew large-area monolayer antimonene on a freshly cleaved $PdTe_2$ substrate in an ultrahigh vacuum (UHV) MBE chamber (2×10^{-10} mbar), and the schematic of growth process was shown in Figure 2.6a [29]. 2D $PdTe_2$ was chosen as a suitable substrate in the growth of antimonene, because of its small mismatch of crystal periodicity with free-standing antimonene and chemically stable surface. From the topological scanning tunneling microscopy (STM) image shown in Figure 2.6b, it is observed that an atomically smooth antimonene film was obtained with no obvious defects or domain boundaries. According to the

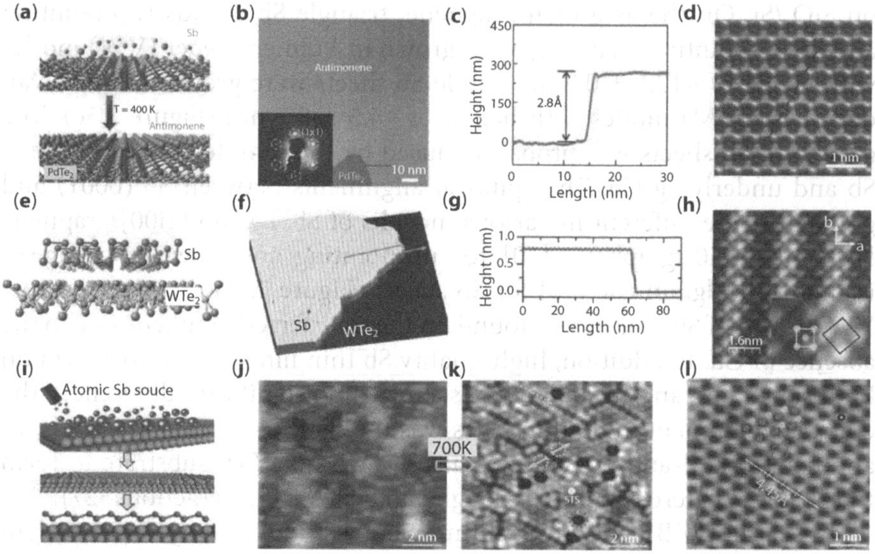

Figure 2.6 (a) Schematic of growth process of monolayer antimonene on 2D $PdTe_2$ substrate. (b) Topographic STM image of large-area antimonene on $PdTe_2$. Inset: LEED pattern of monolayer antimonene on $PdTe_2$ substrate. (c) A height profile of antimonene taken along the red line in panel (b). (d) Atomic-resolution STM image of antimonene. (e) Schematic diagram of epitaxial α-antimonene on 2D Td-WTe_2 substrate. (f) Topographic STM image of monolayer α-antimonene on WTe_2. The area is 120 × 120 nm^2. (g) A line-scan profile of antimonene along the green line in panel (f). (h) Atomic-resolution STM image taken at U = +300 mV, I_t = 100 pA. The area is 8 × 8 nm^2. Inset: the unit cells of $\sqrt{2} \times \sqrt{2}$ lattice and original 1 × 1 lattice. (i) Schematic of the evolution processes from initial Sb atoms to final antimonene on Cu (111) substrates. Topographic STM images of 0.5 monolayer antimonene deposition on Cu (111) surface before (j) and (k) after annealing at 700 K. (l) High-resolution STM image of antimonene. (a–d) Reproduced with permission [29]. Copyright 2017, Wiley-VCH. (e–h) Reproduced with permission [30]. Copyright 2019, Wiley-VCH. (i–l) Reproduced with permission [36]. Copyright 2019, Wiley-VCH.

LEED pattern in the inset of Figure 2.6b, this film is a well-ordered single crystalline with a commensurate (1 × 1) lattice periodicity. The height of this film is about 2.8 Å, which is consistent with that of a monolayer antimonene (Figure 2.6c). Figure 2.6d shows an atomic-resolution STM image of antimonene, it is clearly seen that this epitaxial film has a buckled honeycomb lattice with a periodicity of 4.13 ± 0.02 Å. It is noted that monolayer antimonene grown on $PdTe_2$ was chemically inert to the air. This work first unveils the atomic-scale morphology of epitaxial monolayer β-antimonene.

Interestingly, besides the usual β-antimonene, α-phase antimonene can also be obtained on the 2D substrate by using MBE. Shi *et al.* controllably synthesized micro-sized monolayer α-antimonene on the 2D T_d-WTe_2 substrate [30]. Due to the inert surface and the compatibility with α-antimonene, T_d-WTe_2 was considered as a suitable substrate to achieve the freestanding epitaxial growth of monolayer α-antimonene. As illustrated in Figure 2.6e, α-antimonene takes a BP-like puckered honeycomb structure, which is different from a buckled honeycomb structure of β-antimonene. Figure 2.6f shows the topographic STM image of α-antimonene, it can be seen that Sb forms atomically flat islands on the T_d-WTe_2 with no apparent defects, and the height of the Sb island is around 0.78 nm (Figure 2.6g). The atomic-resolution STM image (Figure 2.6h) revealed a reconstructed √2 √2 lattice of α-antimonene with lattice constants of 0.62 × 0.63 nm. Monolayer α-antimonene possesses a linearly dispersed metallic band crossing the Fermi level, resulting in a high electrical conductivity. Large-scale multilayer antimonene was also grown by controlling the amount of Sb, which showed ultrahigh stability in air.

In the meantime, MBE was employed to grow antimonene on different metal substrates as well, including Ge (111), Ag (111), Pd (111), Cu (110), and Cu (111) [33–36]. Niu *et al.* fabricated single-crystalline monolayer antimonene on Cu (111) and Cu (110) substrates *via* MBE [36]. When Sb atoms were deposited on two substrates at low coverage, surface alloys were first formed, including √3 × √3 R30°-Sb on Cu (111) and c (2 × 2)-Sb on Cu (110). After increasing the coverage to 1 monolayer followed by postannealing, two atomic types of monolayer antimonene were then formed. Schematic of epitaxial growth of Sb atoms on Cu (111) substrates was shown in Figure 2.6i, where Sb atoms were evaporated from a Knudsen cell onto clean Cu (111) surfaces held at room temperature in an UHV chamber (2×10^{-10} mbar). The topographic STM image of 0.5 monolayer antimonene illustrated that Cu (111) surface was covered by short-range ordered single Sb atoms and their small clusters (Figure 2.6j). After annealing at 700 K, a variety of patches with dark holes (missing atoms) and black lines (misalignment of neighboring domains) were generated (Figure 2.6k). When increasing the coverage and postannealing, well-ordered monolayer antimonene was formed, showing a buckled honeycomb lattice and a larger lattice constant than the free-standing antimonene because of a 7.5% tensile strain (Figure 2.6l). But unlike the result of Cu (111) surface, threefold symmetric antimonene was formed on the twofold symmetric c (2 × 2) Sb-Cu (110) surface, inducing a 6.8% compressive strain. The generation of strain also caused the changes of band gaps of antimonene on two substrates.

2.3.4 Other Methods

In addition to the above three common synthesis methods, there are other methods can also be used to synthesize few- or multilayer antimonene nanosheets, nanoribbons, flakes, or Sb films [37–40]. Few-layer β-antimonene nanosheets were successfully prepared by the solution-phase synthesis method, where $SbCl_3$ solutions as the Sb source were reduced by oleylamine [37]. In this anisotropic growth, both dodecylthiol (DDT) and halide ions played an important role in the formation of β-antimonene. The synthesized β-antimonene nanosheets were single-crystalline with lateral size of 0.5–1.5 μm and thickness of 5–30 nm. Besides, multilayer antimonene nanoribbons were synthesized by the plasma-assisted process, where InSb (001) was selected as the substrates and provided Sb source for the formation of antimonene [38]. During the growth process, the indium at the surface of InSb occurred preferentially the nitridation under the action of N_2 plasma, which induced simultaneously the condensation of Sb atoms to form multilayer antimonene. The band gap of multilayer antimonene was opened because of the quantum confinement effect and the turbostratic stacking, then producing the orange light emission (610 nm). Like other 2D materials, antimonene flakes can also be grown through the CVD on SiO_2 substrates [39]. Sb powders were used as the Sb source, and Ar gas was the carrier gas. The growth of antimonene flakes began at 600°C and maintained for 10 min. The obtained flakes showed various shapes, including nanoribbon, hexagon, and trapezoid. The antimonene flakes were very stable in air even when heated by a hot plate below 250°C. Moreover, ultrathin Sb films were alternatively grown on the topological insulators (such as Bi_2Te_2Se and Bi_2Se_3) by a thermal effusion cell [40, 41].

2.4 Applications of Antimonene

2.4.1 Nonlinear Optics

As an excellent nonlinear absorption material, antimonene has broadband nonlinear optical response, high photothermal efficiency, strong Kerr nonlinearity, low saturable intensity, high two-photon absorption coefficient, and large cross-section, offering the potential nonlinear optical applications in all-solid-state lasers, fiber lasers, optical switchers, optical modulators, and optical thresholders [42–47].

Wang *et al.* first reported the passively Q-switched Nd^{3+} solid-state lasers using few-layer antimonene peeled off by the LPE as the saturable absorber

(SA) [42]. With antimonene/quartz as the passively Q-switcher, the laser operations were realized at 946, 1,064 nm in Nd:YAG crystal with pulse widths of 209, 129 ns and peak powers of 1.48, 1.77 W, while the laser emission of Nd:YVO_4 crystal generated at 1,342 nm with pulse width and peak power of 48 ns and 28.17 W, respectively. By fitting the saturable absorption results of nanosecond laser pulses, the saturable intensities of antimonene were obtained to be 0.43 MW/cm^2 at 532 nm and 0.53 MW/cm^2 at 1064 nm. In this work, the 1,342 nm Nd:YVO_4 laser achieved the best results, including the shortest pulse width (48 ns), the highest peak power (28.17 W), and the largest single pulse energy (1.36 μJ). The following drawbacks often occurred in passive Q-switching, including the degradation, failed saturable absorber, and inaccurate modification of repetition rate. To overcome the drawbacks in passive Q-switching, Wang *et al.* devised an actively Q-switched fiber laser with an antimonene-based all-optical modulator (AOM) as a Q-switcher [43]. The antimonene nanosheets fabricated by the LPE were deposited onto a 10 μm microfiber, which was then employed to devise an AOM with a fiber-type Michelsom interferometer (MI) due to the obvious photothermal effect of antimonene. The antimonene-based AOM exhibited a large phase modulation capacity with a conversion efficiency of 0.049 π mW^{-1}, and at the same time, it achieved the intensity modulation with large modulation depth of 25 dB. Also, this AOM showed a long-term stability with only 8.2% decline of phase conversion efficiency after 1 month (Figure 2.7a). The obtained active Q-switching pulses possessed a microsecond duration (the rise time constant of 3.2 ms) and tunable repetition rate ranging from 0.96 to 6.64 kHz (Figure 2.7b).

Using the strong Kerr nonlinearity of antimonene, Song *et al.* devised a new type of optical device based on few-layer antimonene (FLA)-decorated microfiber, which was operated both as an all-optical Kerr switcher and an all-optical wavelength converter [46]. The FLA-based Kerr switcher featured a high extinction ratio of 12 dB with a long-time stability, which could be used to realize the process of controlling light by light in optical communication systems (Figure 2.7c). In addition, by taking advantage of the four-wave mixing (FWM) effect, the designed FWM-based wavelength converter achieved a high conversion efficiency of 63 dB and converted efficiently the modulated radio frequency (RF) signals to the sidebands with a maximum frequency of ~18 GHz, which was a vital part in the optical signal processing (Figure 2.7d). In another work, Song *et al.* also employed FLA-decorated microfiber as an all-optical pulse thresholder to effectively suppress the noise in the transmission system, by which the signal to noise ratio (SNR) was largely improved (~10 dB) [47].

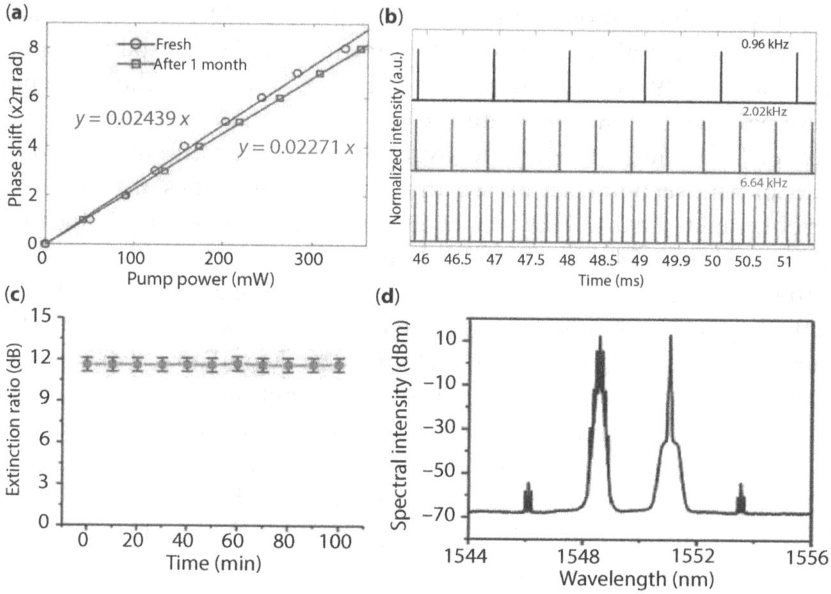

Figure 2.7 (a) Phase shift of the antimonene-based AOM as a function of pump power before and after 1 month. (b) Calculated active Q-switching pulse trains at different repetition rates (0.96, 2.02, 6.64 kHz). (c) The calculated extinction ratio of the FLA-based Kerr switcher. (d) FWM output spectra of the FLA-decorated microfiber with RF modulation (10 GHz). (a, b) Reproduced with permission [43]. Copyright 2019, Wiley-VCH. (c, d) Reproduced with permission [46]. Copyright 2018, Wiley-VCH.

2.4.2 Optoelectronic Device

For optoelectronic applications, materials are usually required to have proper band gaps in the visible region, high mobility, as well as excellent stability. By reducing bulk antimony to only one atomic layer, one can obtain a tunable band gap ranging from 0 to 2.28 eV together with high carrier mobility, which makes antimonene a suitable semiconductor in the field of optoelectronic device [11, 12].

Wang *et al.* first applied antimonene prepared by the modified LPE as a HTL in the perovskite solar cells (PVSCs) [21]. The small differences of the energy levels between antimonene and $CH_3NH_3PbI_3$ perovskite provided a driving force for the hole extraction. As shown in Figures 2.8a, b, the antimonene-HTL device exhibited a higher power conversion efficiency (J_{sc} = 14.6 mA cm^{-2}, EQE = 55%–60%) than the HTL-free device (J_{sc} = 11.2 mA cm^{-2}, EQE < 45%). Besides, the EQE-based integrated J_{sc} in the antimonene-HTL device (14.8 mA cm^{-2}) was also higher than that in the

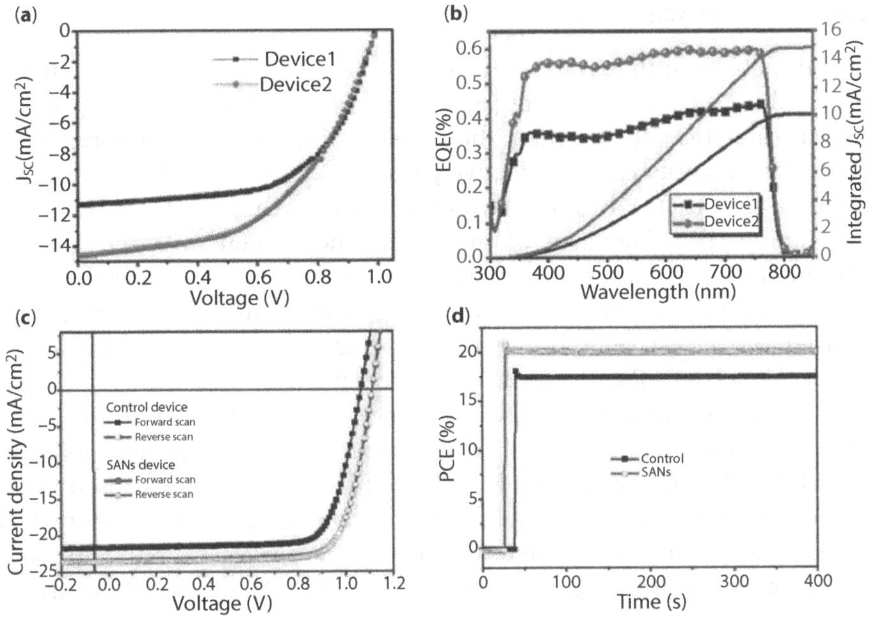

Figure 2.8 (a) Current-density-voltage (*J-V*) curves of devices without (Device 1) and with (Device 2) antimonene HTL. (b) EQE spectra and EQE-based integrated J_{sc} for Device 1 and Device 2. (c) Forward and reverse *J-V* scans of the best-performing control and SANs devices measured at different scan rates. (d) EQE spectra at maximum power output point for best-performing control and SANs devices. (a, b) Reproduced with permission [21]. Copyright 2018, Wiley-VCH. (c, d) Reproduced with permission [48]. Copyright 2018, Wiley-VCH.

HTL-free device (10.1 mA cm^{-2}). Afterwards, Zhang *et al.* also used LPE-exfoliated semiconductive antimonene nanosheets (SANs) as an ideal hole transfer and extraction layer to fabricate the planar inverted PVSCs [48]. In the constructed PVSCs, poly[bis(4-phenyl)(2,4,6-trimethylphenyl) amine] (PTAA) modified SANs were very compatible with $CH_3NH_3PbI_3$ perovskite, meanwhile SANs increased the grain size of $CH_3NH_3PbI_3$ and inhibited the ion migration. All the above factors indicated that SANs were expected to achieve efficient hole transfer from $CH_3NH_3PbI_3$ to the HTL. Compared with the control devices (without SANs layer), the SANs devices obtained higher power conversion efficiencies (PCEs) of 20.05%, V_{oc} of 1.115 V, J_{sc} of 23.46 mA cm^{-2}, and fill factor (FF) of 0.767 in forward scan direction (Figures 2.8c, d). In addition, antimonene quantum dots and sheets were employed as an effective photoactive material in the solar cells and organic photovoltaics as well [49, 50].

2.4.3 Electrocatalysis

Two-dimensional materials with atomic-scale thickness have more active sites and defects, as well as larger surface areas, which can achieve much higher catalytic activity than their bulk counterpart. Therefore, these 2D electrocatalysts can be effectively used to catalyze the CO_2 reduction reaction (CO_2RR), HER, and oxygen evolution reaction (OER) [51–53]. Combining the advantages of metals and non-metals, 2D semi-metals with high carrier concentration and short charge transfer channel exhibit enhanced catalytic activity for low catalytic threshold and efficient charge transfer [54]. Few-layer antimonene is expected to be employed as an active 2D electrocatalyst due to its semi-metallic nature.

Li *et al.* produced few-layer antimonene nanosheets (SbNSs) by cathodic exfoliation, then applied them into the electrocatalytic process of CO_2RR [55]. Compared with bulk Sb, few-layer SbNSs exposed more active edge sites, thus could obtain higher catalytic efficiency. Figure 2.9a shows the linear-sweep voltammetric (LSV) curves of bulk Sb- and SbNSs-modified working electrodes recorded in N_2- and CO_2-saturated

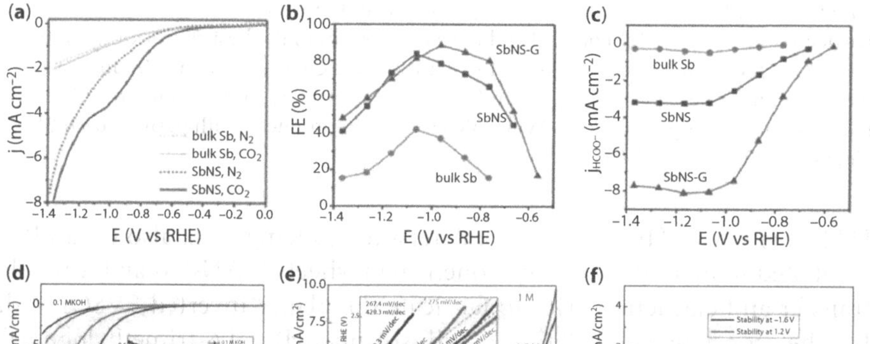

Figure 2.9 (a) LSV curves of bulk Sb- and SbNSs-modified glassy carbon electrodes in N_2- and CO_2-saturated 0.5 M $NaHCO_3$ solutions. Scan rate is 50 mVs^{-1}. FE (b) and partial current density (c) for formate of bulk Sb, SbNSs, and SbNS-G. Polarization curves and Tafel plots of antimonene nanosheets for HER (d) and OER (e) in KOH solutions with different concentrations (0.1, 0.2, 0.5, and 1 M). (f) Long-term stability of antimonene nanosheets in 0.5 M KOH. The bias potentials for HER and OER are −1.6 and 1.2 V, respectively. (a–c) Reproduced with permission [55]. Copyright 2017, Wiley-VCH. (d–f) Reproduced with permission [54]. Copyright 2019, American Chemical Society.

0.5 M $NaHCO_3$ solutions, it is observed that the SbNSs electrode exhibits a lower onset potential and a significantly enhanced catalytic current density compared to bulk Sb electrode. At the same time, the SbNSs electrode shows a broad peak with the center position at −1.06 V in the CO_2-saturated $NaHCO_3$ solutions. The reduction products are composed of H_2, CO, and formate. Furthermore, they also prepared few-layer antimonene nanosheet-graphene (SbNS-G) composites, then compared their catalytic activity with SbNSs, concluding that SbNS-G-modified electrode possessed a higher I_{mass} than SbNSs electrode. As illustrated in Figure 2.9b, the SbNS-G also presented an enhanced selectivity towards product formate at lower overpotentials, where the maximum faradaic efficiency (FE) was 88.5% at an overpotential of 0.87 V (potential of −0.96 V) higher than 84% of SbNSs (potential of −1.06 V). Moreover, the partial current density for formate of SbNS-G at −1.07 V is 1.5 and 16 times larger than that of SbNSs and bulk Sb, respectively (Figure 2.9c). The enhanced catalytic performance of SbNS-G is attributed to the strong electronic interaction between SbNS and graphene.

After that, Ren *et al.* adopted LPE-produced few-layer antimonene nanosheet as a metal-free electrocatalyst for full water splitting in alkaline condition [54]. The antimonene nanosheets showed great potentials in catalyzing both HER and OER, because of the semi-metallic nature together with increased active sites and larger surface area compared to bulk Sb. The lowest Tafel slope of antimonene nanosheets was 217.2 mV dec^{-1} for HER in 0.5 M KOH and 261.3 mV dec^{-1} for OER in 1 M KOH (Figures 2.9d, e). Additionally, bifunctional antimonene nanosheets presented a long-term stability in KOH solution, benefiting for durable and efficient electrocatalysis of full water splitting (Figure 2.9f).

2.4.4 Energy Storage

On the strength of high theoretical capacity of 660 mAh g^{-1}, increased active sites, and fast ion diffusion, few-layer antimonene is considered to be a promising electrode material in the energy storage applications, such as sodium-ion batteries (SIBs), lithium-ion batteries (LIBs), and supercapacitors [56–59].

Gu *et al.* obtained free-standing metallic Sb nanosheets *via* the LPE method, then fabricated SbNS-G hybrid films with tunable densities, which were subsequently applied to the SIBs [56]. The volumetric capacity of SbNS-G film (1.6 mg cm^{-2}) reached to 1,226 mAh cm^{-3} in the initial cycle (Figure 2.10a), which was higher than the value of previously reported Sb/C nanocomposites (100–300 mAh cm^{-3}) and graphene film (80 mAh cm^{-3}).

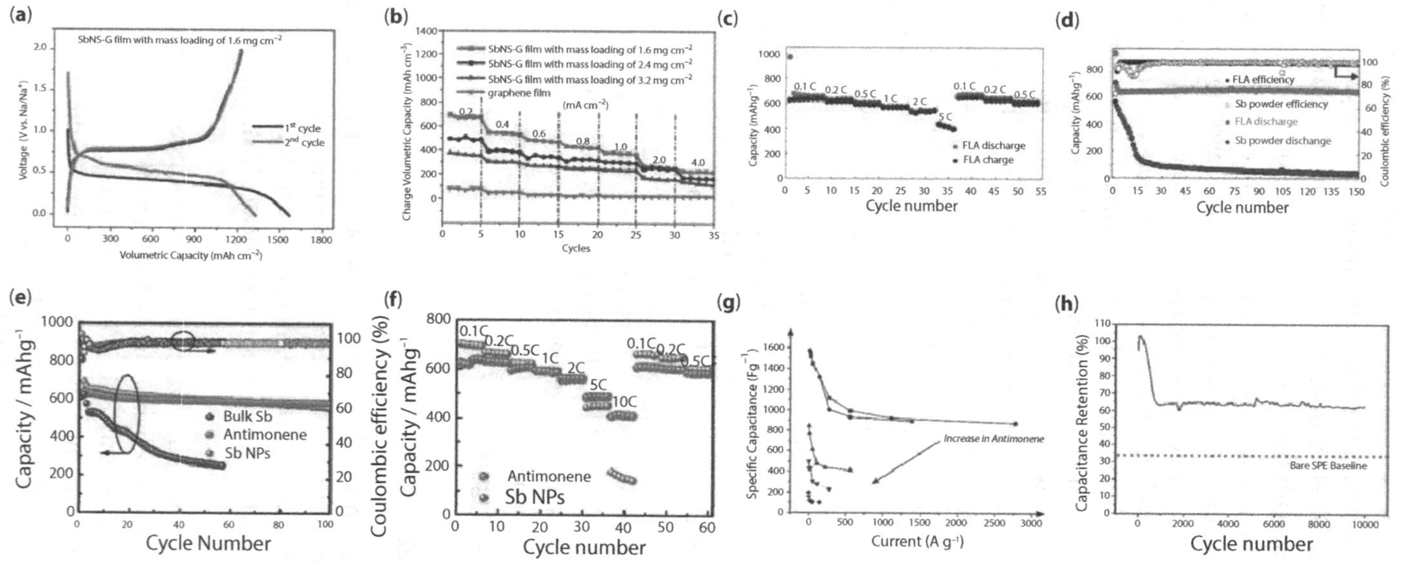

Figure 2.10 (a) The first and two charge/discharge cycles of the SbNS-G film at a current density of 0.1 mA cm^{-2}. (b) Rate capacities of SbNS-G films with different mass loading (1.6 to 3.2 mg cm^{-2}) and graphene film measured at various current densities (0.2 to 4.0 mA cm^{-2}). (c) Rate performances of FLA electrode. (d) Long-term charging/discharging cycles and Coulombic efficiency (CE) of FLA and bulk Sb at 0.5 C. (e) Cycling performances of bulk Sb, FLA and Sb NPs. (f) Rate capacities of FLA and Sb NPs. (g) Specific capacitance of FLA-modified SPE with different amounts of FLA. (h) Capacitance retention ratio versus cycle number of FLA-modified SPE. (a, b) Reproduced with permission [56]. Copyright 2017, Wiley-VCH. (c, d) Reproduced with permission [57]. Copyright 2018, American Chemical Society. (e, f) Reproduced with permission [58]. Copyright 2019, The Royal Society of Chemistry. (g, h) Reproduced with permission [59]. Copyright 2018, Wiley-VCH.

Furthermore, SbNS-G film also exhibited a high-rate capacity with both charge and discharge volumetric capacity of 216 mAh cm^{-3} at a current density of 4.0 mA cm^{-2} (Figure 2.10b), this result was superior to that of graphene film (13 mAh cm^{-3}). SbNS-G film showed simultaneously a good cycle performance for sodium (Na) storage, where the capacity was still stabilized at 110 mAh cm^{-3} after 100 cycles at high current density (4.0 mA cm^{-2}). Afterwards, Tian *et al.* studied the Na storage performance and mechanism of 2D FLA from experiments to theoretical calculations [57]. The FLA electrode delivered a charging capacity of 642 mAh g^{-1} at a current density of 0.1 C and a rate capacity of 429 mAh g^{-1} at 5 C, as well as a stable capacity of 620 mAh g^{-1} at 0.5 C after long-term charging/discharging cycles, which meant the utilization ratio of Sb atoms was as high as 93.9 % (Figures 2.10c, d). For the Na storage mechanism, they thought that FLA experienced a sodiation/desodiation process, where Na first reacted with Sb to form crystalline NaSb intermediate phase then to crystalline Na_3Sb in the sodiation process, conversely during the desodiation process, Na_3Sb first transformed to NaSb then to Sb with a small number of NaSb. This reversible crystalline-phase evolution was also accompanied by anisotropic volume expansion to achieve structural stability and electrochemical redox kinetics. In addition, FLA also exhibited high-rate lithium (Li) storage capabilities, even without the help of additives [58]. As shown in Figure 2.10e, the stable capacity of FLA after 100 cycles at 0.5 C was 584.1 mAh g^{-1} with CE of ~100%, which was higher than that of Sb nanoparticles (NPs) (552.3 mAh g^{-1}). FLA displayed a rate capacity of 488 mAh g^{-1} at 5 C, corresponding to 74% capacity retention of the value at 0.1 C, while the rate capacity of Sb NPs was 443 mAh g^{-1} at 5 C, and this value was only 67% of the initial capacity at 0.1 C (Figure 2.10f). The higher rate performance of FLA than Sb NPs was attributed to the preferential Li ion dispersion along in-plane channel in FLA.

Apart from the applications in Na/Li ion batteries, FLA can also be used as an electrode material in supercapacitors [59]. Compared to the bare screen-printed electrode (SPE), FLA-modified SPE showed improved energy storage capabilities. The specific capacitance was 1,578 F g^{-1} at a current of 14 A g^{-1}, and it retained 63%–65% of the initial capacitance value over 10,000 galvanostatic cycles (Figures 2.10g, h). Moreover, FLA-modified SPE demonstrated a low energy density of 20 mW h kg^{-1} and a high power density of 4.8 kW kg^{-1}.

2.4.5 Biomedicine

Due to the excellent optical, electronic, optoelectronic, physical, and chemical properties, antimonene can be well used in biomedicines, such as photothermal therapy (PTT), photodynamic therapy (PDT), photoacoustic imaging (PAI), photonic drug-delivery platform, and biosensor, etc. [60].

Superior photothermal performance and biocompatibility enable antimonene to be employed as effective photothermal agents (PTAs) for cancer therapy. Tao *et al.* synthesized highly-dispersed 2D antimonene quantum dots (AMQDs) by the LPE and explored their ability of PTT as novel PTAs [61]. To enhance the biocompatibility and stability of AMQDs in physiological medium, 1,2-Distearoyl-sn-glycero-3-phosphoethanolamine-N-[methoxy(poly ethylene glycol)] (DSPE-PEG) was then modified on their surface. From the photothermal heating curves shown in Figure 2.11a, PEG-modified AMQDs exhibited a concentration-dependent photothermal effect with temperature increment as high as 50°C. The obtained photothermal conversion efficacy (PTCE) (η) was 45.5%, which was higher than most of nanomaterials-based PTAs. The near-infrared (NIR) induced degradation of PEG-modified AMQDs in water was very rapid with only several minutes; this feature was superior

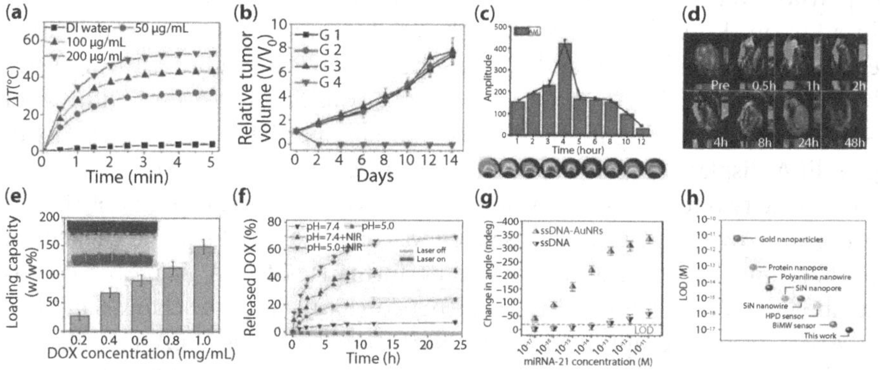

Figure 2.11 (a) Time-dependent photothermal heating curves of PEG-modified AMQDs with different concentrations under irradiation of NIR laser (808 nm, 2 Wcm^{-2}). (b) Relative tumor volumes after treatment with saline (G1), NIR (G2), PEG-modified AMQDs (G3), PEG-coated AMQDs +NIR (G4). (c) Intracellular PA signal versus time. (d) Photoacoustic tomography (PAT) images of tumor at various time points after injection of BSA-AMNFs. (e) DOX concentration-dependent loading capacities on AM-PEG NSs. (f) *In vitro* release profiles of DOX from AM-PEG/DOX NSs under different treatments (pH and pH+NIR). (g) The SPR angle change *versus* miRNA concentration of ssDNA-AuNRs and ssDNA. (h) Comparison of the LOD of antimonene-based SPR sensor with other reported miRNA sensors. (a, b) Reproduced with permission [61]. Copyright 2017, Wiley-VCH. (c, d) Reproduced with permission [63]. Copyright 2019, Wiley-VCH. (e, f) Reproduced with permission [64]. Copyright 2018, Wiley-VCH. (g, h) Reproduced with permission [65]. Copyright 2019, Nature Publishing Group.

to that of BP. Both *in vitro* and *in vivo* therapeutic studies, PEG-modified AMQDs showed potential biocompatibility. When combined with NIR irradiation, PEG-modified AMQDs achieved effective killing of cancer cells and complete ablation of tumors by PTT (Figure 2.11b). In another work, antimonene nanoparticles (~55 nm) cloaked with cell membrane (CmNPs) was used as PTAs in PTT and PDT for high-performance cancer theranostics, benefiting from the enhanced permeability and retention (EPR) effect, superior stability, and good tumor targeting capacity [62]. *In vitro* phototherapy, the viability of 4T1 cells in the CmNP group was lower than that in the antimonene nanoparticle (mNP) group. Upon NIR irradiation, the $^1O_2$ generation and the dead cells were more in the CmNPs group, confirming an effective phototherapy effect. *In vivo* phototherapy, the temperature in the CmNP group (57.1°C) was higher than that in the mNP group (48.3°C). The $^1O_2$ signal was brighter and the tumor was completely inhibited with few abnormalities in the CmNP group under NIR irradiation, again confirming the superior theranostic performance.

Furthermore, LPE-produced antimonene nanoflakes (AMNFs) were used as ideal contrast agents for PAI, due to the excellent photoacoustic (PA) performance [63]. AMNFs showed high molar extinction coefficient of 2.24×10^9 in the NIR window (800 nm), high photothermal conversion capability with temperature increment up to 207.9°C after 30 min NIR irradiation, and high PTCE of 42.36%, indicating superior performance of AMNFs to other widely used PAI contrast agents, such as organic small molecule dyes, gold nanomaterials. AMNFs also exhibited a dose-dependent and wavelength-independent PA signal with the lowest detection limit of 0.6986 μg mL^{-1}. Besides, the low thermal conductivity (15.1 $Wm^{-1}K^{-1}$) and high interfacial thermal conductivity of AMNFs resulted in a stronger PA signal. The photon emission was absent in AMNFs, ensuring the maximum photothermal conversion. To increase the biocompatibility, bovine serum albumin (BSA) was coated on the surface of AMNFs. The above excellent PA performance of BSA-AMNFs allowed sensitively monitoring the cellular uptake process *in vitro* and clearly imaging the small volume tumor *in vivo* (Figures 2.11c, d).

Tao *et al.* adopted PEG-modified antimonene (AM-PEG) nanosheets (NSs) as a robust photonic drug-delivery platform for multimodal-imaging-guided cancer theranostics [64]. AM-PEG NSs first showed considerable PTCE of 41.8% and high loading capacity of doxorubicin (DOX) molecules with the largest value of 150% (Figure 2.11e). Then, the release of DOX molecules presented both pH-sensitive and NIR-sensitive characteristics, which enabled the platform to be designed simpler (Figure 2.11f). *In vitro*, AM-PEG/DOX NSs exhibited photo-induced deep penetrating ability into tumor spheroids and enhanced intracellular uptake. Therefore, combined with NIR irradiation, AM-PEG/DOX NSs killed almost all the tumor cells, indicating an efficient therapeutic effect. *In vivo*, AM-PEG/DOX NSs continued to accumulate and retain in the tumor sites, then completely eliminated the tumors and inhibited their growth, attributing to a synergistic effect of PTT and chemotherapy. With good biocompatibility and potential degradability, AM-PEG NSs were very promising drug-delivery platform for cancer theranostics.

Having more delocalized 5s/5p orbitals, antimonene can interact strongly with single-stranded DNA (ssDNA), allowing for ultrasensitive detection of cancer-associated MicroRNA (miRNA) molecules. Xue *et al.* developed an antimonene-based surface plasmon resonance (SPR) sensor and performed trace attomolar-level quantitative detection of miRNA-21 and miRNA-155 [65]. Benefiting from the signal amplification of gold nanorods (AuNRs), this antimonene-based SPR biosensor showed an extremely low limit of detection (LOD) of 10 aM to miRNA-21, which was the highest sensitivity comparing to other reported miRNA sensors (Figures 2.11g, h).

2.4.6 Magneto-Optic Storage

Nonmagnetic doping in antimonene can induce the sp-electron ferromagnetic order with a high Curie temperature (T_c). Tang *et al.* synthesized fluorinated antimonene (F-antimonene) *via* an electrochemical exfoliation and synchronous fluorination (EESF) method and explored its magnetic properties [66]. The T_c value of F-antimonene was obtained as high as 717 K, and it was one of the highest T_c value determined experimentally for 2D ferromagnetic materials. Moreover, the saturation magnetization of F-antimonene (~0.1 emu g^{-1}) was obtained at the F/Sb ratio of 0.20. Based on the first-principles calculations considering the spin-orbit coupling (SOC) effect of heavy Sb element, the long-range ferromagnetic order of F-antimonene resulted from the F adatom and its low-carrier-density sp-electron-polarized impurity subbands. Robust high-T_c ferromagnetism in few-layer F-antimonene offers potentials in the magneto-optic storage and magnetic transport applications.

2.5 Conclusion and Outlook

Antimonene, a newly discovered 2D materials, has remarkable optical, electronic, optoelectronic, physical, and chemical properties, including environmental stability, wideband saturable adsorption, tunable band gap, high carrier mobility, generous active sites, superior photothermal performance, and biocompatibility, which enable it to have potential applications in many fields. In traditional 2D materials, changes in the number of layers usually affect only the size of band gap. What makes 2D antimonene unique is its strong spin-orbit coupling and the semi-metallic nature of its bulk form, allowing for adjustable electronic behaviors by varying the layer numbers. These significant features drive researchers to prepare antimonene to better understand its properties and develop its practical applications. Like other 2D materials, mechanical exfoliation was the first method to peel off high-quality mono- or few-layer antimonene at ambient conditions, and then LPE was used to realize the scalable production of antimonene. The epitaxial growth technique was further explored to deposit large-area few- or even single-layer antimonene on different substrates. Other methods were also developed to prepare antimonene with various morphs, such as solution-phase synthesis, plasma-assisted growth, and CVD. Future research can focus on achieving more effective exfoliation, since the successful preparation of antimonene was the premise of verifying the theoretically predictive properties and exploring its practical applications.

Recent works have unveiled some appealing properties of antimonene in experiments and further demonstrated its practical applications. Few-layer antimonene showed broadband nonlinear optical response and high photothermal efficiency (48%), allowing for the applications in lasers, optical switchers, optical modulators, and optical thresholders, which were important in the next-generation optical communication. With a tunable band gap (0–2.28 eV) and high carrier mobility, antimonene was employed as an effective photoactive material in the PVSCs and organic photovoltaics. Generous active sites and large surface area in antimonene made it a suitable catalyst in the catalytic process of CO_2RR, HER, and OER. On the basis of high theoretical capacity (660 mAh g^{-1}) along with fast ion diffusion, few-layer antimonene was considered to be an excellent electrode material in SIBs, LIBs, and supercapacitors. As an effective photothermal and contrast agent, antimonene can be used in the PTT, PDT, and PAI for high-performance cancer theranostics. Based on the current status of research on the properties and applications of antimonene, it is expected to enhance the performance by functionalization or modification and then develop its potential applications in more other fields.

To summarize, it is particularly important to predict the properties of antimonene more accurately in theory. Moreover, effective methods are needed to exfoliate or synthesize 2D antimonene, and then, high performance and diverse applications are expected to be realized in future.

References

1. Novoselov, K.S., Geim, A.K., Morozov, S.V., Jiang, D., Zhang, Y., Dubonos, S.V., Grigorieva, I.V., Firsov, A.A., Electric field effect in atomically thin carbon films. *Science*, 306, 666, 2004.
2. Son, Y.W., Cohen, M.L., Louie, S.G., Half-metallic graphene nanoribbons. *Nature*, 444, 347, 2006.
3. Radisavljevic, B., Radenovic, A., Brivio, J., Giacometti, V., Kis, A., Single-layer MoS_2 transistors. *Nat. Nanotechnol.*, 6, 147, 2011.
4. Li, L., Yu, Y., Ye, G.J., Ge, Q., Ou, X., Wu, H., Feng, D., Chen, X., Zhang, Y., Black phosphorus field-effect transistors. *Nat. Nanotechnol.*, 9, 372, 2014.
5. Tran, V., Soklaski, R., Liang, Y., Yang, L., Layer-controlled band gap and anisotropic excitons in few-layer black phosphorus. *Phys. Rev. B*, 89, 235319, 2014.
6. Wang, X., Jones, A.M., Seyler, K.L., Tran, V., Jia, Y., Zhao, H., Wang, H., Yang, L., Xu, X., Xia, F., Highly anisotropic and robust excitons in monolayer black phosphorus. *Nat. Nanotechnol.*, 10, 517, 2015.
7. Favron, A., Gaufres, E., Fossard, F., Phaneuf-L'Heureux, A.L., Tang, N.Y., Levesque, P.L., Loiseau, A., Leonelli, R., Francoeur, S., Martel, R., Photooxidation and quantum confinement effects in exfoliated black phosphorus. *Nat. Mater.*, 14, 826, 2015.
8. Zhang, S., Yan, Z., Li, Y., Chen, Z., Zeng, H., Atomically thin arsenene and antimonene: Semimetal-semiconductor and indirect-direct band-gap transitions. *Angew. Chem. Int. Ed.*, 54, 3112, 2015.
9. Pizzi, G., Gibertini, M., Dib, E., Marzari, N., Iannaccone, G., Fiori, G., Performance of arsenene and antimonene double-gate MOSFETs from first principles. *Nat. Commun.*, 7, 12585, 2016.
10. Chen, X., Yang, Q., Meng, R., Jiang, J., Liang, Q., Tan, C., Sun, X., The electronic and optical properties of novel germanene and antimonene heterostructures. *J. Mater. Chem. C*, 4, 5434, 2016.
11. Zhang, S., Xie, M., Li, F., Yan, Z., Li, Y., Kan, E., Liu, W., Chen, Z., Zeng, H., Semiconducting group 15 monolayers: A broad range of band gaps and high carrier mobilities. *Angew. Chem. Int. Ed.*, 55, 1666, 2016.
12. Wang, X., Song, J., Qu, J., Antimonene: From experimental preparation to practical application. *Angew. Chem. Int. Ed.*, 58, 1574, 2019.
13. Butler, S.Z., Hollen, S.M., Cao, L., Cui, Y., Gupta, J.A., Gutiérrez, H.R., Heinz, T.F., Hong, S.S., Huang, J., Ismach, A.F., Johnston-Halperin, E., Kuno, M., Plashnitsa, V.V., Robinson, R.D., Ruoff, R.S., Salahuddin, S., Shan, J., Shi, L., Spencer, M.G., Terrones, M., Windl, W., Goldberger, J.E., Progress, challenges, and opportunities in two-dimensional materials beyond graphene. *ACS Nano*, 7, 2898, 2013.
14. Ares, P., Aguilar-Galindo, F., Rodriguez-San-Miguel, D., Aldave, D.A., Diaz-Tendero, S., Alcami, M., Martin, F., Gomez-Herrero, J., Zamora, F.,

Mechanical isolation of highly stable antimonene under ambient conditions. *Adv. Mater.*, 28, 6332, 2016.
15. Castellanos-Gomez, A., Buscema, M., Molenaar, R., Singh, V., Janssen, L., van der Zant, H.S.J., Steele, G.A., Deterministic transfer of two-dimensional materials by all-dry viscoelastic stamping. *2D Mater.*, 1, 011002, 2014.
16. Ares, P., Zamora, F., Gomez-Herrero, J., Optical identification of few-layer antimonene crystals. *ACS Photonics*, 4, 600, 2017.
17. Abellan, G., Ares, P., Wild, S., Nuin, E., Neiss, C., Miguel, D.R., Segovia, P., Gibaja, C., Michel, E.G., Gorling, A., Hauke, F., Gomez-Herrero, J., Hirsch, A., Zamora, F., Noncovalent functionalization and charge transfer in antimonene. *Angew. Chem. Int. Ed.*, 56, 14389, 2017.
18. Ciesielski, A. and Samori, P., Graphene via sonication assisted liquid-phase exfoliation. *Chem. Soc. Rev.*, 43, 381, 2014.
19. Gibaja, C., Rodriguez-San-Miguel, D., Ares, P., Gomez-Herrero, J., Varela, M., Gillen, R., Maultzsch, J., Hauke, F., Hirsch, A., Abellan, G., Zamora, F., Few-layer antimonene by liquid-phase exfoliation. *Angew. Chem. Int. Ed.*, 55, 14345, 2016.
20. Lin, W., Lian, Y., Zeng, G., Chen, Y., Wen, Z., Yang, H., A fast synthetic strategy for high-quality atomically thin antimonene with ultrahigh sonication power. *Nano Res.*, 11, 5968, 2018.
21. Wang, X., He, J., Zhou, B., Zhang, Y., Wu, J., Hu, R., Liu, L., Song, J., Qu, J., Bandgap-tunable preparation of smooth and large two-dimensional antimonene. *Angew. Chem. Int. Ed.*, 57, 8668, 2018.
22. Gusmao, R., Sofer, Z., Bousa, D., Pumera, M., Pnictogen (As, Sb, Bi) nanosheets for electrochemical applications are produced by shear exfoliation using kitchen blenders. *Angew. Chem. Int. Ed.*, 56, 14417, 2017.
23. Ares, P., Palacios, J.J., Abellán, G., Gómez-errero, J., Zamora, F., Recent progress on antimonene: A new bidimensional material. *Adv. Mater.*, 30, 1703771, 2018.
24. Koma, A., van der Waals epitaxy—A new epitaxial growth method for a highly lattice-mismatched system. *Thin Solid Films*, 216, 72, 1992.
25. Ji, J., Song, X., Liu, J., Yan, Z., Huo, C., Zhang, S., Su, M., Liao, L., Wang, W., Ni, Z., Hao, Y., Zeng, H., Two-dimensional antimonene single crystals grown by van der Waals epitaxy. *Nat. Commun.*, 7, 13352, 2016.
26. Sun, X., Lu, Z., Xiang, Y., Wang, Y., Shi, J., Wang, G.C., Washington, M.A., Lu, T.M., van der Waals epitaxy of antimony islands, sheets, and thin films on single-crystalline graphene. *ACS Nano*, 12, 6100, 2018.
27. Liu, X., Sun, G., Chen, P., Liu, J., Zhang, Z., Li, J., Ma, H., Zhao, B., Wu, R., Dang, W., Yang, X., Dai, C., Tang, X., Chen, Z., Miao, L., Liu, X., Li, B., Liu, Y., Duan, X., High-performance asymmetric electrodes photodiode based on Sb/WSe_2 heterostructure. *Nano Res.*, 12, 339, 2018.
28. Lei, T., Liu, C., Zhao, J., Li, J., Li, Y., Wang, J., Wu, R., Qian, H., Wang, H., Ibrahim, K., Electronic structure of antimonene grown on Sb_2Te_3 (111) and Bi_2Te_3 substrates. *J. Appl. Phys.*, 119, 015302, 2016.

29. Wu, X., Shao, Y., Liu, H., Feng, Z., Wang, Y., Sun, J., Liu, C., Wang, J., Liu, Z., Zhu, S., Wang, Y., Du, S., Shi, Y., Ibrahim, K., Gao, H., Epitaxial growth and air-stability of monolayer antimonene on $PdTe_2$. *Adv. Mater.*, 29, 1605407, 2017.
30. Shi, Z., Li, H., Yuan, Q., Song, Y., Lv, Y., Shi, W., Jia, Z., Gao, L., Chen, Y., Zhu, W., Li, S., Van der Waals heteroepitaxial growth of monolayer Sb in a puckered honeycomb structure. *Adv. Mater.*, 31, 1806130, 2019.
31. Chen, K., Lee, L.M., Chen, H.A., Sun, H., Wu, C., Chen, H., Lin, K., Tseng, Y.C., Kaun, C.C., Pao, C.W., Lin, S.Y., Multi-layer elemental 2D materials antimonene, germanene and stanene grown directly on molybdenum disulfides. *Semicond. Sci. Technol.*, 34, 105020, 2019.
32. Fortin-Deschenes, M., Jacobberger, R.M., Deslauriers, C.A., Waller, O., Bouthillier, E., Arnold, M.S., Moutanabbir, O., Dynamics of antimonene-graphene van der Waals growth. *Adv. Mater.*, 31, 1900569, 2019.
33. Fortin-Deschenes, M., Waller, O., Mentes, T.O., Locatelli, A., Mukherjee, S., Genuzio, F., Levesque, P.L., Hebert, A., Martel, R., Moutanabbir, O., Synthesis of antimonene on germanium. *Nano Lett.*, 17, 4970, 2017.
34. Shao, Y., Liu, Z.L., Cheng, C., Wu, X., Liu, H., Liu, C., Wang, J.O., Zhu, S.Y., Wang, Y.Q., Shi, D.X., Ibrahim, K., Sun, J.T., Wang, Y.L., Gao, H.J., Epitaxial growth of flat antimonene monolayer: A new honeycomb analogue of graphene. *Nano Lett.*, 18, 2133, 2018.
35. Jałochowski, M. and Krawiec, M., Antimonene on Pb quantum wells. *2D Mater.*, 6, 045028, 2019.
36. Niu, T., Zhou, W., Zhou, D., Hu, X., Zhang, S., Zhang, K., Zhou, M., Fuchs, H., Zeng, H., Modulating epitaxial atomic structure of antimonene through interface design. *Adv. Mater.*, 31, 1902606, 2019.
37. Peng, L., Ye, S., Song, J., Qu, J., Solution-phase synthesis of few-layer hexagonal antimonene nanosheets via anisotropic growth. *Angew. Chem. Int. Ed.*, 58, 9891, 2019.
38. Tsai, H.S., Chen, C.W., Hsiao, C.H., Ouyang, H., Liang, J.H., The advent of multilayer antimonene nanoribbons with room temperature orange light emission. *Chem. Commun.*, 52, 8409, 2016.
39. Wu, Q. and Song, Y.J., The environmental stability of large-size and single-crystalline antimony flakes grown by chemical vapor deposition on SiO_2 substrates. *Chem. Commun.*, 54, 9671, 2018.
40. Kim, S.H., Jin, K.H., Kho, B.W., Park, B.G., Liu, F., Kim, J.S., Yeom, H.W., Atomically abrupt topological p-n junction. *ACS Nano*, 11, 9671, 2017.
41. Kim, S.H., Jin, K.H., Park, J., Kim, J.S., Jhi, S.H., Yeom, H.W., Topological phase transition and quantum spin hall edge states of antimony few layers. *Sci. Rep.*, 6, 33193, 2016.
42. Wang, M., Zhang, F., Wang, Z., Wu, Z., Xu, X., Passively Q-switched Nd^{3+} solid-state lasers with antimonene as saturable absorber. *Opt. Express*, 26, 4085, 2018.

43. Wang, Y., Huang, W., Wang, C., Guo, J., Zhang, F., Song, Y., Ge, Y., Wu, L., Liu, J., Li, J., Zhang, H., An all-optical, actively Q-switched fiber laser by an antimonene-based optical modulator. *Laser Photonics Rev.*, 13, 1800313, 2019.
44. Zhang, F., Wang, M., Wang, Z., Han, K., Liu, X., Xu, X., Excellent nonlinear absorption properties of β-antimonene nanosheets. *J. Mater. Chem. C*, 6, 2848, 2018.
45. Liu, G., Zhang, F., Wu, T., Li, Z., Zhang, W., Han, K., Xing, F., Man, Z., Ge, X., Fu, S., Single-and dual-wavelength passively mode-locked erbium-doped fiber laser based on antimonene saturable absorber. *IEEE Photonics J.*, 11, 1503011, 2019.
46. Song, Y., Chen, Y., Jiang, X., Liang, W., Wang, K., Liang, Z., Ge, Y., Zhang, F., Wu, L., Zheng, J., Ji, J., Zhang, H., Nonlinear few-layer antimonene-based all-optical signal processing: Ultrafast optical switching and high-speed wavelength conversion. *Adv. Opt. Mater.*, 6, 1701287, 2018.
47. Song, Y., Liang, Z., Jiang, X., Chen, Y., Li, Z., Lu, L., Ge, Y., Wang, K., Zheng, J., Lu, S., Ji, J., Zhang, H., Few-layer antimonene decorated microfiber: Ultra-short pulse generation and all-optical thresholding with enhanced long term stability. *2D Mater.*, 4, 045010, 2017.
48. Zhang, F., He, J., Xiang, Y., Zheng, K., Xue, B., Ye, S., Peng, X., Hao, Y., Lian, J., Zeng, P., Qu, J., Song, J., Semimetal-semiconductor transitions for monolayer antimonene nanosheets and their application in perovskite solar cells. *Adv. Mater.*, 30, 1803244, 2018.
49. Zhang, C., Li, Y., Zhang, P., Qiu, M., Jiang, X., Zhang, H., Antimonene quantum dot-based solid-state solar cells with enhanced performance and high stability. *Sol. Energy Mater. Sol. Cells*, 189, 11, 2019.
50. Wang, Z., Zhang, R., Zhao, M., Wang, Z., Wei, B., Zhang, X., Feng, S., Cao, H., Liu, P., Hao, Y., Wang, H., Xu, B., Pennycook, S.J., Guo, J., High-yield production of stable antimonene quantum sheets for highly efficient organic photovoltaics. *J. Mater. Chem. A*, 6, 23773, 2018.
51. Voiry, D., Salehi, M., Silva, R., Fujita, T., Chen, M., Asefa, T., Shenoy, V.B., Eda, G., Chhowalla, M., Conducting MoS_2 nanosheets as catalysts for hydrogen evolution reaction. *Nano Lett.*, 13, 6222, 2013.
52. Yu, X., Prevot, M.S., Guijarro, N., Sivula, K., Self-assembled 2D WSe_2 thin films for photoelectrochemical hydrogen production. *Nat. Commun.*, 6, 7596, 2015.
53. Zhang, K., Jin, B., Park, C., Cho, Y., Song, X., Shi, X., Zhang, S., Kim, W., Zeng, H., Park, J.H., Black phosphorene as a hole extraction layer boosting solar water splitting of oxygen evolution catalysts. *Nat. Commun.*, 10, 2001, 2019.
54. Ren, X., Li, Z., Qiao, H., Liang, W., Liu, H., Zhang, F., Qi, X., Liu, Y., Huang, Z., Zhang, D., Li, J., Zhong, J., Zhang, H., Few-layer antimonene nanosheet: A metal-free bifunctional electrocatalyst for effective water splitting. *ACS Appl. Energy Mater.*, 2, 4774, 2019.
55. Li, F., Xue, M., Li, J., Ma, X., Chen, L., Zhang, X., MacFarlane, D.R., Zhang, J., Unlocking the electrocatalytic activity of antimony for CO_2 reduction by

two-dimensional engineering of the bulk material. *Angew. Chem. Int. Ed.*, 56, 14718, 2017.
56. Gu, J., Du, Z., Zhang, C., Ma, J., Li, B., Yang, S., Liquid-phase exfoliated metallic antimony nanosheets toward high volumetric sodium storage. *Adv. Energy Mater.*, 7, 1700447, 2017.
57. Tian, W., Zhang, S., Huo, C., Zhu, D., Li, Q., Wang, L., Ren, X., Xie, L., Guo, S., Chu, P.K., Zeng, H., Huo, K., Few-layer antimonene: Anisotropic expansion and reversible crystalline-phase evolution enable large-capacity and long-life Na-ion batteries. *ACS Nano*, 12, 1887, 2018.
58. Gao, Y., Tian, W., Huo, C., Zhang, K., Guo, S., Zhang, S., Song, X., Jiang, L., Huo, K., Zeng, H., Tailoring natural layered β-phase antimony into few layer antimonene for Li storage with high rate capabilities. *J. Mater. Chem. A*, 7, 3238, 2019.
59. Martínez-Periñán, E., Down, M.P., Gibaja, C., Lorenzo, E., Zamora, F., Banks, C.E., Antimonene: A novel 2D nanomaterial for supercapacitor applications. *Adv. Energy Mater.*, 8, 1702606, 2018.
60. Tao, W., Kong, N., Ji, X., Zhang, Y., Sharma, A., Ouyang, J., Qi, B., Wang, J., Xie, N., Kang, C., Zhang, H., Farokhzad, O.C., Kim, J.S., Emerging two-dimensional monoelemental materials (Xenes) for biomedical applications. *Chem. Soc. Rev.*, 48, 2891, 2019.
61. Tao, W., Ji, X., Xu, X., Islam, M.A., Li, Z., Chen, S., Saw, P.E., Zhang, H., Bharwani, Z., Guo, Z., Shi, J., Farokhzad, O.C., Antimonene quantum dots: Synthesis and application as near-infrared photothermal aagents for effective cancer therapy. *Angew. Chem. Int. Ed.*, 56, 11896, 2017.
62. Lu, G., Lv, C., Bao, W., Li, F., Zhang, F., Zhang, L., Wang, S., Gao, X., Zhao, D., Wei, W., Xie, H.Y., Antimonene with two-orders-of-magnitude improved stability for high-performance cancer theranostics. *Chem. Sci.*, 10, 4847, 2019.
63. Yu, J., Wang, X.H., Feng, J., Meng, X., Bu, X., Li, Y., Zhang, N., Wang, P., Antimonene nanoflakes: Extraordinary photoacoustic performance for high-contrast imaging of small volume tumors. *Adv. Healthcare Mater.*, 8, 1900378, 2019.
64. Tao, W., Ji, X., Zhu, X., Li, L., Wang, J., Zhang, Y., Saw, P.E., Li, W., Kong, N., Islam, M.A., Gan, T., Zeng, X., Zhang, H., Mahmoudi, M., Tearney, G.J., Farokhzad, O.C., Two-dimensional antimonene-based photonic nanomedicine for cancer theranostics. *Adv. Mater.*, 30, 1802061, 2018.
65. Xue, T., Liang, W., Li, Y., Sun, Y., Xiang, Y., Zhang, Y., Dai, Z., Duo, Y., Wu, L., Qi, K., Shivananju, B.N., Zhang, L., Cui, X., Zhang, H., Bao, Q., Ultrasensitive detection of miRNA with an antimonene-based surface plasmon resonance sensor. *Nat. Commun.*, 10, 28, 2019.
66. Tang, X., Hu, L., Fan, T., Zhang, L., Zhu, L., Li, H., Liu, H., Liang, J., Wang, K., Li, Z., Ruan, S., Zhang, Y., Fan, D., Chen, W., Zeng, Y., Zhang, H., Robust above-room-temperature ferromagnetism in few-layer antimonene triggered by nonmagnetic adatoms. *Adv. Funct. Mater.*, 29, 1808746, 2019.

3

Synthesis and Properties of Graphene-Based Materials

U. Naresh[1], N. Suresh Kumar[2], D. Baba Basha[3], Prasun Benerjee[4], K. Chandra Babu Naidu[4]*, R. Jeevan Kumar[1], Ramyakrishna Pothu[5] and Rajender Boddula[6]

[1]*Department of Physics, Sri Krishnadevaraya University, Anantapur, India*
[2]*Department of Physics, JNTUA, Anantapur, India*
[3]*Department of Physics, College of Computer and Information Sciences, Majmaah University, Al'Majmaah, Saudi Arabia*
[4]*Department of Physics, GITAM Deemed to be University, Bangalore, India*
[5]*College of Chemistry and Chemical Engineering, Hunan University, Changsha, China*
[6]*CAS Center for Excellence in Nanoscience, National Center for Nanoscience and Technology, Beijing, China*

Abstract

The graphene is a highly preferable material among all carbon materials and it acquired considerable significance owing to its extraordinary properties. However, this chapter offers a special focus on the preparation methods, structure, properties and applications of graphene, graphene oxide (GO), and graphene-related materials (GRM). In particular, the GRMs like carbon nanotubes provided lot of benefits to medicinal fields. The importance of this aspect is described in brief. Apart from these, the thermal stability (physical property) is illustrated via the phonon behavior after applying the thermal energy to the GRM system. Furthermore, the electronic properties are discussed at length due to having an exposure of outstanding electronic and transport behavior of charges. In addition, an aspect like how these graphene-based materials are equipped as advanced transistors, FETS, inverters, and integrated circuits is explained.

Keywords: Graphene, synthesis, GRMS, electrical properties, physical properties

**Corresponding author*: chandrababu954@gmail.com

Inamuddin, Rajender Boddula, Mohd Imran Ahamed and Abdullah M. Asiri (eds.)
Monoelements: Properties and Applications, (57–72) © 2020 Scrivener Publishing LLC

3.1 Introduction

In last decades, the researchers focused on their attention on the carbon-based nanomaterials for the applications in energy storage and transformation of energy from one form to another form owing to its significant physical properties such as structural, mechanical, thermal, magneto-electrical, optical, and physicochemical properties [1]. In addition to these properties, the carbon materials have several advantages such as feasibility and structural variation by adding the organic and inorganic elements [2]. Among these carbon materials, few carbon derivatives like carbon nanotubes and graphene are the most studied and reliable materials being this kind of materials has the distinct applications in the field of solar cells, Li-ion batteries supercapacitors and fuel cells which are the part of the energy conversion and storage devices [3].

In particular, the graphene (Figure 3.1) and its related derivatives are the most impacted materials in the material science due to its unique properties. This is due to the fact that graphene is the only material which has excessive mechanical strength and electrical properties as two- and three-dimensional orientations [4]. The graphene structure oriented as monolayer, two-dimensional layer, and mixer of metals and graphene layers were studied extensively. In addition to this layered structure, the graphene exists as sheets, flakes which are derived from the bulk graphite and by depositing some inorganic substrates [5]. On the other hand, the materials which are matrix of the graphene are called Graphene-related materials (GRMs), which were used in biosensors, medical applications

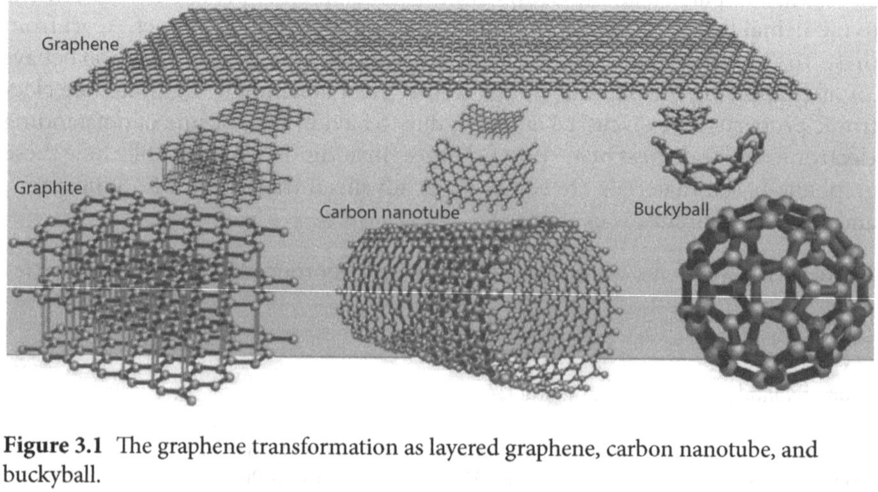

Figure 3.1 The graphene transformation as layered graphene, carbon nanotube, and buckyball.

and various industrial, technological applications due to remarkable properties [6].The remarkable properties means that the graphene has shows well biocompatibility with the surrounding natural resources, safety in the uses in various biological applications. The invention of novel GRMs is a complicated phenomena towards synthesis and characterization techniques to check physical properties such as stiffness, hydrophobicity, morphological sensing properties, etc., with doping of graphene in GRMs [7]. Graphene oxide is one of most studied material in the researchers in the materials science because this kind of materials offers much more chemistries with stability; hence, these materials used widely in electronic field, memory storage devices, energy-storing devices, etc. [8].

Generally, the graphene is a family of carbon. Graphene is the thinnest material although it conducts electricity and heat. Also, the graphene can also be acts as very good transparent material. However, the development of graphene happens in several steps [9]. The three most significant carbon-based materials are found such as zero-dimensional fullerenes of carbon, one-dimensional carbon nanotubes, and two-dimensional graphene. Its applications have become extremely larger from the past two decades, after successful production of graphene in 2004 by Andre Geem and his colleagues as it became the "rising star" in material science [10].

The history of graphene dates from 1859, when the scientist Benjamin Bordie synthesized a lamellar structure of graphene oxide by reacting graphite with concentrated nitric acid and Potassium chlorate. Afterward, the graphene was prepared by Brodie's method in 1919 from graphene oxide. When the band theory of graphite was given by the Wallace at 1947, then by based on band theory, the researchers Ruess and Vogt investigated conductivity of the graphene thin layer in 1948. Steps directed to prepare graphene using similar adopted procedure for carbon nanotubes yielded thicker graphite films. The researchers Lewis and Ubbeloholde isolated carbon atom which had maximum conductivity from the graphene plane which is extracted from graphite in 1960. In 1961, Boehm Hanns-Peter, group of scientists, synthesized graphene sheets and recorded structural analysis with the help of the XRD and morphological studies by TEM first time accurately [11].

Even though, the importance of the graphene came in 2010 when the researchers Konstantin Novoselov and Andre Geim from Manchester won the prestigious Noble prize in the year 2010 for the study on graphene for their innovative work on 2D graphene material. In this investigation, the graphene extracted from a portion of graphite which is the part of ordinary pencils. They obtained atom size carbon array using common strip with just size one atom thickness [12].

The graphene gets importance as it exhibits better physical and chemical properties. For example, the graphene can be a stronger even it has size in nano size. The electronic properties of graphene are remarkable as it performs very good conductions than the other materials, and the heat conductions are well [13]. Whereas, the optical properties are extreme, as showing the complete transparent nature although it exhibits high density in itself. The researchers found that it is impossible to have many important properties in one material with stability like graphene.

However, the two-dimensional graphene materials become the new group of materials which have the exclusive properties targeted by researchers. The quantum mechanics principles are possible with the results of experiments of the graphene. The different kind of applications is enhanced by creating new graphene materials. Suppose the electronic properties of graphene that is in manufacture of electronic system of graphene transistors are effective than the usual silicon transistors [14].

The practical applications such as touch screens, touché pads, solar cells, and other electronic systems can easily developed with the graphene well transparent and electrical properties [15]. On the other hand, the graphene mixed polymers can turn new functional hybrid materials into perfect conductors of electricity with thermal and mechanical strength. Hence, the graphene mixed materials are found as in the good peripherals of newly fabricated electrical devices and systems because the graphene attains thinness and elasticity.

3.2 Applications

The high thermal conduction and high current density enables larger energy creation, less optical absorption offered by its maximum transparent nature. Low noise offers application in electrical communications but it has disadvantage that it is not semiconductor as it is not having energy bandgap. As of course from last decades, the commonly used materiel in electronic application is silicon at room temperature by exhibiting narrow energy bandgap [16]. Usual silicon for electron devices has lower electron mobility than the graphene electron mobility. However, silicon is the most used material in electronic application in place of memory card, motherboard fabrication, and various electronic fabrication systems, etc., due to its unique properties. On the other hand, the graphene high mechanical strength makes the requirement to replace the silicon. The flexibility and stiffness of the graphene make it a good choice for the wearable, comfortable electronic technology. The high transmittance of graphene sheets makes the user in touch screens, touch pads, and similar devices. The transistor is the other important and backbone part of electronic system. The graphene mixed or graphene-related transistors are the new trend in these days due to its remarkable properties [17]. As maintained previous discussion, the high electron-hole mobility and zero bandgap make the fabrication special and logic field effective transistors (FETs); this type of FETS is called as graphene field-effect transistor (GFETS).

In optoelectronic field, considerable attention has attained by the graphene because of its extreme optical properties with wide electron and vacancies ratio and has good absorptive properties of white light. These properties were helped to made advanced optoelectronic devices like photodetector optical waveguide, light-emitting diode (LED), frequency modulator, and photonic devices [18]. The graphene materials become the choice of the materials science which can be used in antenna fetch between low-frequency ranges to high-frequency range. The appropriate dielectric material is crucial for antenna because the thickness and dielectric constant of the material strongly effect on antennas resonant frequency. From literature, there are many materials including silicon and aluminum, and its mixtures are used as substrates of graphene patch in the direction of the antenna-based materials. For example, silicon nitride with graphene patch is synthesized as smooth surface and reflective and its electric insulate behavior makes the effective material to antenna and communication engineering applications. Also, aluminum oxide with graphene substrate enhanced the dielectric resistance and decrease mechanical friction makes a broad range of applications.

In the field of biomedical applications, graphene-GRMs show remarkable properties which makes possible to create new advances by exploring biosensors, drug delivery, and engineering devices of tissue engineering technology. The GRMs' main properties such as safety, biocompatibility, and suitable mechanical properties make the GRM the most suitable material for biological application. Nevertheless, the biocompatibility, dose rate, surface properties, and utilization of the GRMs are dependent on the size, shape, mechanical strength, purity, etc. On the other hand, the material graphene oxide (GO) is used a widely material due to its reactivity with the stability at room temperature. GRMs have been widely reported for the use of induce reactive oxygen species (ROS) injected into cells, resulting in lipid and DNA injure, activating cell death. GRMs such as hydrophilic oxidized and nano-sized are diffused into biological cells and go through membranes. In addition to GRMs and GOs, the layered structure graphene is the one of the choice for distinguished material to study human lung cells detections and also several nano-size GRMs platelets used to put down on the lung surface to treat the lunge-related injuries [19].

3.3 Structure

The graphene turns into "wonder material" due its remarkable properties. The structure of the material (Figures 3.2 and 3.3) is made of carbon atoms and oriented as flat shape. The atomic number is six for carbon and the configuration is fallowed as $1s^2\ 2s^2\ 2p^2$ so as it has four electrons valance in the p-orbital. All the carbon atoms in graphene are arranged in a standard hexagon pattern C-C bond length 1.42 A⁰. As it connecting with sp^2 bonding of each atom, it forms the honeycomb crystal lattice [54, 38]. This two-dimensional graphene consists of sp^2 bonds. The surprising properties

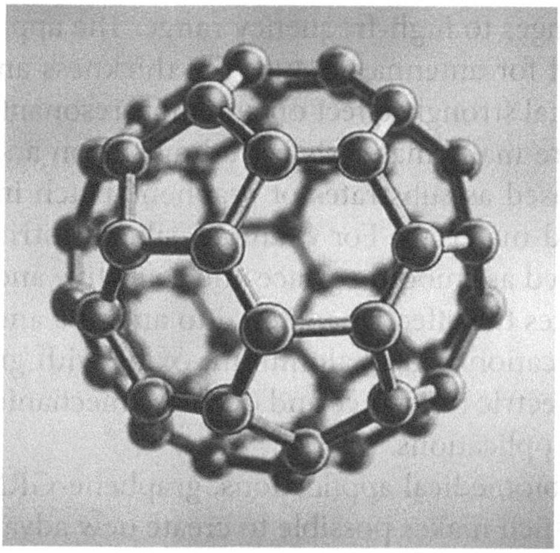

Figure 3.2 Chemical structure.

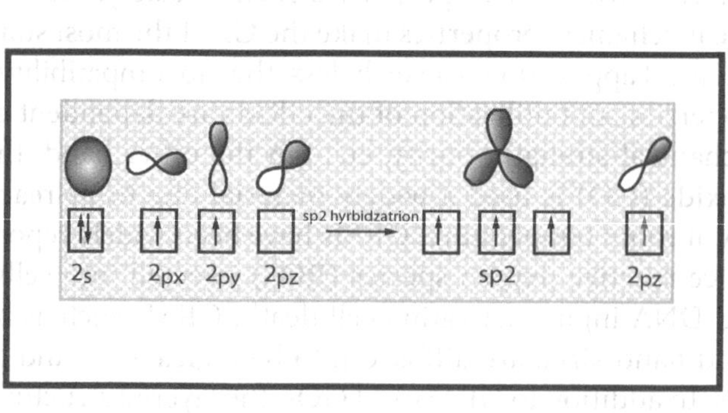

Figure 3.3 Orbital structure.

of graphene are initiated from the 2p orbital; the structure forms p state band that delocalize on a sheet of carbons containing graphene. Hence, the graphene obtains high thermal stability and lower effective mass and carries high mobility even it shows high transparent nature. The graphene also shows advanced properties than the other common materials. The need of high-quality graphene in bulk and nano form, because of this fact, several researchers in literature proposed different types of synthesis techniques to made graphene and its derivatives with distinguished physical and chemical properties. Graphene has great chemistry because its electronic structure reacts as an oxidizer or a reducing agent. Graphene arranges carbon atoms. Electron conductivity, weak interactions between graphene layers are due to the adsorption of atoms are caused by sp^2-hybridized bonds of the carbon.

3.3.1 Graphene-Related Materials

There are several polymers and nano-particle (metal, metal oxide, semiconductor) alloys that depend on the exclusive properties of graphene. The properties of graphene have a similar mechanism of CNT (Carbon nanotubes) which are those have high thermal and electrical properties and large surface area (2,620 m^2/g) [20] due to its dimensional variation. Mechanical exfoliation of graphite not suitable for large-scale production, the chemical-oxidation of graphene-oxide from the graphite provides an easier technique to attain large quantities of GO that can be chemically and electrochemically reduced or thermally into graphene. The given GOs and rGOs have huge production opportunities to explore the flat structure of carbon by incorporating polymers and nanoparticles in graphene mixtures.

The graphene sheets are fabricated by exfoliation of the graphene layers as single or group of carbon atoms. Plenty of the GO layers were formed by the oxidation of carbon, hydrogen, and oxygen molecules. The removal of oxygen groups might have responsible for the formation of layered shape of the GO sheets. These chemically acquired graphene sheets are called as reduced graphene oxide sheets, functionalized graphene [21]. The structural changes happened because of the covalent bonded functional groups. The morphological properties studied by the researchers founded as some destructed areas which were arises oxygen derivatives. One report suggests that the honeycomb lattice is preserved in graphene in GO is displacement of carbon atoms. Thus, GO can be described as a random distribution of oxidized regions, with most carbon atoms and oxygen were functionalized groups and non-oxidized regions preserving as sp^2 hybridization. From the last decade, GO and rGO are hot topics in the research and development of graphene, particularly in relation to mass applications of graphene [22].

3.3.2 Synthesis Techniques

The fabrication or synthesis of graphene depends on the process, extraction of graphene layers with respect to required size and application. Various methods have been invented for early-stage production thin graphic films. In 1970–1971, the carbon atoms were precipitated as thin graphite layers in the metallic surface [23]. Then, in 1975, some layers of graphite were synthesized by chemical methods of decomposition on a single crystal platinum surface, but due to lack of graphene characterization techniques or limited possible applications. In the late 90s, Roof and colleagues attempted to separate thin graphic flakes on SiO_2 surfaces by mechanically rubbing the sample in the HOPG (Highly Oriented Pyrolytic Graphite). This was followed by the use of a similar synthesis technique that was introduced by the Kim *et al.* and reported electrical properties at 2005 [24–28].

But, it is true that the rapid progress in graphene research began after Geum and colleagues first published their partitioning work on SiO_2 graphene and its flow on the surface. Since the invention of graphene in 2004, the synthesis methods for fabricating various thin graphic films have been developed and some are membrane graphene. The micro- and nano-size graphene layers formation from the graphite depend on the synthesis type. The mechanical exfoliation of graphite as graphene is the commonly used synthesis type; this method yields high-quality graphene and mass production as to synthesize layer-level graphene (Figure 3.4).

In recent years, various methods for graphene synthesis have been established; nevertheless, mechanical exfoliation, chemical exfoliation, and thermal chemical vapor synthesis of deposition (CVD) are the most conventional techniques. There are other methods of nanotube and microwave unzipping, such as synthesis. AFM cantilever is capable of gin harmonization in spite of mechanical exfoliation a few layers of graphene, the process thickness limit varies from 10 g of graphene to comparable 30-layer graphene [29, 30].

3.3.3 Mechanical Exfoliation of Graphene Layers

The graphene in the form of sheets cannot change their ultimate properties. The graphene structurally contains vendor-wall bonds and interactions between the layers. The methods of exfoliation by mechanical process

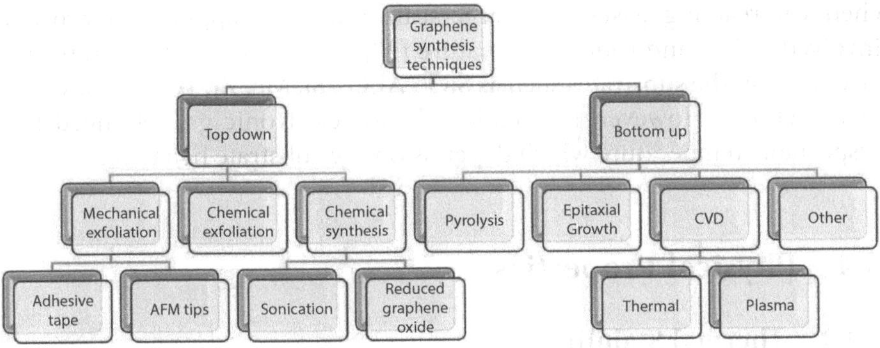

Figure 3.4 Synthesis methods of graphene.

break the weak interaction bands. The liquid phase exfoliation is another famous method to exfoliate graphene layers from graphite [31].

3.3.4 Chemical Vapor Deposition of Graphene Layers

Chemical Vapor Deposition (CVD) is another alternative technique of producing layers of graphene. The graphene layers are extracted from the graphite by catalytic or targeting of carbon-atoms on surface of the metal. The CVD is one of the best methods to produce single-layer graphene from the last decade.

3.3.5 Hummer Method of Graphene

The Hummer technique is the one of the chemical-based method for the preparation of graphene. The process of this method includes as the graphite as carbon precursor and it was oxidized through the sulfuric acid. The water and oxygen peroxide cause the divide the GO sheets and removal of impurities done by cleaning the HCl till it gets natural pH value [32]. Graphite sheets are also synthesized in the electrolytic solution of ionic liquids such as deionized water and imidazolium using a constant voltage to the graphite anode and cathode (electrochemical method).

3.3.6 Plasma-Enhanced Chemical Vapor Deposition of Graphene Layers

In this method, the graphene layers were created in the void chamber through plasma utilization on the substrate surface by compressing chemical reacting gases. The large-scale industrial application process involved at low and room temperature [33, 34]. The depositing graphene layers are on the substrates such as SiO_2, Al_2O_3, Si, Mo, Zr, W, Ta, Cu, stainless steel, etc. However, the semiconductor electronic gadgets need the respectable temperature which depends on the substrate behavior.

3.4 Physical Properties

3.4.1 Thermal Stability

It has gained immense consideration of graphene in recent years due to its thermal conductivity and stability. It is well-known fact that the thermal energy activated from the phonon vibration and transportation of phonons in the lattice of unit cell of crystallite. It has been found that phonons can be controlled in the same way as controlled of electrons, and researchers proposed the concepts for phonon devices such as rectifiers, logic gates, and thermal transistors. The thermal rectifiers have low thermal resistance in a particular direction but have high thermal resistance in the opposite direction. Researchers have committed much effort to explain the theoretical phenomena and to initialize the theoretical studies as the thermal stability devices of graphene. The thermal energy can also attribute due to the impurities and structural defects occurred in the graphene substrate or composite solid crystals. Engineering of the graphene-related devices requires good heat transportations and thermal stability; these properties are arranged by the graphene [35, 36].

An effective strategy for achieving high calibration of graphene nanoparticles (GNS) in GNS/Cu alloys follows spark plasma sintering by vacuum filtration method. The coated graphene nanosheets of the graphene copper alloys possess the essential anisotropic thermal properties. Yang *et al.* [37] investigated the thermal properties of graphene which incorporated by the nitrogen nano-ribbons at different temperatures at different concentrations of nitrogen (0.11% to 2.31%) using molecular equilibrium dynamics. The results achieved as nitrogen defective nature inhibited because of the graphene incorporation results a significant reduction in thermal conductivity; this fact expressed that the atoms of nitrogen reduce the thermal conductivity with respect to temperature. In addition, thermal recirculation is detected, then the ration doped with nitrogen increases with increasing. Finally, to confirm the accuracy of thermal rectification, they computed power-spectra of the phonons of TNGN at 300 K with doped nitrogen concentrations of 0.11% and 2.31% [37]. The nitrogen-doped graphene nanoribbons show thermal consistency as increasing from low temperature to high temperature, it shows as increases first and then decreases. The temperature sensitivity is smaller than pure graphene that influences the consistency of the graphene devices which are doped with nitrogen that can make a large difference in temperature. High frequency is often used to generate phonons and reduce the small thermal correction caused by point defects in small amounts of nitrogen atoms (with a rating of 0.11%). In addition, the enlargement in the asymmetric interface area caused by growth of nitrogen-atoms prevents high-frequency tangential phonons and increases variability in the flow of heat in two opposite directions, indicating the strong upward drift in the correction.

Zhang *et al.* [41] investigated for functional thermal stable materials for energy storage devices such as Li-ion batteries. In graphene/polymer composites, the polymer chain and phonon dispersion take longer to transfer heat energy through the polymer chain than graphene. They prepared polydimethylsiloxane (PDMS)/graphene tubes combination synthesized and thermal properties were investigated. The heat transformation in the prepared composite is possible with graphene ordered structure. Therefore, the alloys show the largest 10%–80 % increase heat transfer functionality of alloys that is satisfactory for application.

3.4.2 Electronic Properties

The electronic properties of the graphene composite depend on the structure of the graphene. The graphene structure looks as honeycomb net-type structure as shown Figure 3.5. Each of the graphene nets has combined through the three s- and p-bonds of carbon atoms stand for the sp^2-hybridization plane and then forms a diagonal p-bond with the shoulder, which passes through the entire graphene whereas p-orbital electron is free to move on the electron of graphene. This behavior causes to the

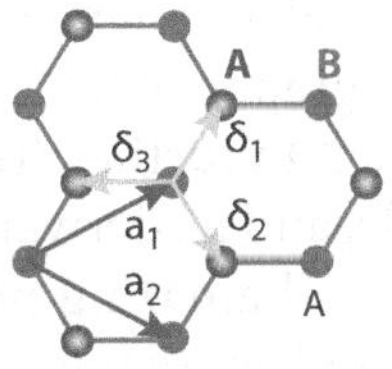

Figure 3.5 Organic structure of graphene.

well conductance at room temperature and the theoretical background can be explained through the quantum mechanics. We know that the free electron in conduction bond is responsible for the electron conductivity, whereas the absence of the electrons makes hole concentration in valence bond. Hence, the electron-hole pairs formed through the conduction band electrons and valance band holes which can move freely are also called free charge carriers [42]. The graphene enhances the electrical properties for electronic devices such as Li-ion batteries, supercapacitors, solar cells, etc. Theoretically, electrons in graphene are obeyed Fermi-Dirac statistics, which makes graphene investigation results obeyed principles of quantum electrodynamics [43]. Meanwhile, from the last decade, the carbon mixers or graphene dependent transistors like FET (field effect transistor) gained much attention by the researchers [44]. However, as we compared to the mobility of charge carriers in carbon nanotubes, the charge carriers in graphene show large mobility. The controller of electron conductance that FET is fabricated through the graphene composite by the researchers Merrick *et al.* [45–50]. The researchers found that saturated current and wide conductance make it ideal for electronic application (figure 7 in reference [45]) that exhibits the device outcomes after testing. In 2011, the earlier graphene integrated circuit was engineered. The bipolar behavior of graphene composite is demonstrated successfully. They reported that estimation of signal power is directly proportionate to input RF-signal through PRF ¼ 12 dBm for the LO-signal greater than PLO 20 dBm. Further, the small amplitude values arise due to the attraction of high-frequency noise by the canal inductance [50–59]. The graphene has outstanding chemical stability as well as thermal stability. The formed graphene heterostructures and H-BN complex show excellent electrical properties, which allows them to have broader scope in the fabrication of nano-electronic devices.

3.5 Conclusions

In summary, the wonder material graphene remarkable properties are discussed in brief. The history of graphene that is how it is born and how the graphene develops from ancient to further. The synthesis methods to prepare graphene, graphene oxide, and its derivatives discussed into the point. The graphene derivatives showed amazing properties like thermal stability with its blender (organic and inorganic elements) as showing perfect mechanical strength by having honeycomb structure. The electronic properties are conferred from the graphene family. The electronic properties of graphene and the derivatives of graphene equipped as advanced transistors, FETS, inverters, and integrated circuits discussed by various researchers are communicated briefly.

References

1. Shang, N.G., Papakonstantinou, P., McMullan, M., Catalyst-free efficient growth, orientation and bio sensing properties of multilayer graphene nano flake films with sharp edge planes. *Adv. Funct. Mater.*, 18, 21, 3506–3514, 2008.
2. Obraztsov, A.N., Zolotukhin, A.A., Ustinov, A.O., Volkov, A.P., Svirko, Y., Jefimovs, K., DC discharge plasma studies for nano structured carbon CVD. *Diam. Relat. Mater.*, 12, 3–7, 917–920, 2003.
3. Chhowalla, M., Teo, K.B.K., Ducati, C., Rupesinghe, N.L., Amaratunga, G.A.J., Ferrari, A.C., Roy, D., Robertson, J., Milne, W.I., Growth process conditions of vertically aligned carbon nanotubes using plasma enhanced chemical vapor deposition. *J. Appl. Phys.*, 90, 5308, 2001.
4. Wang, J.J., Zhu, M.Y., Outlaw, R.A., Zhao, X., Manos, D.M., Holloway, B.C., Synthesis of carbon nanosheets by inductively coupled radio-frequency plasma enhanced chemical vapor deposition. *Carbon*, 42, 14, 2867–2872, 2004.
5. Moseler, M., Gumbsch, P., Casiraghi, C., Ferrari, A.C., Robertson, J., The ultrasmoothness of diamond-like carbon surfaces. *Science*, 309, 5740, 1545–1548, 2005.
6. Lee, J., RF performance of pre-patterned locally embedded back- gate graphene device, in: *2010 International Electron Devices Meeting–Technical Digest*, IEEE, New York, 2010.
7. Van Bommel, A.J., Crombeen, J.E., Van Tooren, A., LEED and Auger electron observations of the SiC(0001) *surface. Surf. Sci.*, 48, 2, 463–472, 1975.
8. Dresselhaus, M.S., Jorio, A., Saito, R., Characterizing graphene, graphite, and carbon nanotubes by raman spectroscopy. *Condens. Matter Phys.*, 1, 1, 89–108, 2010.
9. Saha, S.K., Waghmare, U.V., Krishnamurthy, H.R., Sood, A.K., Phonons in few-layer graphene and interplanar interaction: A first-principles study. *Phys. Rev. B.*, 78, 16, 165421, 2008.
10. Casiraghi, C., Ferrari, A.C., Ohr, R., Flewitt, A.J., Chu, D.P., Robertson, J., Dynamic roughening of tetrahedral amorphous carbon. *Phys. Rev. Lett.*, 91, 226104, 2003.
11. Forbeaux, I., Themlin, J.-M., Charrier, A., Thibaudau, F., Debever, J.-M., Solid-state graphitization mechanisms of silicon carbide 6H SiC polar faces. *Appl. Surf. Sci.*, 162, 406, 2000.
12. Hass, J., Jeffrey, C.A., Feng, R., Li, T., Li, X., Song, Z., Berger, C., de Heer, W.A., First, P.N., Conrad, E.H., Highly-ordered graphene for two dimensional electronics. *J. Appl. Phys.*, 92, 2479, 2002.

13. Emtsev, K.V., Speck, F., Seyller, T., Ley, L., Riley, J.D., Interaction, growth, and ordering of epitaxial graphene on SiC surfaces: A comparative photoelectron spectroscopy study. *Phys. Rev. B*, 77, 155303, 2008.
14. Kedzierski, J., Hsu, P.-L., Healey, P., Wyatt, P., Keast, C., Sprinkle, M., Berger, C., Heer, W.D., Epitaxial graphene transistors on SiC substrates. *IEEE Trans. Electron Devices*, 55, 2078, 2008.
15. Schwierz, F., Graphene transistors. *Nat. Nanotechnol.*, 5, 487, 2010.
16. Bhuyan, M.S.A., Uddin, M.N., Islam, M.M., Bipasha, F.A., Hossain, S.S., Synthesis of graphene. *Int. Nano Lett.*, 6, 2, 65–83, 2016.
17. Zhang, Y.B., Small, J.P., Pontius, W.V., Kim, P., Fabrication and electric-field dependent transport measurements of meso scopic graphite devices. *Appl. Phys. Lett.*, 86, 073104, 2005.
18. Reina, A., Jia, X.T., Ho, J., Nezich, D., Son, H., Bulovic, V., Mildred Dresselhaus, S., Kong, J., Large area, few-layer graphene films on arbitrary substrates by chemical vapor deposition. *Nano Lett.*, 9, 1, 30–35, 2009.
19. Jiao, L.Y., Wang, X.R., Diankov, G., Wang, H.L., Dai, H.J., Facile synthesis of high-quality graphene nanoribbons. *Nat. Nanotechnol.*, 5, 5, 321–325, 2010.
20. Kosynkin, D.V., Higginbotham, A.L., Sinitskii, A., Lomeda, J.R., Dimiev, A., Price, B.K., Tour, J.M., Longitudinal unzipping of carbon nanotubes to form graphene nanoribbons. *Nature*, 458, 7240, 872–876, 2009.
21. Zhu, Y., Murali, S., Cai, W., Li, X., Suk, J.W., Potts, J.R., Ruoff, R.S., Graphene and Graphene Oxide: Synthesis, Properties, and Applications. *Adv. Mater.*, 22, 35, 3906–3924, 2010.
22. Jiao, L.Y., Zhang, L., Wang, X.R., Diankov, G., Dai, H.J., Narrow graphene nano ribbons from carbon nanotubes. *Nature*, 458, 7240, 877–880, 2009.
23. Xin, G., Hwang, W., Kim, N., Cho, S.M., Chae, H., A graphene sheet exfoliated with microwave irradiation and interlinked by carbon nanotubes for high-performance transparent flexible electrodes. *Nanotechnology*, 21, 40, 405201, 2010.
24. Zhang, W., Cui, J., Tao, C.-A., Wu, Y., Li, Z., Ma, L., Wen, Y., Li, G., A Strategy for producing pure single-layer graphene sheets based on a confined self-assembly approach. *Angew. Chem. Int. Ed.*, 48, 32, 5864–5868, 2009.
25. Wang, X., You, H., Liu, F., Li, M., Wan, L., Li, S., Li, Q., Xu, Y., Tian, R., Yu, Z., Xiang, D., Cheng, J., Large-scale synthesis of few-layered graphene using CVD. *J. Chem. Vap. Depos.*, 15, 1–3, 53–56, 2009.
26. Wang, Y., Chen, X., Zhong, Y., Zhu, F., Loh, K.P., Large area, continuous, few-layered graphene as anodes in organic photovoltaic devices. *Appl. Phys. Lett.*, 95, 063302, 2009.
27. Pandey, D., Scanning probe microscopy study of exfoliated oxidized graphene sheets. *Surf. Sci.*, 602, 9, 1607–1613, 2008.
28. Paredes, J.I., Atomic force and scanning tunneling microscopy imaging of graphene nanosheets derived from graphite oxide. *Langmuir*, 25, 10, 5957–5968, 2009.

29. Pei, S., Direct reduction of graphene oxide films into highly conductive and flexible graphene films by hydrophilic acids. *Carbon*, 48, 4466–4474, 2010.
30. Saha, S.K., Bhaumik, S., Maji, T., Mandal, T.K., Pal, A.J., Solution-processed reduced graphene oxide in light-emitting diodes and photovoltaic devices with the same pair of active materials. *RSC Adv.*, 4, 67, 35493–35499, 2014.
31. Schniepp, H.C., Functionalized single graphene sheets derived from splitting graphite oxide. *J. Phys. Chem. B*, 110, 17, 8535–8539, 2006.
32. Shin, H.-J., Efficient reduction of graphite oxide by sodium boro hydride and its effect on electrical conductance. *Adv. Funct. Mater.*, 19, 1987–1992, 2009.
33. Chen, Z.H., Lin, Y.M., Rooks, M.J., Avouris, P., Graphene nano-ribbon electronics. Phys. E-Low-Dimens. *Syst. Nanostruct.*, 40, 2, 228–232, 2007.
34. Dresselhaus, M.S., Jorio, A., Saito, R., Characterizing graphene, graphite, and carbon nanotubes by raman spectroscopy. *Condens. Matter Phys.*, 1, 1, 89–108, 2010.
35. Wang, L., Meric, I., Huang, P.Y., Gao, Q., Gao, Y., Tran, H., Taniguchi, T., Watanabe, K., Campos, L.M., Muller, D.A., One-dimensional electrical contact to a two-dimensional material. *Science*, 342, 614–617, 2013.
36. Saha, S.K., Waghmare, U.V., Krishnamurthy, H.R., Sood, A.K., Phonons in few layer graphene and inter planar interaction: A first-principles study. *Phys. Rev. B Condens. Matter*, 78, 16, 165421, 2008.
37. Ponomarenko, L.A., Geim, A.K., Zhukov, A.A., Jalil, R., Morozov, S.V., Novoselov, K.S., Grigorieva, I.V., Hill, E.H., Cheianov, V.V., Fal Ko, V.I., Tunable metal-insulator transition in double-layer graphene hetero structures. *Nat. Phys.*, 7, 12, 958–961, 2011.
38. Evans, D.A., Mcglynn, A.G., Towlson, B.M., Gunn, M., Jones, D., Jenkins, T.E., Winter, R., Poolton, N.R.J., Determination of the optical band-gap energy of cubic and hexagonal boron nitride using luminescence excitation spectroscopy. *J. Phys. Condens Matter*, 20, 7, 075233, 2008.
39. Watanabe, K., Taniguchi, T., Kanda, H., Direct-bandgap properties and evidence for ultraviolet lasing of hexagonal boron nitride single crystal. *Nat. Mater*, 3, 6, 404–409, 2004.
40. Yang, B., Li, D., Qi, L., Li, T., Yang, P., Thermal properties of triangle nitrogen-doped graphene nanoribbons. *J. Phys. Lett. A*, 383, 12, 1306–1311, 2019.
41. Zhang, Y.-F., Ren, Y.-J., Guo, H.-C., Bai, S-I., Enhanced thermal properties of PDMS composites containing vertically aligned graphene tubes. *Appl. Therm. Eng.*, 150, 840–848, 2019.
42. Javey, A., Kim, H., Brink, M., Wang, Q., Ural, A., Guo, J., Mcintyre, P., Mceuen, P., Lundstrom, M., Dai, H., Javey, A., High- dielectrics for advanced carbon nanotubes transistors and logic gates. *Nat. Mater*, 1, 4, 241–246, 2002.
43. Javey, A., Jing, G., Farmer, D.B., Qian, W., Yenilmez, E., Gordon, R.G., Lundstrom, M., Dai, H., Self-aligned ballistic molecular transistors and electrically parallel nanotube arrays. *Nano Lett.*, 4, 7, 1319–1322, 2004.
44. Klinke, C., Chen, J., Afzali, A., Avouris, P., Charge transfer induced polarity switching in carbon nano tube transistors. *Nano Lett.*, 5, 5, 555–558, 2006.

45. Kang, S.J., Kocabas, C., Ozel, T., Shim, M., Pimparkar, N., Alam, M.A., Rotkin, S.V., Rogers, J.A., High-performance electronics using dense, perfectly aligned arrays of single-walled carbon nanotubes. *Nat. Nanotechnol.*, 2, 4, 230–236, 2007.
46. Meerick, I., Han, M.Y., Young, A.F., Ozyilmaz, B., Kim, P., Shepard, K.L., Current saturation in zero-band gap, top-gated graphene field-effect transistors. *Nat. Nanotechnol.*, 3, 11, 654–659, 2008.
47. Avouris, P., Chen, Z., Perebeinos, V., Carbon-based electronics. *Nat. Nanotechnol.*, 2, 10, 605–615, 2014.
48. Akinwande, D., Close, G.F., Wong, H.S.P., Analysis of the frequency response of carbon nanotubes transistors. *IEEE Trans. Nanotechnol.*, 5, 5, 599–605, 2006.
49. Schedin, F., Geim, A.K., Morozov, S.V., Hill, E.W., Blake, P., Katsnelson, M.I., Novoselov, K.S., Detection of individual gas molecules adsorbed on graphene. *Nat. Mater*, 6, 9, 652–655, 2007.
50. Li, S.L., Miyazaki, H., Kumatani, A., Kanda, A., Tsukagoshi, K., Low operating bias and matched input-output characteristics in graphene logic inverters. *Nano Lett.*, 10, 10, 2357–2362, 2010.
51. Guerriero, E., Polloni, L., Rizzi, L.G., Bianchi, M., Mondello, G., Sordan, R., Graphene audio voltage amplifier. *Small*, 8, 3, 356, 2012.
52. Kim, W., Riikonen, J., Li, C., Chen, Y., Lipsanen, H., Highly tunable local gate controlled complementary graphene device performing as inverter and voltage controlled resistor. *Nanotechnology*, 24, 39, 395202, 2013.
53. Liao, A.D., Wu, J.Z., Wang, X., Tahy, K., Jena, D., Dai, H., Pop, E., Thermally limited current carrying ability of graphene nano ribbons. *Phys. Rev. Lett.*, 106, 25, 256801, 2011.
54. Liu, L., Ryu, S., Tomasik, M.R., Stolyarova, E., Jung, N., Hybertsen, M.S., Steigerwald, M.L., Brus, L.E., Flynn, G.W., Graphene oxidation: Thickness-dependent etching and strong chemical doping. *Nano Lett.*, 8, 7, 1965–1970, 2008.
55. Xue, Y., Qian, L., He, G., Xu, K., Lin, J., Hu, X., Hu, J., Excellent electrical conductivity of the exfoliated and fluorinated hexagonal boron nitride nanosheets. *Nanoscale Res. Lett.*, 8, 1, 1–7, 2013.
56. Xia, F., Pere beinos, V., Lin, Y.M., Wu, Y., Avouris, P., The origins and limits of metal-graphene junction resistance. *Nat. Nanotechnol.*, 6, 3, 179–184, 2011.
57. Yang, X., Liu, G., Balandin, A.A., Mohanram, K., Triple-mode single-transistor graphene amplifier and its applications. *ACS Nano*, 4, 4, 5532–5538, 2010.
58. Rizzi, L.G., Bianchi, M., Behnam, A., Carrion, E., Guerriero, E., Polloni, L., Pop, E., Sordan, R., Cascading wafer-scale integrated graphene complementary inverters under ambient conditions. *Nano Lett.*, 12, 8, 3948, 2012.
59. Lin, Y.M., Valdesgarcia, A., Han, S.J., Farmer, D.B., Meric, I., Sun, Y., Wu, Y., Dimitrakopoulos, C., Grill, A., Avouris, P., Wafer-scale graphene integrated circuit. *Science*, 332, 6035, 1294–1297, 2011.

4

Theoretical Study on Graphene Oxide as a Cancer Drug Carrier

Satya Narayan Sahu[1], Saraswati Soren[1], Shanta Chakrabarty[2*] and Rojalin Sahu[1†]

[1]*School of Applied Sciences, Kalinga Institute of Industrial Technology, Deemed to be University, Bhubaneswar, India*
[2]*School of Mechanical Engineering, Kalinga Institute of Industrial Technology, Deemed to be University, Bhubaneswar, India*

Abstract

In the current scenario, cancer is one of the widely spread disease next to AIDS. The treatment of this disease is still a big challenge for the oncologist. During the treatment of cancer, patient suffers severely because of chemotherapy and radiations. Delivering the optimal mixture of drugs to the targeted cell is very crucial to avoid the adverse effect of drugs on other cells and tissues of the body. In this context, carbon-based 2D graphene oxide has brought a new hope for effective targeted drug delivery to the cancerous cell. The objective of our study here is to analyze the structural stability of graphene oxide (GO) and cancer drugs complexes and the binding mechanism of GO as a cancer drug carrier. To accomplish the study, we have carried out computational analysis by using different computational tools and techniques. Visualization and image preparation were performed by using Discovery Studio. The computational study, which covered in this manuscript, is bio-compatibility of graphene oxide and cancer drugs. The molecular docking simulation study was taken to analyze the binding mechanisms of the graphene oxide as a cancer drug carrier. We have shown here how the graphene sheet is use for cancer drug carrier and deliver at low P^H of the affected cell and bio-compatibility of graphene by *in silico* method. The essential cancer drug compounds like 5-flourouracil, Ibuprofen, Camptothecine, and Doxorubicin are taken for the study. We have observed the carrier and deliver mechanism through binding performance with the carrier.

**Corresponding author*: chakrabarty.shanta@gmail.com
†*Corresponding author*: rsahufch@kiit.ac.in

Inamuddin, Rajender Boddula, Mohd Imran Ahamed and Abdullah M. Asiri (eds.)
Monoelements: Properties and Applications, (73–86) © 2020 Scrivener Publishing LLC

Keywords: Graphene oxide, cancer, molecular docking, molecular interaction, drug carrier, *In silico* study, binding energy, drug delivery

4.1 Introduction

Graphene (G) or graphene oxide (GO) is considered as carbon-based 2D material and has emerged as a potentially promising area of research in the field of nano medicine [1]. In comparison to graphene, GO shows more interesting properties owing to its built-in functionalization sites. It is typically produced from the graphite by a modified Hummer's method or by the chemical treatment such as subsequent dispersion and exfoliation in possible organic solvents or water and oxidation [2]. It is noticed that the oxygen containing in GO are mostly in the form of hydroxyl and epoxy functional group (basal plane) with smaller amount of carbonyl, ohenol, carboxyl, quinine, and lactone (sheet edges) [3]. The extensive range of oxygen containing functional group on sheet edges and basal plane of GO makes it suitable to yield well dispersed solution of GO sheet in both organic solvents as well as in water [4]. Typical structure of G and GO are shown in Figure 4.1.

The presence of various functional groups on the surface of GO offers an adjustable flake-size distribution with outstanding water solubility and excellent physiochemical properties [5]. In the last decade, GO has been incessantly attracting the researchers with its imminent application in the field of biomedical as gene/drug delivery, biomedical imaging, cancer therapy, tissue engineering, etc. [6]. The applicability of GO in the biomedical research demands the confirmation and essentialness of its biocompatibility, biodegradability, and toxicity nature through extensive *in vivo* and *in vitro* studies by using either animals model, specific cell lines, or theoretical analysis [7]. Though there are the vast applications of GO in the medical field in various manners, but its use is predominantly beneficial for targeted drug delivery in cancerous cell. Cancer is caused by uncontrolled cell growth, which is leads to death and affects the millions of human being across the world population. In the current scenario, the number of cancer cases increases with the modern life style [8]. The treatment of this disease is still a big challenge for the oncologist. During the

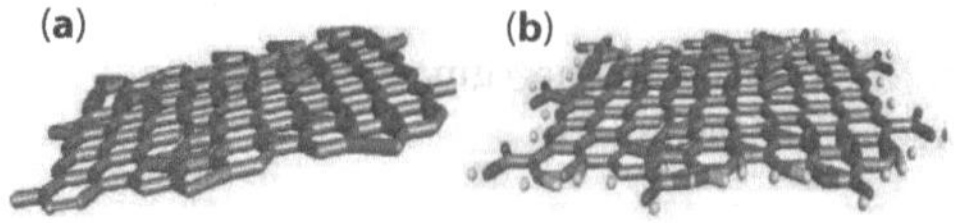

Figure 4.1 Schematic representation of (a) graphene and (b) graphene oxide.

treatment of cancer, patient suffers a lot because of chemotherapy and radiations [9]. Delivering the optimal mixture of drugs to the targeted cell is very crucial to avoid the adverse effect of drugs on other cells and tissues of the body. In this context, carbon-based 2D graphene oxide has brought a new hope for effective targeted drug delivery to the cancerous cell [10]. The fictionalization of GO enhances its bio-compatibility, hydrophilicity, and excellent stability in cell medium which leads to availability of enough basal planes of graphene oxide for loading of drugs [11].

The presence of large number of OH and COOH group on the surface of GO increases the possibility of covalent bonds [12]. The graphene oxide and its functional derivatives exhibit excellent material characteristics for targeted drug delivery for cancer. The drugs can easily bind to the 2D layer of GO due to availability of different functional group on its surface, which in turn reduces the mobility of drug molecule significantly. The decrease of mobility facilitate the effective drug delivery to the cancerous cell by GO. The efficiency of drug delivery can be controlled by customizing the lateral dimensions, layer number, and surface area of GO. The alteration of thickness and size of GO sheets significantly influence the drug load and drug release properties [13]. The controlled way of drug delivery system affords more efficient therapies for cancer treatment with minimal toxicity by delivery directly to temporal and spatial requirements for various applications [14, 15]. The controlled drug delivery system may be advantageous in different *in vitro* analysis, such as exploratory cell biology experimentation or high-throughput drug screening analysis. Due to the adjustable properties like stability and fine control on over doses of the drugs, the novel GO is recommended by several researchers as advance drug delivery technologies by enabling tailored drug-release profiles [16]. It has already been demonstrated in the literature that by fine tuning the flake size of GO, its internalization into cells can be regulated [17].

The objective of our study is to analyze the structural stability of graphene oxide and cancer drugs complexes and the binding mechanism of GO as a cancer drug carrier. To accomplish our research insight, we carried out a computational analysis by using different computational tools and techniques. The computational study, which covered in this manuscript are bio-compatibility of graphene oxide and cancer drugs. The molecular docking simulation study was taken to analyze the binding mechanisms of the graphene oxide as a cancer drug carrier. In Section 4.2, we have discussed the molecular level interactions of graphene oxide with biomolecules, whereas in Section 4.3 we used some computational technique to study the interaction between graphene oxide and different cancer drugs. The last section of the manuscript presents the overall results and discussion of the present study.

4.2 Molecular Interaction of Biomolecules and Graphene Oxide

Graphene oxide is a molecule of carbon and oxygen and it is an aromatic network, which provides the open surface for covalent and non-covalent interactions with biomolecules [18]. The derivatives of graphenes or graphene oxide (GO) have carboxyl, epoxide, and hydroxyl functional groups on its edges and basal plane. Nowadays, graphene is used as a cancer drug carrier, whereas the derivatives of graphene like graphene oxide and redox graphene oxide (RGO) are bio-compatible in nature. The interaction of biomolecules with graphene oxide and its derivatives are described below.

4.2.1 Molecular Interaction of DNA With Graphene Oxide

From the different kinds of biomolecules, DNA is an important biomolecule. Functionalize graphenes are used as drug carriers due to its bio-compatibility, renewability, and helix structures [19]. The π-π interaction or non-covalent interaction between the hexagonal cells of graphene oxide and the ring structure of nucleolus bases of single stranded DNA is adsorb onto the graphene oxide. Zhang *et al.* modeled a fluorescent biosensing system to detect the biotin, which have high binding affinity between graphene oxide and single stranded DNA. At first, the modification of single stranded DNA with biotin at 3' end and 6-carboxyfluorescin at the 5' end of DNA. Streptavidin binds with biotinylated single stranded DNA to protect the single stranded DNA by hydrolysis by exo-nuclease I enzyme. Then, π-π interaction between streptavidin and single stranded DNA was absorbed by graphene oxide. This can provide fluorescent sensing of small biomolecules [20]. According to Li *et al.*, the single stranded DNA and RGO/Pt nanocomposite are highly catalytic for methanol oxide [21]. Mixing of graphene oxide with single stranded DNA and the mixture was reduced by using sodium citrate. At that time, single stranded DNA was paralyzed by the RGO, due to π-π interaction. The electrostatic force of attraction between the positive charged graphene and negative charged nucleotides enables the DNA to immobilize on graphene surface. Zhang *et al.* composed amino functionalized graphene (G-NH_2) by using 1-octadecylamine (OTA) [20]. The DNA strands were attached to the G-NH_2 by electrostatic binding interactions between negatively charged phosphates groups of DNA strands and positively charged amino groups of G-NH_2. For interaction between DNA and GO, the hydrogen bonding is most

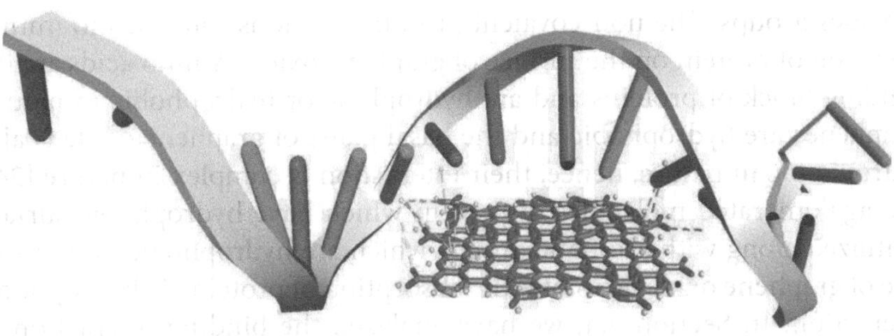

Figure 4.2 Schematic representation of interaction between graphene oxide with ssDNA.

important. The H-bonding is the primary forces for interaction of single stranded DNA and GO. According to Xue *et al.*, the Plasmon resonance investigates the interaction strength between DNA and GO. The non-covalent interaction of DNA and GO was showing unsatisfactory results to stabilize G-DNA nano hybrids. Therefore, the covalent interaction of DNA and graphene has been adopted to immobilize the DNA by graphene surface [22]. Zhang *et al.* composed a stable graphene and pyrene butyric acid nano composites, which is used as immobilization platform for DNA. In Figure 4.2, we have shown a schematic representation of binding interaction of graphene oxide and single stranded DNA.

4.2.2 Molecular Interaction of Protein With Graphene Oxide

On combining graphene and protein, a new hybrid fabrication having biological activity is observed. Immobilization of protein onto graphene takes place via non-covalent and covalent interaction. 1-pyrenebutyric acid N-hydroxysucinimide (PHNS) ester was first used to functionalize nano sheets of graphene oxide in order to form a binding site for protein on the surface of RGO. Bi-functionalized PHNS interacts with graphene basal plane pi-pi interaction between RGO and pyrenyl and also provides succinimidyl ester groups, which are highly reactive to nucleophilic substitution by primary and secondary amines that exist on the ferritin surface. Ferritin and RGO nano sheet on covalent interaction forms RGO-ferritin nanohybrids. Nano fibers of protein get bounded onto surface of graphene by covalent bond. It was demonstrated by Li and his coworkers that graphene oxide sheets via various interactions can absorb protein without the need of any additional ligands [23]. The presence of hydrophobic aromatic group on the graphene oxide surface favors the hydrpobhically driven protein and the covalent reaction with lysine units was favored by epoxides groups. The non-covalent pi-pi interactions can leads to immobilization of protein on the surface of graphene oxide. Amino acids are the building block of proteins and are hydrophilic or hydrophobic in nature. Graphenes are hydrophobic and the basal plane of graphene oxide is also hydrophobic in nature; hence, their interaction is complex in nature [24]. The agglomerated molecules of protein which have hydrophobic surface is utilized along with aromatic groups which are hydrophobic on the surface of graphene oxide facilitates the absorption of protein via hydrophobic interaction. In Section 4.4, we have analyzed the binding interaction of graphene oxide with cancer drugs and its behavior as cancer drug carrier by using computational method.

4.3 Computational Method

In this study, we have shown how the graphene sheet is use for cancer drug carrier and deliver at low P^H of the affected cell and bio-compatibility of graphene by *in silico* method. The essential cancer drug compounds like 5-flourouracil, Ibuprofen, Camptothecine, and Doxorubicin are taken for the study. We have observed the carrier and deliver mechanism through binding performance with the carrier. Here, graphene is taken as a drug carrier, which is experimentally proved and previously studied, but in our study, this process is employed to prove it theoretically. The study was performed by using only computational tools and algorithm. In this study, molecular docking approach was used to find out the binding performance of graphene sheet with cancer drug compounds. Molecular docking is frequently used to analyze the binding performance of small molecules [25–27]. Molecular modeling involves the process of docking by which we can analyze or predict the favored binding coordination/performance of a molecule to the other and they bind to each other to form a stable complex model. The aim of molecular docking simulation is to analyze the molecular orientation by computational method [28–31]. The molecular docking analysis was performed by using Autodock 4.2 tool in which Autodock vina software is used [32]. The docking study was performed with the flexible ligand and the rigid receptor by tacking grid size 40, 56, and 40 points in x, y, and z directions was created with center_x = 9.244, center_y = 0.046, and center_z = 0.0 for graphene oxide sheet. For docking study, the input file should be in .pdb file format, and to converted the structure in to .pdbqt file format, we have to remove bad contacts and added hydrogens on polar contacts of the receptor by using Autodock tool where both the internal geometry of the receptor and ligand is kept fixed and docking is performed or may be flexible docking (induced fit) which allows conformational changes. Two approaches are particularly popular within the molecular docking. One approach uses for searching algorithms that describes the receptor and the ligand binding modes. Another approach is based on to calculate the interaction energies between ligand-receptor in pairwise. Both approaches are based on mathematical methods to find out the different interactions, including non-covalent interactions. During docking process, the intermolecular interactions were evaluated by scoring functions [33–36]. The predicted grid size and coordinates of graphene oxide was investigated by molecular docking method. The initial coordinates of graphene oxide and studied cancer drug molecules are taken from the structures which are taken from Pubchem database. The visualization and image representation were performed by using Discovery Studio [37]. The molecular interaction between graphene oxide and cancer drugs would be obtained from inter molecular interactions. Hence, docking plays an important role in the molecular recognition process.

4.4 Results and Discussion

Molecular docking approach shows the binding coordination between graphene oxide with existing drug molecules for cancer treatment. In this study, we have chosen the frequently used drug molecules which are use for cancer treatment. These are 5-flourouracil, Ibuprofen, Camptothecine, and Doxorubicin. We have used this docking study to analyze the binding and adsorption performance over the graphene oxide.

4.4.1 Binding Behavior Between Graphene Oxide With Cancer Drugs (5-Flourouracil, Ibuprofen, Camptothecine, and Doxorubicin)

We have found the binding energy value, −7.2 kcal/mol, when the graphene oxide binds with 5-Fluorouracil. The $\pi...\pi$ stacking interactions are found within the range of 4.67 to 4.84 Å and this interactions are present between π orbitals of both the molecules. The pictorial representation of docking between graphene oxide with 5-Fluorouracil was shown in Figure 4.3a. The binding energy with graphene oxide and Camptothecine, Doxorubicin, and Ibuprofen were −21.5, −21.0, and −14.8 kcal/mol, respectively, and it is shown in Table 4.1.

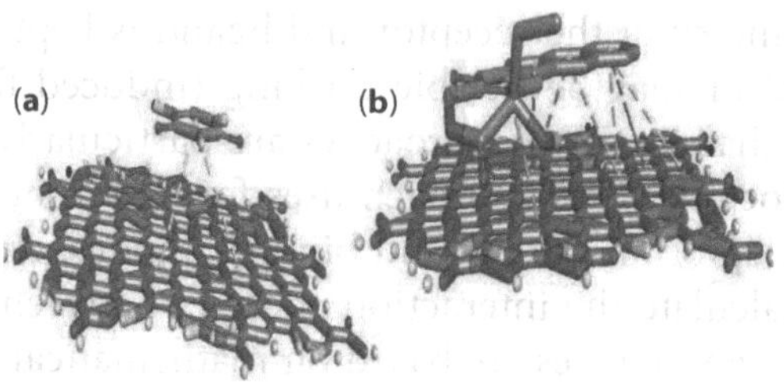

Figure 4.3 Schematic representation of binding coordinate between graphene oxide with (a) 5-Fluorouracil and (b) Camptothecin. The black colored stick model represented as graphene oxide sheet.

In case of Camptothecin, there is one electrostatic interaction present. The bond length of O...π is 4.47 Å. There are four numbers of π...π stacking interaction are present with bond lengths in the ranges 3.65 to 5.41 Å. Whereas three numbers of π... alkyl interactions are formed. The pictorial representation of docking between graphene oxide with Camptothecin is shown in Figure 4.3b.

When we have analyzed the binding performance between graphene oxide and Doxorubicin, we have found that there are four numbers of π...π interactions having bond lengths in the ranges 4.78, 3.59, 4.20, 4.31, 5.19, and 4.83 Å, where as two numbers of π...alkyl interactions are present. The binding representation is shown in Figure 4.4a. In case of Ibuprofen, there are four numbers of π...alkyl interactions are present between graphene oxide and Ibuprofen where bond lengths are differed from 3.82 to 4.64 Å and two numbers of π-stacked interactions are formed with varying bond lengths. The schematic representation of binding performance between graphene oxide and Ibuprofen is shown in Figure 4.4b. The binding behavior of graphen oxide with Camptothecin and Doxorubicin has shown best binding performance due to high binding energy.

Overall, in this study, we have discussed about graphene oxide as the cancer drug carrier for the drugs like 5-Flourouracil, Ibuprofen, Camptothecine, and Doxorubicin. Here, graphene oxide sheets have very high binding energy with Camptothecin and Doxorubicin. The negative value of binding energy of graphene oxide with discussed drug molecules signifies the stability in aqueous solutions. In the present study, the drug molecules Camptothecin and Doxorubicin are showing best performance due to the highest number of π...stacked interactions, π...alkly and electrostatic interactions. The π....π stacking interactions are the non-covalent interactions, where as in case of π-system, it could be diffuse electrons to exhibit attractive forces. Dougherty *et al.* studied that the non-covalent

Table 4.1 Binding and dominant mechanisms of graphene oxide with various cancer drugs.

Drug molecules	Binding energy (kcal/mol)	Binding atoms	Nature of the bond	Bond length (Å)
5-Fluorouracil	−7.2	π orbitals - π orbitals	π- π stacked	4.67
		π orbitals - π orbitals	π- π stacked	4.84
Camptothecin	−21.5	O - π orbitals	electrostatic	4.47
		π orbitals - π orbitals	π- π stacked	4.42
		π orbitals - π orbitals	π- π stacked	5.32
		π orbitals - π orbitals	π- π stacked	4.97
		π orbitals - π orbitals	π- π stacked	4.78
		π orbitals - π orbitals	π - alkyl	4.79
		π orbitals - π orbitals	π - alkyl	3.65
		π orbitals - π orbitals	π - alkyl	5.41
Doxorubicin	−21.0	π orbitals - π orbitals	π - π stacked	4.78
		π orbitals - π orbitals	π - π stacked	3.59
		π orbitals - π orbitals	π - π stacked	4.20
		π orbitals - π orbitals	π - π stacked	4.31
		π orbitals - π orbitals	π - alkyl	5.19
		π orbitals - π orbitals	π - alkyl	4.83

(Continued)

Table 4.1 Binding and dominant mechanisms of graphene oxide with various cancer drugs. (*Continued*)

Drug molecules	Binding energy (kcal/mol)	Binding atoms	Nature of the bond	Bond length (Å)
Ibuprofen	−14.8	C - π orbitals	π - sigma	3.68
		C - π orbitals	π - alkyl	3.82
		C - π orbitals	π - alkyl	3.93
		C - π orbitals	π - alkyl	4.64
		C - π orbitals	π - alkyl	4.34
		π orbitals - π orbitals	π - π stacked	4.68
		π orbitals - π orbitals	π - π stacked	5.23

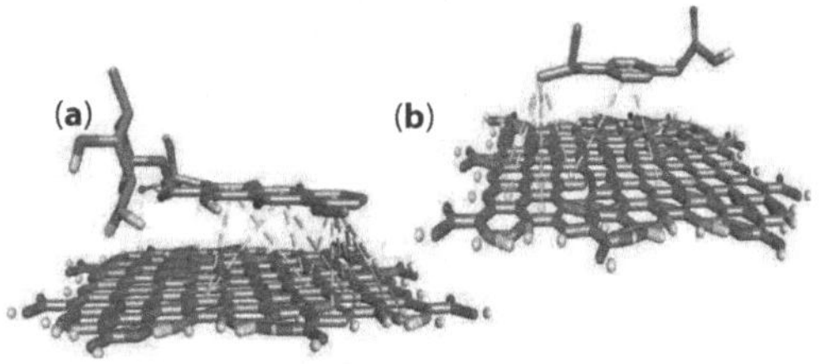

Figure 4.4 Schematic representation of binding coordinate between graphene oxide with (a) Doxorubicin and (b) Ibuprofen. The black colored stick model represented as graphene oxide sheet.

π...π stacking interactions of chemical substance with graphene oxide shows the sufficient stability in case of aqueous solutions, where as our results also reflect the same [38]. Depan *et al.* have reported that use of polymer which is positively charged in order to confine Doxorubicin-loaded graphene oxide, which finally resulted in Doxorubicin-graphene oxide-folate (a nano carrier system) having the property of releasing P^H response drug [39]. Pure graphene and graphene oxide possess very high conjugated system, i.e., pi-pi delocalized system which enables in binding with aromatic compounds like cancer drugs, fluroscein, and other hybrid molecules via pi-pi stacking [40]. The details of binding energy by docking analysis of graphene oxide with cancer drug molecules are given in the Table 4.1. Visualization and image preparation were performed by using Discovery Studio.

4.5 Conclusion

In this study, we have studied graphene oxide as a suitable drug carrier for anticancer drugs 5-Flourouracil, Ibuprofen, Camptothecine, and Doxorubicin computationally. The negative binding energy values of GO with these drug molecules signify the stability of these molecules in aqueous solutions. Among the four drug molecules, Camptothecin and Doxorubicin have shown best loading performance due to the highest number of π...π stacking interactions, π...alkly and electrostatic interactions. Overall, here, both the covalent and non-covalent interactions are present. We got the binding energy value −7.2 kcal/mol, when the graphene oxide binds with 5-Fluorouracil. The π...π stacking interactions are found within the range of 4.67 to 4.84 Å and this interactions are present between π orbitals of both the molecules. There is only one electrostatic interaction present in case of Camptothecin, and four numbers of π...π stacking interactions present with bond lengths 4.47 Å and in the range of 3.65 to 5.41 Å, respectively. Whereas three numbers of π...alkyl interactions are formed. There are four numbers of π...π interactions having bond lengths in the ranges 4.78, 3.59, 4.20, 4.31, 5.19, and 4.83 Å, and two numbers of π...alkyl interactions are present in case of graphene oxide and Doxorubicin. In case of Ibuprofen, there are four numbers of π...alkyl interactions that are present between graphene oxide and Ibuprofen where the bond lengths are differed from 3.82 to 4.64 Å and two numbers of π-stacked interactions are formed with varying bond lengths. The binding behavior of graphen oxide with Camptothecin and Doxorubicin has shown best binding performance due to high binding energy.

References

1. Mao, H.Y., Laurent, S., Chen, W., Akhavan, O., Imani, M., Ashkarran, A.A., Mahmoudi, M., Graphene: Promises, facts, opportunities, and challenges in nanomedicine. *Chem. Rev.*, 113, 5, 3407–3424, 2013.
2. Botas, C., Álvarez, P., Blanco, P., Granda, M., Blanco, C., Santamaría, R., Romasanta, L.J., Verdejo, R., López-Manchado, M.A., Menéndez, R., Graphene materials with different structures prepared from the same graphite by the Hummers and Brodie methods. *Carbon*, 65, 156 164, 2013.
3. Gupta, A., Shaw, B.K., Saha, S.K., Bright green photoluminescence in aminoazobenzene-functionalized graphene oxide. *J. Phys. Chem. C*, 118, 13, 6972–6979, 2014.

4. Yang, H., Shan, C., Li, F., Han, D., Zhang, Q., Niu, L., Covalent functionalization of polydisperse chemically-converted graphene sheets with amine-terminated ionic liquid. *Chem. Commun.*, 26, 3880–38825, 2009.
5. Khan, Z.U., Kausar, A., Ullah, H., Badshah, A., Khan, W.U., A review of graphene oxide, graphene buckypaper, and polymer/graphene composites: Properties and fabrication techniques. *J. Plast. Film Sheet.*, 32, 4, 336–379, 2016.
6. Chung, C., Kim, Y.-K., Shin, D., Ryoo, S.-R., Hong, B.H., Min, D.-H., Biomedical applications of graphene and graphene oxide. *Acc. Chem. Res.*, 46, 10, 2211–2224, 2013.
7. Tabish, T.A., Pranjol, M.Z.I., Hayat, H., Rahat, A.A.M., Abdullah, T.M., Whatmore, J.L., Zhang, S., *In vitro* toxic effects of reduced graphene oxide nanosheets on lung cancer cells. *Nanotechnology*, 28, 50, 504001, 2017.
8. Zhang, L., Xia, J., Zhao, Q., Liu, L., Zhang, Z., Functional graphene oxide as a nanocarrier for controlled loading and targeted delivery of mixed anticancer drugs. *Small*, 6, 4, 537–544, 2010.
9. McQuestion, M., Fitch, M., Howell, D., The changed meaning of food: Physical, social and emotional loss for patients having received radiation treatment for head and neck cancer. *Eur. J. Oncol. Nurs.*, 15, 2, 145–151, 2011.
10. Feng, L., Wu, L., Qu, X., New horizons for diagnostics and therapeutic applications of graphene and graphene oxide. *Adv. Mater.*, 25, 2, 168–186, 2013.
11. Liu, J., Cui, L., Losic, D., Graphene and graphene oxide as new nanocarriers for drug delivery applications. *Acta Biomater.*, 9, 12, 9243–9257, 2013.
12. Chen, D., Feng, H., Li, J., Graphene oxide: Preparation, functionalization, and electrochemical applications. *Chem. Rev.*, 112, 11, 6027–6053, 2012.
13. Kim, H., Lee, D., Kim, J., Kim, T., Kim, W.J., Photothermally triggered cytosolic drug delivery via endosome disruption using a functionalized reduced graphene oxide. *ACS Nano*, 7, 8, 6735–6746, 2013.
14. Bharti, C., Nagaich, U., Pal, A.K., Gulati, N., Mesoporous silica nanoparticles in target drug delivery system: A review. *Int. J. Pharm. Invest.*, 5, 3, 124, 2015.
15. Weaver, C.L., LaRosa, J.M., Luo, X., Cui, X.T., Electrically controlled drug delivery from graphene oxide nanocomposite films. *ACS Nano*, 8, 2, 1834–1843, 2014.
16. Vasconcelos, T., Sarmento, B., Costa, P., Solid dispersions as strategy to improve oral bioavailability of poor water soluble drugs. *Drug Discovery today*, 12, 23–241068–1075, 2007.
17. Zheng, X.T., Than, A., Ananthanaraya, A., Kim, D.-H., Chen, P., Graphene quantum dots as universal fluorophores and their use in revealing regulated trafficking of insulin receptors in adipocytes. *ACS Nano*, 7, 7, 6278–6286, 2013.
18. Li, D., Zhang, W., Yu, X., Wang, Z., Su, Z., Wei, G., When biomolecules meet graphene: From molecular level interactions to material design and applications. *Nanoscale*, 8, 47, 19491–19509, 2016.

19. Premkumar, T. and Geckeler, K.E., Graphene–DNA hybrid materials: Assembly, applications, and prospects. *Progress Polym. Sci.*, 37, 4, 515–529, 2012.
20. Zhang, H., Li, Y., Su, X., A small-molecule-linked DNA–graphene oxide-based fluorescence-sensing system for detection of biotin. *Anal. Biochem.*, 442, 2, 172–177, 2013.
21. Li, M., Pan, Y., Guo, X., Liang, Y., Wu, Y., Wen, Y., Yang, H., Pt/single-stranded DNA/graphene nanocomposite with improved catalytic activity and CO tolerance. *J. Mater. Chem. A*, 3, 19, 10353–10359, 2015.
22. Xue, T., Cui, X., Guan, W., Wang, Q., Liu, C., Wang, H., Qi, K., Singh, D.J., Zheng, W., Surface plasmon resonance technique for directly probing the interaction of DNA and graphene oxide and ultra-sensitive biosensing. *Biosens. Bioelectron.*, 58, 374–3795, 2014.
23. Li, D., Zhang, W., Yu, X., Wang, Z., Su, Z., Wei, G., When biomolecules meet graphene: From molecular level interactions to material design and applications. *Nanoscale*, 8, 47, 19491–19509, 2016.
24. Wang, G., Shen, X., Wang, B., Yao, J., Park, J., Synthesis and characterisation of hydrophilic and organophilic graphene nanosheets. *Carbon*, 47, 5, 1359–1364, 2009.
25. Pattanayak, S.K. and Chowdhuri, S., Effect of water on solvation structure and dynamics of ions in the peptide bond environment: Importance of hydrogen bonding and dynamics of the solvents. *J. Phys. Chem. B*, 115, 13241–13252, 2011.
26. Pattanayak, S.K., Prashar, N., Chowdhuri, S., Effect of temperature and pressure on the structure, dynamics, and hydrogen bond properties of liquid N-methylacetamide: A molecular dynamics study. *J. Chem. Phys.*, 134, 154506, 2011.
27. Pattanayak, S.K. and Chowdhuri, S., A molecular dynamics simulations study on the behavior of liquid N-methylacetamide in presence of NaCl: Structure, dynamics and H-bond properties. *J. Mol. Liq.*, 172, 102–109, 2012.
28. Pattanayak, S.K. and Chowdhuri, S., Size dependence of solvation structure and dynamics of ions in liquid n-methylacetamide: A molecular dynamics simulation study. *J. Theor. Comput. Chem.*, 11, 361–377, 2012.
29. Pattanayak, S.K. and Chowdhuri, S., Effects of concentrated NaCl and KCl solutions on the behaviour of aqueous peptide bond environment: Single-particle dynamics and H-bond structural relaxation. *Mol. Phys.*, 111, 3297–3310, 2013.
30. Sahu, S.N., Moharana, M., Sahu, R., Pattanayak, S.K., Impact of mutation on podocin protein involved in type 2 nephrotic syndrome: Insights into docking and molecular dynamics simulation study. *J. Mol. Liq.*, 281, 549–562, 2019.
31. Panda, J., Sahoo, J.K., Panda, P.K., Sahu, S.N., Samal, M., Pattanayak, S.K., Sahu, R., Adsorptive behavior of zeolitic imidazolate framework-8 towards

anionic dye in aqueous media: Combined experimental and molecular docking study. *J. Mol. Liq.*, 278, 536–5455, 2019.
32. Norgan, A.P., Coffman, P.K., Kocher, J.-P.A., Katzmann, D.J., Sosa, C.P., Multilevel parallelization of AutoDock 4.2. *J. Cheminf.*, 3, 1, 12, 2011.
33. Pattanayak, S.K. and Chowdhuri, S., Effects of methanol on the hydrogen bonding structure and dynamics in aqueous N-methylacetamide solution. *J. Mol. Liq.*, 194, 141–148, 2014.
34. Bissoyi, A., Singh, A.K., Pattanayak, S.K., Bit, A., Sinha, S.K., Patel, A., Jain, V., Patra, P.K., Understanding the molecular mechanism of improved proliferation and osteogenic potential of human mesenchymal stem cells grown on a polyelectrolyte complex derived from non-mulberry silk fibroin and chitosan. *Biomed. Mater.*, 13, 015011, 2017.
35. Bissoyi, A., Pattanayak, S.K., Bit, A., Patel, A., Singh, A.K., Behera, S.S., Satpathy, D., Alphavirus nonstructural proteases and their inhibitors, in: *Viral Proteases Their Inhib*, pp. 77–104, 2017.
36. Hiremath, S.M., Suvitha, A., Patil, N.R., Hiremath, C.S., Khemalapure, S.S., Pattanayak, S.K., Negalurmath, V.S., Obelannavar, K., Armakovic, S.J., Armaković, S., Synthesis of 5-(5-methyl-benzofuran-3-ylmethyl)-3H-[1, 3, 4] oxadiazole-2-thione and investigation of its spectroscopic, reactivity, optoelectronic and drug likeness properties by combined computational and experimental approach. *Spectrochim. Acta A*, 205, 95–110, 2018.
37. Heiat, M., Najafi, A., Ranjbar, R., Latifi, A.M. and Rasaee, M.J., Computational approach to analyze isolated ssDNA aptamers against angiotensin II. *J. Biotechnol*, 230, 34–39, 2016.
38. Dougherty, D.A., The cation– π interaction. *Acc. Chem. Res.*, 46, 4, 885–893, 2012.
39. Depan, D., Shah, J., Misra, R.D.K., Controlled release of drug from folate-decorated and graphene mediated drug delivery system: Synthesis, loading efficiency, and drug release response. *Mater. Sci. Eng.: C*, 31, 7, 1305–1312, 2011.
40. Kavitha, T., Abdi, S.I.H., Park, S.-Y., pH-sensitive nanocargo based on smart polymer functionalized graphene oxide for site-specific drug delivery. *Phys. Chem. Chem. Phys.*, 15, 14, 5176–5185, 2013.

5

High-Quality Carbon Nanotubes and Graphene Produced from MOFs for Supercapacitor Application

Mandira Majumder[1,3*], Ram B. Choudhary[1], Anukul K. Thakur[2], Rabah Boukherroub[3] and Sabine Szunerits[3†]

[1]*Nanostructured Composite Materials Laboratory, Department of Physics, Indian Institute of Technology (Indian School of Mines), Dhanbad, Dhanbad, India*
[2]*Department of Physical Sciences, Indian Institute of Science Education and Research, Berhampur, Berhampur, India*
[3]*Univ. Lille, CNRS, Centrale Lille, ISEN, Valenciennes, UMR 8520 - IEMN, Lille, France*

Abstract
Metal-organic frameworks (MOFs) are a family of porous materials consisting of unique structural features such as large surface area, wide physicochemical tunability, and substantial stability. These properties make MOFs of wide interest for numerous applications ranging from energy storage devices and batteries, to sensors, and catalysis. These materials not only possess unique properties but can act as promising precursors for the synthesis of other carbon forms. This book chapter aims to group together the significant work reported in the last years on MOFs derived carbon (graphene and carbon nanotubes) and MOF-carbon composite materials with a special emphasis on the use of these nanostructures for energy storage devices (supercapacitors).

Keywords: MOFs, MOF derived carbon, supercapacitors, CNT, graphene

**Corresponding author*: mandira.16dr000220@ap.ism.ac.in; mandiramajumder92@gmail.com
†*Corresponding author*: sabine.szunerits@univ-lille.fr

Inamuddin, Rajender Boddula, Mohd Imran Ahamed and Abdullah M. Asiri (eds.)
Monoelements: Properties and Applications, (87–118) © 2020 Scrivener Publishing LLC

5.1 Introduction

Carbon has engrossed a huge interest due to the discovery of various allotropes [i.e., carbon nanotubes (CNTs), fullerenes, nanofibers, graphene, etc.] exhibiting extraordinary physicochemical properties such as high strength, outstanding resistance to corrosion, as well as good electrical and thermal conductivities [1–3]. Currently, half of all the commercial lithium ion batteries incorporate carbon nanofibers, obtained from carbon nanotubes, to enhance the energy capacity [3]; CNTs have found their use in increasing the strength of tennis rackets and reducing their weight [2]. The ability of CNTs to store hydrogen up to 65% of their mass makes their use as a cheap and efficient alternative to fossil fuels. These examples show that advancements in carbon-based nanotechnology is one of the most dynamic research field currently centered around the synthesis, characterization, and application of CNTs [4, 5]. The synthetic approaches of CNTs are based on arc discharge, laser ablation, flame synthesis, saline solution method, spray pyrolysis, etc. [7–9]. However, there are several technical difficulties in developing highly ordered and well organized CNTs [10]. Also, CNTs synthesized through these methods display strong agglomeration tendency, reducing their effective surface and electrical conductivity. CNTs, grown in a porous silicon template and/or aluminum-based anodic oxide pores using the Chemical Vapor Deposition (CVD) technique, exhibit some directional growth, but post-synthesis treatment with HF to dissolve the anodic oxide leads often to the damage of the CNTs [11]. However, while progress has been made, the controllable and oriented synthesis of CNTs using a low-cost, controllable, and eco-friendly method remains a key challenge [12].

Graphene, defined as a single-atom wide sheet of carbon atoms continuously arranged in a honeycomb-like hexagonal pattern, is the wonder material of the 20th century revolutionizing all research fields. Since graphene is the thinnest material in the world, it also possesses the highest surface to volume ratio. This property makes graphene suitable to be incorporated as electrode material in solar cells, fuel cells, batteries, and supercapacitors, the four major energy-related areas [13–17]. Graphene and graphene derived materials (notably graphene oxide and reduced graphene oxide) can be synthesized using various approaches, the most common one being exfoliation of graphite in liquid media and chemical vapor deposition (CVD), followed by micromechanical cleavage, epitaxial growth on SiC substrates, or unzipping of CNTs [18]. However, graphene oxide (GO) production by chemical means globally suffers from some serious quality problems and reproducibility. Eigler *et al.* reported recently on the chemical route for the synthesis of high-quality defect-free graphene nanosheets [19] showing charge mobility equivalent to 250 $cm^2 V^{-1} s^{-1}$. Nevertheless, wet chemical synthesis suffers from the involvement of toxic solvents, which calls for a need for greener synthetic methods.

Tuning the morphology through the formation of ordered porous structures in graphene has been demonstrated to be an effective way to improve the active surface area of the materials. When it comes to porosity, graphene oxide (GO), synthesized by the Hummers' method, was reported to possess a surface area of 332 $m^2 g^{-1}$, which further increased to 617 $m^2 g^{-1}$ upon reduction to reduced graphene oxide (rGO) using hydrazine [20]. A surface area as large as 3,100 $m^2 g^{-1}$ was achieved upon activation of microwave exfoliated GO with KOH to form nanopores of 1 nm and mesopores of 4 nm in size [21]. Indeed, in comparison with a graphene layer, porous graphene exhibits an increased number of active sites, more exposed edges and shortcuts between different layers of graphene for improved ion diffusion, which are favorable for electrochemical applications [22]. The various types of porous carbon are shown in Figure 5.1 [23–25].

Pore structure and pore size distribution are indeed important aspects of any synthesized carbon material and become a determining factor for performance. The International Union of Pure and Applied Chemistry

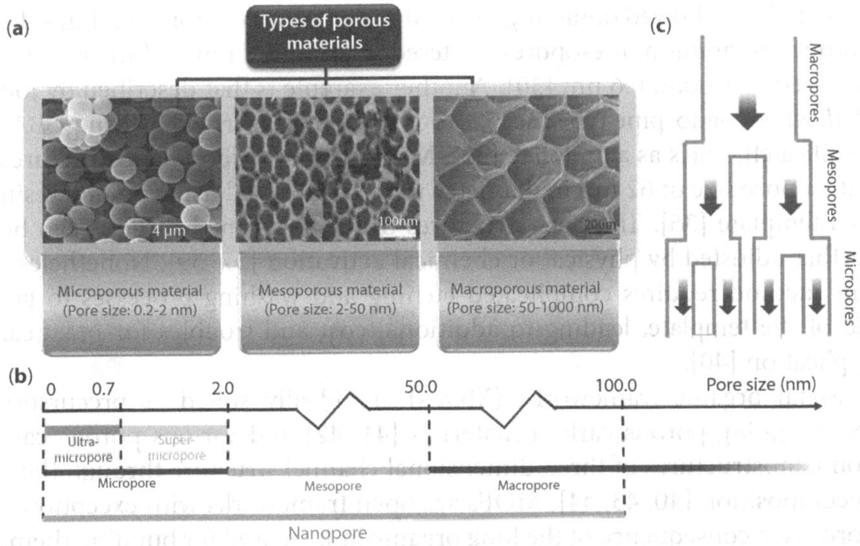

Figure 5.1 (a) Schematic diagram showing various types of porous materials [23–25]. (b) Scale bar of the various kinds of pores. (c) The formation path of various pores within a material @Copyright permission from references [23–25].

(IUPAC) recommends classification of porous carbon materials into three classes: macroporous (>50 nm), mesoporous (more than 2 nm to less than 50 nm), and microporous (<2 nm). Carbon particles with bimodal porosity (consisting of both meso- and micropores) possessing a surface area of 2,488 $m^2 g^{-1}$ were recently produced through the use of rape pollen as a carbon precursor and $CuCl_2$ as the activating agent [26]. Creation of a porous carbon macroporous network through carbonization of potassium stearate in an inert atmosphere was reported by Luo *et al.* [27]. The recycling of waste polyurethane by means of mechanochemistry and activation with the help of potassium carbonate produced nitrogen-doped porous carbon material [28]. Hydrothermal carbonization of waste sugarcane bagasse followed by chemical activation using KOH resulted in high specific surface area (3,151 $m^2 g^{-1}$) bimodal pore structured (micro and mesoporous) carbon material [29].

One of the major difficulties remaining is the synthesis of nano-carbons with controlled and uniformly distributed pore size. To mitigate the problem of uncontrolled and non-uniform pore size distribution, template carbonization was adopted [30]. Template carbonization [31–34] is a rather recent approach allowing the implementation of micro-, meso-, and macropores [30]. Ordered microporous carbon with a pore size below 2 nm and a specific surface area of 3,600 $m^2 g^{-1}$ was also reported using zeolite template with a pore size >0.6 nm –0.7 nm [30]. Hexagonal mesoporous silica templates allowed obtaining ordered mesoporous carbon with bimodal porous distribution: mesopores centered at around 2 nm and micropores centered at around 0.6 nm [30]. Another example is that described by Joo *et al.* where nano-pipe type mesoporous carbon was formed by using SBA-15 silica channels as a template [35]. Macroporous carbon nanostructures with a pore size of 62 nm in diameter were synthesized using phenol resin as a template [36]. The pore structures of the so-formed carbons can be further adjusted by physical or chemical activation [37–39]. Nonetheless, this method requires complicated etching and washing processes to get rid of the template, leading to additional cost and troubles for practical application [40].

Metal organic frameworks (MOFs) are ideally suited as precursors for designing porous carbon materials [41, 42] and various porous carbon nanostructures of three-dimensional channel structure through their decomposition [40, 43, 44]. MOFs are open frameworks with exceptional porosity, a consequence of the long organic linkers used for building them, conferring MOFs with large storage space and numerous adsorption sites. Until today, over 20,000 different MOFs have been synthesized and their formation is only limited by imagination. One current research activity is the integration in a controlled way of MOFs with other carbon-related materials [45, 46]. It is expected that this will result in single-phase MOFs with increased stability and electrical conductivity, while preserving the porous structure and template influence. These synergistic effects will lead to the applications of MOF-carbon composites outside the areas of single-phase MOFs [47–51] and interdisciplinary areas.

This book chapter aims to group together the significant work reported in the last years on MOFs derived carbon (graphene and carbon nanotubes) and MOF-carbon composites. A special focal point is on the use of MOF derived carbons as supercapacitive electrode materials. Indeed, the ever-increasing demand of energy causes a serious concern about the rapid depletion of various non-renewable and conventional sources of energy along with increased environmental pollution. For making efficient use of some non-conventional/renewable sources of energy like solar light or wind, the associated energy storage device plays a key role in determining the performance and efficiency. Among the various energy storage devices, supercapacitors are emerging as a promising candidate bridging the gap between conventional batteries and the capacitors in terms of maintenance, rate of charge and discharge, power, and energy density. The interest of using MOFs is thus timely and will be demonstrated on some selected examples here.

5.1.1 The Basics of Metal Organic Frameworks (MOFs)

MOFs are defined by Yaghi and Hailian as structures consisting of pores mainly constructed from coordinate bonding of organic linkers (or other bridging ligands) and metal ions, as illustrated in (Figure 5.2a) [52]. The organic units are the bridging ligands that include carboxylates, sulfonates, phosphonates, and various other compounds, which are heterocyclic in nature (Figure 5.2b). The inorganic part consists of metal ions or secondary building units (SBU). A huge number of SBU structures with varied number of sites of extension can be observed in MOF structure: triangle type (three sites), square paddle-wheel type (four sites), trigonal prism type (six sites), and octahedron type (six sites) (Figure 5.2c) [53, 54]. Their high surface area and tunable pore size make them appropriate for various applications including gas storage and separation, catalysis, sensor-based applications, drug loading and delivery, as well as scaffolds for energy driven applications [55, 57]. Some of the benchmarks of MOFs are MOF-5 ($Zn_4O(bdc)_3$, bdc = terephthalate), HKUST-1, MIL-53(Fe), chromium (III)

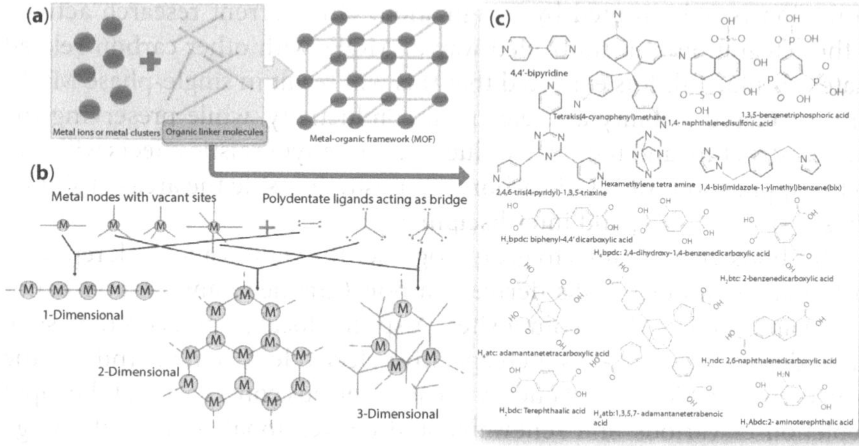

Figure 5.2 (a) Schematic representation of the structure of metal organic frameworks consisting of inorganic and organic units (MOFs). (b) A variety of secondary building units (SBU) geometries with different number of points of extension. (c) Chemical structures of various well-known organic ligands used for the synthesis of MOFs containing carboxylic functionalities, nitrogen, sulfur, phosphorous, and heterocycles.

terephthalate (MIL-101), or SBA-15 with high porosity and low pressure gas sorption, as well as with high chemical stability for application in drug delivery [58–60].

MOFs display a pore size up to 2 nm, which can well accommodate molecules with a small size. However, these small pores are not suitable for the incorporation of bigger molecules like enzymes and proteins. It has been tried to grow the pore size towards the meso-porous regime (2–50 nm) and reduce the size of the crystal towards the nanometer regime [61]. MOFs prove to be advantageous over other carbon-based materials largely due to their tunable porosity and ease of synthesis. Furthermore, these porous materials can act as a template for engineering other nanoporous carbon [62–64]: heating MOFs up to 1,000°C results in carbonization and the development of porous carbon nanomaterials [65].

5.2 Carbonization of MOFs

Carbonization of MOFs is an easy and efficient approach to synthesize porous carbon nanostructures. As shown in Figure 5.3, various reports are available, where MOFs are pyrolyzed at high temperature to obtain carbon materials such as nanoporous carbon (NC), highly porous carbon (HPC),

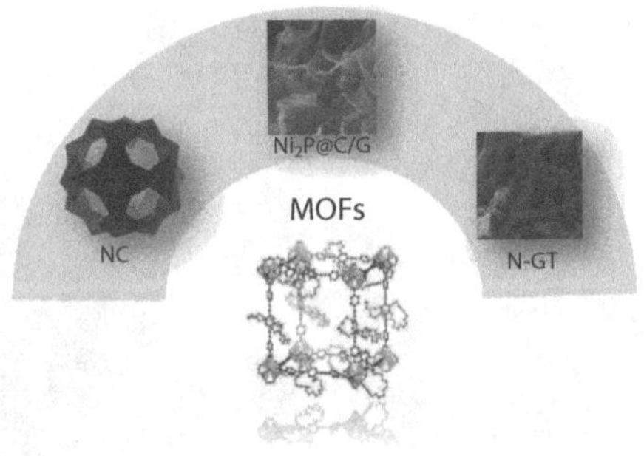

Figure 5.3 MOFs used as templates and/or precursors for the fabrication of porous nanocarbons and composite materials [68, 70, 72, and 73] @Copyright permission from references [68, 70, 72, and 73].

carbon-confined Ni_2P nanocrystals supported on graphene (Ni_2P@C/G), nano graphene tube (N-GT), and MOF derived carbon (MOFC) [66–73]. Apart from these carbon forms, MOFs can also be used as precursors for the synthesis of CNTs and graphene. The following section includes a brief discussion about the synthesis of CNTs and graphene using MOFs.

5.2.1 Conversion of MOFs Into Carbon Nanotubes (CNTs)

MOFs represent suitable precursors for the synthesis of CNTs with adjustable diameter, tunable pore size, and conductivity [74–76]. Nitrogen-enriched carbon tubes (NCNTs) were successfully prepared *via* the carbonization of ZIF-67 MOF in an inert atmosphere (Figure 5.4a). The formation of hollow-structured hierarchical CNTs frameworks resulted in a specific surface area of 513 m^2 g^{-1} and pore volume equal to 1.16 cm^3 g^{-1} with an average pore size of ~7.8 nm. The authors inferred that in this entire process, the ZIF-67 MOF particles not only served as the carbon and nitrogen sources for the growth of NCNTs, but equally acted as a self-template for the development of the framework morphology [74]. Similarly, Su *et al.* reported the synthesis of N-doped carbon nanotubes (NCNTs) by the direct pyrolysis of Zn–Fe-ZIF nanospheres implemented as the precursor (Figure 5.4b), displaying a specific surface area of 130 to 150 m^2 g^{-1}, as estimated from nitrogen adsorption-desorption isotherms [76].

Several other reports are available where apart from CNTs, MOFs were used as precursors to synthesize other nanomaterials like graphene, graphene-based composites, and other nanostructure-based composites [77–81].

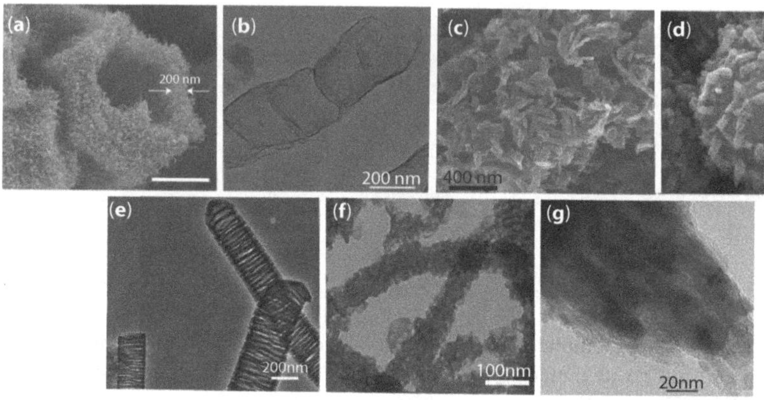

Figure 5.4 (a) Nitrogen-enriched carbon tubes (NCNTs) derived from ZIF-67 MOF @ Copyright permission from reference [74]. (b) N-doped carbon nanotubes obtained by carbonization of Zn–Fe-ZIF @Copyright permission from reference [76]. (c) Carbonized MOF-74-Rods @Copyright permission from reference [82]. (d) Carbonized Ni-MOFs @ Copyright permission from reference [84]. (e) Nano tubular structure of the carbonized Al-PCP MOF @Copyright permission from reference [75]. (f) Nanowires obtained from the carbonization of ZIF-8 MOF @Copyright permission from reference [79]. (g) Carbonized Fe-MOF @Copyright permission from reference [81].

5.2.2 MOFs Derived Graphene Like Carbon and Graphene-Based Composites

As discussed earlier, the synthesis of graphene and graphene composites remains challenging with respect to yield, cost, use of toxic solvents, etc. Several studies demonstrated the application of MOFs as a starting material for designing graphene-like carbon. Pachfule *et al.* proposed an interesting strategy for the synthesis of graphene-like nanoribbons of a few layer thickness using a sonochemical method, followed by thermal activation of the so-obtained carbon nanorods by the carbonization of MOF-74-rods (30–60 nm of diameter and 200–500 nm in length) [82] (Figure 5.4c).

Apart from using MOFs as precursors to synthesize porous carbons, MOFs can be integrated with other carbon-based nanomaterials to form porous graphene-based composite materials. An N-doped porous carbon@graphene (N-PC@G) composite with a sandwich-like structure was synthesized from GO and Zn^{2+}-based imidazolate framework (ZIF-8 MOF) [83]. A high specific surface area of 1,094.3 $m^2\ g^{-1}$ was reported for the obtained sandwiched carbon composite material due to a bimodal-pore structure (micropores and mesopores) [83]. Ni-based MOFs (Ni-MOFs), showing exclusive graded layered ball-within-ball nanoscale structure, have also been reported (Figure 5.4d) [84]. As a consequence of the thermal annealing of the Ni-MOF, NiO/Ni nanocrystals were encaptivated within a graphene hollow structure to result in NiO/Ni/graphene nanostructures. The ball-within-ball structure that was exhibited by the parent Ni-MOF before thermal annealing remained intact in the so formed NiO/Ni/graphene nanostructures. The resulting NiO/Ni/graphene nanostructures displayed fair conductivity [84]. Synthesis of Ni nanoparticles (Ni NPs) encapsulated few-layered nitrogen-doped graphene (Ni@NC) by high temperature annealing of Ni-MOF in nitrogen atmosphere has been reported [85]. The work demonstrated that the annealing temperature can have serious implications on the structure and the properties of the resulting Ni NP incorporated nitrogen-doped graphene [85]. Cao *et al.* proposed a facile approach for the synthesis of metal-oxide covered graphene composites. MOFs were first grown on three-dimensional graphene networks (3DGNs) that could act as the starting materials for producing metal oxides. The requisite composites consisting of graphene network and metal oxide were then produced by subsequent annealing under argon gas followed by annealing in air. The annealing of the composites in an argon atmosphere led to the betterment in the quality and surface roughness of the material. Selected-area electron diffraction pattern indicated that amorphous particles of ~10 nm were generated. Raman spectra revealed that after annealing the composite, the carbon part of ZIF-8 was transformed into amorphous carbon. This amorphous carbon acted as a buffer, obstructing further shrinkage of ZIF-8 crystals at the time of annealing [86]. Further, reduced graphene oxide-covered MoO_3 composites were synthesized by mixing Mo-based MOF with graphene oxide sheets followed by high temperature annealing [87]. The reported strategy is also useful to synthesize other graphene/metal oxide-based composites involving other appropriate MOFs [88].

5.2.3 MOFs Precursors for the Preparation of Porous Carbon Nanostructures Other Than Graphene and CNTs

The production of porous nanocarbon by heat treatment of MOFs with three different metal cations (Ga^{3+}, Cr^{3+}, and V^{3+}) at 800°C under a continuous flow of nitrogen was described by Chen *et al.* [89]. The work highlighted that the morphology as well as the obtained pore structures can be tuned by altering the calcination temperature. Carbonizing Al-PCP MOF, formed by first mixing $Al(NO_3)_3{\cdot}9H_2O$, 1,4-Naphthalenedicarboxylic acid and H_2O, using a hydrothermal method at 780°C resulted in a mesoporous nanostructure with a surface area of 1,103 $m^2 g^{-1}$ (Figure 5.4e) [75]. Zhang *et al.* proposed a nanowire-engaged designing of MOF nanofibers, where very thin tellurium nanowires with outstanding dispersity acted as templates, resulting in a controlled growth of ZIF-8 nanofibers (Figure 5.4f). The obtained nanofibers were successfully converted into highly-porous carbon nanofibers by means of doping *via* calcination [79]. Large size (diameter >100 nm) N-doped graphene tubes could be successfully synthesized from MIL-100(Fe) formed by using dicyandiamide (DCDA) and iron (II) acetate *via* high temperature graphitization [80]. Fe-MOF was also implemented as a precursor in order to synthesize porous rod-like carbonaceous electrode materials (Figure 5.4g). Fe-MOF carbonized (MOFC) material was reported assuming rod-like morphology with an average diameter of 80 nm and an average length of 300 nm [81].

5.3 Effect of MOF Pyrolysis Temperature on Porosity and Pore Size Distribution

Pyrolysis is a widely used approach to convert MOFs into porous carbon nanostructures based on the heating of MOFs at high temperatures [90–92]. While the pyrolysis temperature correlates to the specific surface area of the resulting MOF-derived carbon nanostructure [90, 92], there is an absence of any unified regularity between the pore properties of the products and variation of the pyrolysis temperature. This is due to the variation of the pyrolysis method involved for different MOF precursors and related to various other molecular and structural components. Nonetheless, some researchers have synthesized a series of Zn-containing MOFs and pyrolyzed them to infer existence of direct proportionality between the specific surface area of the carbonized products and the carbon content of ligand molecules in MOFs [93, 94]. A detailed study of the effect of the temperature variation during the process of pyrolysis was presented by Li *et al.* for a Zn-based MOF. The work involved the synthesis of Zn-MOF by implementing the solvothermal method with DMF as a solvent, followed by pyrolysis of the obtained Zn-MOF precursor at various temperatures. Pyrolysis resulted in a complete destruction of the long-range arranged morphology of the Zn-MOF precursor together with the complete destruction of the Zn and carbon composites both on the mesoscopic scale and at the molecular level. The resulting carbon nanomaterial exhibited hierarchical pore size distribution (PSD). With an enhancement in the pyrolysis temperature, PSD widens in the microporous range together with extension of both the pore volume and surface area [95].

To study the influence of temperature on the formed pores, the work by Liu *et al.* is taken as example, where nanoporous carbon materials were synthesized through the carbonization of furfuryl alcohol (FA) as a carbon precursor and MOF-5 as the template [96] at various temperatures (530, 650, 800, 900, and 1,000°C). The carbonization temperature merely affected the pore size distribution of the synthesized samples as can be perceived from the N_2 adsorption-desorption curves (Figure 5.5a), obtained by plotting the volume of N_2 absorbed (V_{ads}) as a function of the relative pressure (P/P_0). The pore size distribution curve, obtained by plotting differential pore volume distribution (D_v/D_d) vs. pore diameter (Figure 5.5b), also shows the effect of temperature of synthesis on the pore size distribution. The temperature of carbonization affected significantly the surface area of the obtained porous nanocarbon. The plot of the carbonization temperature vs. volume of N_2 adsorbed (V_{ads}) exhibited a "V" shaped curve

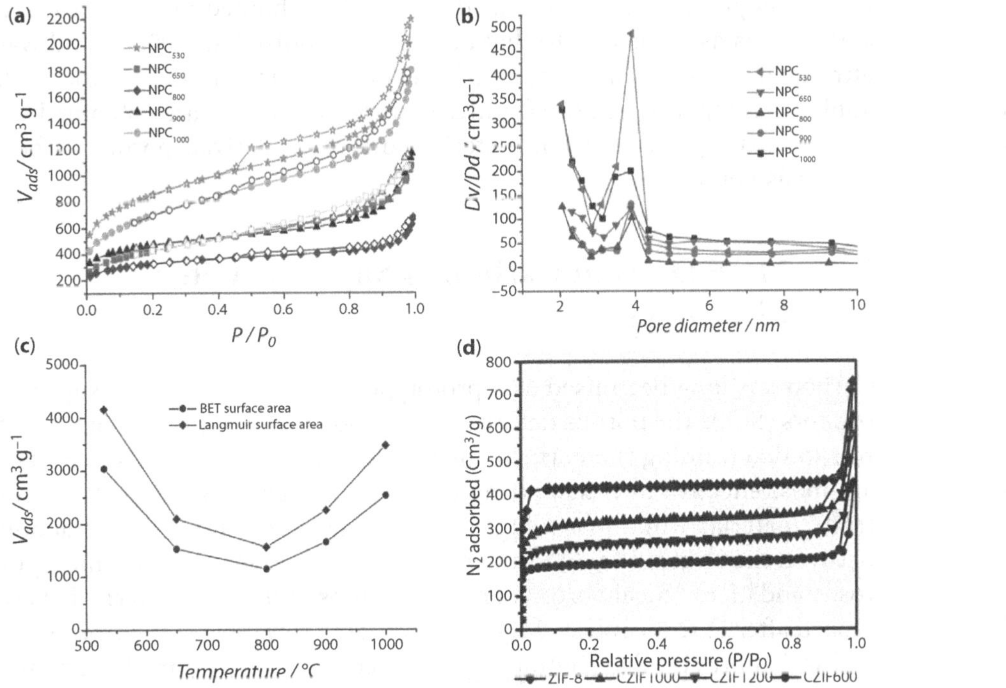

Figure 5.5 (a) N_2 adsorption-desorption isotherms of the N-doped porous carbon (NPC) materials @Copyright permission from reference [96]. (b) BJH pore size distribution of the NPC materials @Copyright permission from reference [96]. (c) BET and Langmuir surface area variation with the carbonization temperature @Copyright permission from reference [96]. (d) The N_2 adsorption-desorption isotherms of the ZIF-8 and the materials obtained from its carbonization @Copyright permission from reference [97].

(Figure 5.5c). MOF-5 decomposed into ZnO at temperatures above 500°C, which could be removed completely by washing with HCl leaving a large number of nanopores, resulting in the highest surface area. However, at temperature greater than 900°C, metallic Zn was formed and vaporized, resulting in increased pores, which in turn led to enhancement in the surface area with further increase in the carbonization temperature [96].

Yet, in another article by Abbasi *et al.* where ZIF-8 was carbonized, analysis of the nitrogen adsorption desorption curves revealed a microporous structure along with the presence of macroporous regions in between the particle voids [97]. The formation of the microporous structure was attributed to the contribution from the ZIF-8 and the evaporation of Zn during the heating process [97]. Unlike the observation made in the previous work, the active area of the sample heat-treated at 600°C decreased as compared to the parent MOF, while it increased when the MOF was treated at 1,000°C, as can be inferred from the N_2 adsorption-desorption curves (Figure 5.5d). ZIF-8 treated at 1,200°C exhibited finally a reduced surface area as compared to that of ZIF-8 carbonized at 1,000°C and was attributed to the densification of the porous carbon framework [97]. Table 5.1 compares the pore features and specific surface area obtained for the various reported carbon materials and MOF derived porous carbon materials/composites.

5.4 MOF Derived Carbon as Supercapacitor Electrodes

Carbon has long been used as a prototype electrode material in supercapacitors (SCs). The porous nature of the carbon material plays a significant role in determining the performance of these SCs. Mesoporous materials are most effective in transporting electrons and charges within the electrode material while providing a large surface area for the interaction between the electrode active material and the electrolyte. Hence, mesoporosity and hierarchical porosity are the two most important criteria that are looked after in this context. MOFs, with a large variation of their pore size and wide pore size distribution, have proved to be very promising precursors for the synthesis of porous carbon electrodes used in SCs.

Supercapacitors differ from conventional capacitors in the sense that they can provide 100 times more capacitance as they do not use dielectric material in between the two oppositely charged electrodes. Instead, charge storage in SC occurs mainly through electrostatic double layer formation and/or fast reversible redox interaction between the electrode and the

Table 5.1 Quantitative and qualitative aspects of the pore features for the previously reported carbon materials and MOF derived carbon materials.

Materials	**Precursors**	**Method**	**Specific surface area ($m^2\ g^{-1}$)**	**Pore structure**	**Ref**
Rape Pollen Carbon (RPC)	Rape pollen	Activation using $CuCl_2$	2,488	Bi-modal (meso- and microporous)	[26]
Porous Carbon (CK800)	Potassium Stearate	Carbonization	576.4	Macroporous network with intervening micropores	[27]
N-doped Carbon	Polyurethane	Ball milling, carbonization and activation using potassium carbonate	2,150	-	[28]
Waste Bagasse-based Highly Porous Carbon (WB_HPC)	Waste Bagasse	Hydrothermal carbonization and chemical activation using KOH	3,151	Bi-modal (micro- and mesopores)	[29]
3D Networked Porous Carbon	Mixture of starch and KOH	Sol gel method and carbonization	1,367.87	Microporous	[98]
MOF-5 derived Carbon	MOF-5 & Furfuryl alcohol	Pyrolysis	2,872	Microporous	[40]

(*Continued*)

Table 5.1 Quantitative and qualitative aspects of the pore features for the previously reported carbon materials and MOF derived carbon materials. (*Continued*)

Materials	**Precursors**	**Method**	**Specific surface area ($m^2 g^{-1}$)**	**Pore structure**	**Ref**
Zn-MOF derived Carbon	Zn-MOF	Pyrolysis with N_2 flow	1,513	Microporous	[95]
MOF derived Carbon	V(OH)(NDC)	Heating at 800°C	43	Bi-modal (micro- and mesoporous)	[99]
	Cr(OH)(NDC)		435		
	Ga(OH)(NDC)		274		
ZIF-8 MOF derived carbon and Graphene composite	ZIF-8 & rGO	Pyrolyzing at 900°C	1,094	Bi-modal (meso- and microporous)	[83]
MOF-5 derived Carbon	MOF-5 & melamine	Pyrolysis	765	Bi-modal (meso- and microporous)	[100]

electrolyte, as shown in the Figure 5.6a. The formation of an Electric Double Layer (EDL) is schematically seen as a charge separation phenomenon consisting of two double parallel layers of accumulated charge occurring at the interface when a material is exposed to a fluid. One layer is mainly the surface charge caused as a result of ions adsorption onto the surface of the electrode or due to chemical interactions. The other charged layer is formed as a result of ions pulled towards the electrode surface charge through Coulomb force. The ions from the second layer are not firmly attached to the electrode surface and hence can move within the fluid due to electrical or thermal interactions and this is termed as "diffuse layer". There are some other materials that can store charge in the bulk by highly reversible redox reactions like metal oxides and polymers and are termed as pseudocapacitors, as shown in the Figure 5.6b.

Since, capacitance (C) is directly proportional to the area of the electrodes (A) and inversely proportional to the distance between the two oppositely charged electrodes (d) as:

$$C \propto A \tag{5.1}$$

$$C \propto \frac{1}{d} \tag{5.2}$$

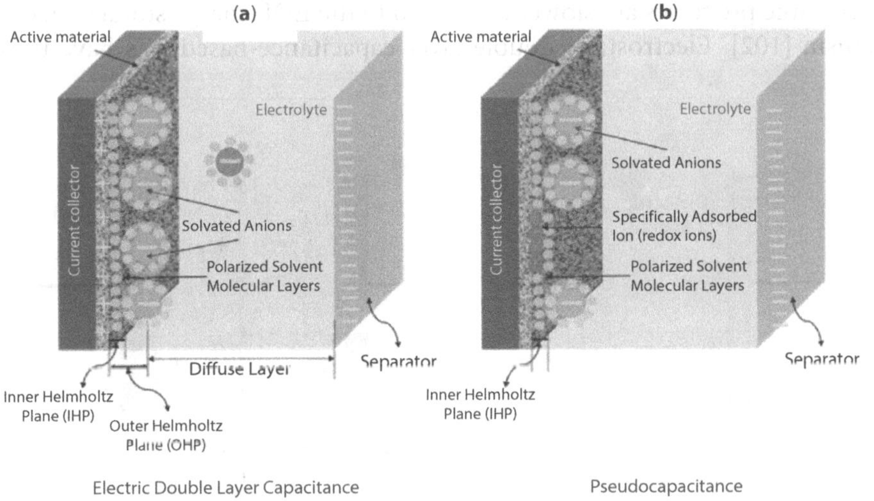

Figure 5.6 Charge storage mechanism in supercapacitors (SC). (a) Electric double layer formation. (b) Pseudocapacitive mechanism of charge storage.

high capacitance is achievable when highly porous material is used as electrode material where the total active surface area for interaction between the electrode and the electrolyte is increased as compared to the conventional capacitors (Figure 5.7a). In addition, the width of the two oppositely charged layers (the Helmholtz layers) is highly reduced (varying between 0.6 and 1 nm) in SC [101], as shown in Figure 5.7b. SCs differ from the ordinary capacitors in the sense that they do not use usual dielectric materials in solid form like ceramic porcelain, glass, mica, plastics, and metal oxides, but rather associate materials with electrostatic double-layer capacitance and/or electrochemical pseudocapacitance such as highly porous carbon materials, conducting polymers, transition metal oxides, etc. [102], contributing to the overall capacitance of the capacitor. Electric double-layer capacitance (C_{DL}) (where the charge is stored electrostatically) is a result of the charge separation caused at the electrode-electrolyte interface and the charge is stored in the so formed two oppositely charged layers (double layers). This type of interaction between the electrode active material and the electrolyte is mainly restricted to the surface of the electrode material and is highly reversible along with fast charge-discharge rates. Mostly, carbon materials store charge through the formation of EDL. In case of materials other than carbon, like metal oxides or polymers, the charge is stored through high reversible redox reaction between the electrolyte and the electrode active material. The faradaic interactions are not only restricted to the surface, but also intrude deep inside the interiors of the electrode material. Since, the charge is stored in the bulk, the charge-discharge processes are slower compared to the EDL charge storage mechanism [102]. Electrostatic double-layer capacitance-based SCs have thus

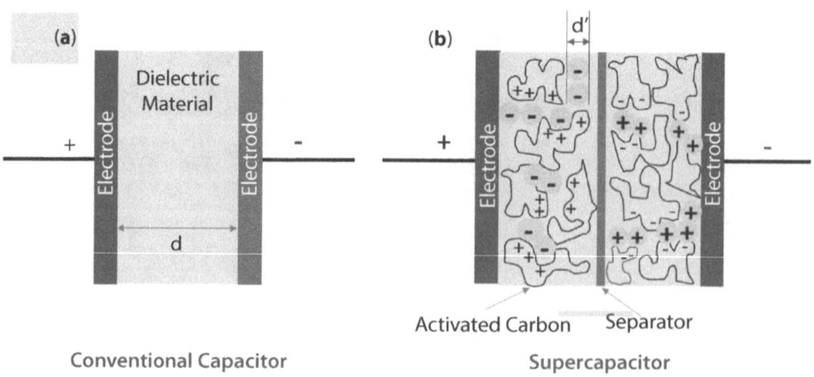

Figure 5.7 Schematic diagram representing the difference between the charge storage mechanism of (a) a conventional capacitor and (b) a supercapacitor.

the unique advantage of fast charging-discharging. In addition, large cycle life, low maintenance, light weightiness, cost-effectiveness, and eco-friendliness are some of the unique features that make these devices so promising for energy storage applications.

Supercapacitor cells can have four configurations depending on the type of active electrode material:

(i) Symmetric supercapacitor cells, consisting of similar kind and type of electrode materials as both the electrodes (e.g., RuO_2//RuO_2)
(ii) Internal Series Hybrids marked by two different active materials coated on the two electrodes, but each electrode consisting of a single active material (e.g., MnO_2//carbon nanotubes)
(iii) Internal Parallel Hybrids, characterized by two identical electrodes consisting of minimum two different active materials (e.g., RuO_2-graphene//RuO_2-graphene)
(iv) Double Hybrids, marked by two different electrodes, each consisting of a minimum of two different active materials (e.g., polyaniline-graphene//RuO_2-activated carbon) [103].

Use of carbon material in various forms such as graphene, carbon nanotubes, carbon nano onions, carbon nanohorn, hemispherical carbon, etc., is considered very promising [104]. Table 5.2 summarizes examples of such carbon materials reported as supercapacitive electrode materials and the corresponding specific capacitance [105–114].

Though carbon materials are the pioneering materials, they face the issue of low scale synthesis and high cost besides the fact that their specific capacitance is limited by the surface charge storage mechanism [104]. Conducting polymers, on the other hand, are synthesized using facile methods on a bulk scale with low cost and can store charge in the bulk [115]. Polyaniline and polypyrrole are the most exclusively studied conducting polymers as supercapacitive electrode materials [115]. However, these conducting polymers face the problem of limited surface area due to extensive agglomeration. Also, due to fragile backbone, the rate capability and cyclic stability showcased by the conducting polymers are not very impressive [116]. Pure polyaniline displays a specific capacitance of 400–500 F g^{-1} in acidic medium [116]. Polypyrrole exhibits a greater capacitance per unit volume (400–500 F cm^{-3}) owing to its dense growth [115]. However, the dense growth leads to the problem of limited access for the electrolytic ions deep into the electrode material, reducing its capacitance

Table 5.2 Comparative table listing specific capacitance obtained for various MOF derived carbon electrode materials and other carbon-based materials.

Material	Specific capacitance ($F\ g^{-1}$)	Condition	Electrolyte	Ref.
Onion like carbon	52	$10\ mV\ s^{-1}$	3.5 M KCl	[110]
Thermolyzed MOF-5 (MC-Zn)	95	$1\ A\ g^{-1}$	30 wt% KOH	[75]
NPC-600	100	$5\ mV\ s^{-1}$	1 M H_2SO_4	[42]
Thermolyzed HKUST-1 (MC-Cu)	105.3	$1\ A\ g^{-1}$	30 wt% KOH	[75]
Nitrogen-enriched nonporous carbon	140	$1\ A\ g^{-1}$	6 M KOH	[108]
N-doped porous carbon material (NPC-900)	149	$0.5\ A\ g^{-1}$	6 M KOH	[106]
Microporous carbon spheres	164	$1\ mA\ cm^{-2}$	30 wt% KOH	[109]
Thermolyzed Al-PCP (MC-Al)	173.6	$1\ A\ g^{-1}$	30 wt% KOH	[75]
Carbonized of zinc-based MOF-5 (CMOF-5-80)	218	$1\ A\ g^{-1}$	1 M H_2SO_4	[105]
KOH-activated carbon aerogels	220	$0.125\ F\ g^{-1}$	1 M H_2SO_4	[112]
Honeycomb like MnO_2/CNT	324	$0.5\ A\ g^{-1}$	Polyvinyl alcohol/ H_2SO_4	[113]
Rod-like activated porous carbon (RAPC)	332	$1\ A\ g^{-1}$	6 M KOH	[107]

(Continued)

Table 5.2 Comparative table listing specific capacitance obtained for various MOF derived carbon electrode materials and other carbon-based materials. (*Continued*)

Material	Specific capacitance ($F\ g^{-1}$)	Condition	Electrolyte	Ref.
Polypyrrole/carbon black	336	$5\ mA\ cm^{-2}$	$1\ M\ NaNO_3$	[111]
Hemispherical N-doped carbon spheres integrated with polyindole (h-NCS/PIn)	397	$1\ A\ g^{-1}$	$1\ M\ H_2SO_4$	[114]
rGO/MoO_3	404	$0.5\ A\ g^{-1}$	H_2SO_4/ polyvinyl alchohol gel	[87]
Carbonized Fe-based MOF (MOFC-600)	972	$1\ A\ g^{-1}$	6 M KOH	[81]

[115]. Metal oxides are another class of materials being studied as supercapacitor electrode materials. Though they can be easily synthesized and can store charges in the bulk of the material, they face the problem of limited surface area, toxicity, and low conductivity [117]. Exceptionally, Hu *et al.* reported nanotubular array of hydrous RuO_2 with a high value of specific capacitance of 1,300 $F\ g^{-1}$ [118].

Apart from the active material, the electrolyte also takes up a major role in providing a stable potential window to the supercapacitive cell, playing a major role in determining the energy and power delivered, and cyclic life of the supercapacitor device. Various kinds of electrolytes generally employed are:

a) Aqueous electrolytes which can provide a large conductivity: these can be acidic (e.g., 1 M H_2SO_4), basic (e.g., 6 M KOH), and/or neutral (Li_2SO_4, Na_2SO_4 etc.)
b) Organic electrolytes which can provide a large potential window, e.g., acetonitrile or poly propylene carbonate

c) Ionic liquids which are advantageous to obtain a potential window as high as 6 V along with good thermal stability and low vapor pressure, e.g., 1-decyl-3-methylimidazolium bis(trifluoro-methylsulfonyl)imide [103].

The present state-of-the-art in SCs is mostly about the use of sustainable, stable, non-toxic, easily synthesizable materials in combination with suitable electrolyte. Pyrolysis of organic waste materials (acting as carbon-rich precursors) is a very attractive and efficient way to obtain porous carbon. For instance, activated carbon derived from coconut shell exhibits a specific capacitance of 246 F g^{-1} in 0.5 M H_2SO_4 electrolyte [119]. Very importantly, surface pyrolyzed MOFs can address the problem of high cost, low scale synthesis, toxicity, agglomeration, low surface area, which are generally encountered by carbon materials for super capacitor application [56].

It is the work by Diaz *et al. via* the use of Co8-MOF-5, which highlighted for the first time the promising aspects of MOFs as supercapacitive electrode materials. Though the specific capacitance obtained for this material was not very impressive (0.49 F g^{-1} at 25 mV s^{-1} determined from cyclic voltammetry, and 0.30 F g^{-1} at the current density of 10 mA g^{-1} calculated from the charging-discharging curves), the electrical conductivity of the MOF was greater than most of the metal oxide-based electrodes reported. Also, Co8-MOF-5 behaved like electrochemical double layer supercapacitor material in acetonitrile mainly subjugated by the impact of the black carbon that has been used as the additive [120, 121]. Right after this breakthrough, the research group of Han reported a pseudocapacitive behavior of a cobalt-based MOF with a specific capacitance of 206.76 F g^{-1} [121, 122]. Inspired by these works, investigation of MOFs as electrode materials for supercapacitors eventually attracted wide interest. In the following years, Wang *et al.* prepared interweaved flexible ZIF-67 MOF with polyaniline electrode using an electrochemical deposition process. The specific areal capacitance (2,146 mF cm^{-2} at a scan rate of 10 mV s^{-1}) for the electrode material is the highest reported till date for any MOF-based electrode material [121, 123]. However, in general, the performance of MOF in pristine form still remained limited due to poor conductivity. Also, the cyclic stability and rate performance could not be enhanced much. Carbonizing MOFs turned out to be a possible way to extract the benefit of the high porosity as well as enhanced conductivity due to carbonization. In the following paragraphs, some of the reported works on carbonized MOFs used as supercapacitor electrode materials have been outlined.

Five carbon materials with pores in the nanoscale regime were produced by the method of polymerization followed by carbonization of furfuryl alcohol, acting as the carbon precursor. This was incorporated within a MOF-5 made of $Zn_4O(bdc)_3$ as the metal salt and 1,4-benzenedicarboxylate as the organic linker. These electrodes resulted in a specific capacitance above 100 F g^{-1} at a scan rate of 5 mV s^{-1} [42]. Li *et al.* converted a Cu-based MOF into N-doped carbon species with hierarchical porosity, exhibiting a regular flower-like morphology with the help of thermal carbonization. The material exhibited a surface area of 32.3 m^{-2} g^{-1} and a capacitance of 149 F g^{-1} at 0.5 A g^{-1} [106]. Carbonization of zinc-based MOF-5 mixed with 80% of polyvinyl fluoride (PVDF) resulted in a specific capacitance of 218 F g^{-1} at the current density of 1 A g^{-1} in 1 M H_2SO_4 [105]. Jiang *et al.* synthesized a rod-like carbon with hierarchical porous structure, derived from MOFs based on aluminum, with a surface area of 1,886 m^2 g^{-1}, corresponding to a high pore volume of 1.29 cm^3 g^{-1} and a uniform pore size of 4.25 nm. This material showcased excellent electrochemical characteristics along with reliability in long run and large value of specific capacitance of 345 F g^{-1} at 1 mV s^{-1} and 332 F g^{-1} at 1 A g^{-1} [107]. Cao *et al.* synthesized reduced graphene oxide (rGO) sheet wrapped around MoO_3 by simply mixing rGO with Mo-MOF, followed by annealing. The material achieved a specific capacitance of 404 F g^{-1} at 0.5 A g^{-1}. This material also exhibited extremely good cyclic stability [87]. Khan *et al.* successfully converted MOF-199 into a porous carbon material by simple pyrolysis at 700°C under inert atmosphere. In alkaline electrolyte (1 M KOH), the electrode material displayed a specific capacitance of 705 F g^{-1} at $j = 2$ mA g^{-1} [124]. Sui *et al.* reported the synthesis of Fe incorporated carbon derived from MOF. The specific capacitance for the calcinated Fe-based MOF electrode was 972 F g^{-1} at 1 A g^{-1} [81]. Table 5.2 summarizes the specific capacitance values obtained for MOF derived carbon-based electrode materials and other forms of carbon-based electrode materials.

5.5 Conclusions and Perspectives

In this chapter, a brief review of the implication of MOFs as supercapacitor electrodes, their formation strategies, and the importance of porosity has been discussed. The discussion has been taken forward by including some of the recent findings, positioning the suitability of MOFs for energy driven applications. With an ever increasing energy demand all over the world, a decrease in the availability of non-renewable energy source in the near future together with environment considerations, alternative and ecofriendly sources of energy are eagerly searched for. Carbon-based supercapacitors have long been considered ideally suited in this respect. The use of MOF has allowed to obtain porous carbon materials with superior supercapacitve behavior in the last years. While the future of MOFs is bright, several challenges related to MOF derived nanomaterials remain. Though several MOFs have been reported, only a very few of them have been practically used for SCs. There is still a need for a systematic study of the various types of MOFs to obtain a clear understanding of the influence of different metals, guest species, and organic linkers on the morphology, chemical structure, and characteristics of the pristine MOFs and derived materials. The influence and role of parameters such as pressure, temperature, and other additives/species on the conversion of MOFs into various carbon forms remain still to be understood. Although the structure of MOFs can be determined by X-ray diffraction technique, hydrogen atoms cannot be precisely located. As a result of this, the local structure about hydrogen is poorly determined. Techniques such as nuclear magnetic resonance spectroscopy need to be employed to get a deeper and complete insight of the local structures of the MOFs [125]. This might help in addition finding the right conditions to convert MOFs into specific carbon forms such as CNTs or graphene, something still rare. The implementation of *in situ* spectroscopic measurements, high-resolution imaging, and theoretical calculations might further help to obtain deeper insight into the complex mechanisms driving MOF conversion and use.

For large-scale applications of MOF, apart from thermal annealing, less energy consuming processes would be of advantage for their carbonization. The direct use of MOFs without undergoing any conversion process to be used as supercapacitor would be the ideal situation. In the present scenario, poor electrical conductivity remains one of the main problems, restricting the direct application of MOFs.

Acknowledgement

The authors acknowledge the support of CNRS and the University of Lille for providing a platform for fruitful discussions on the topic. Also, the authors acknowledge Raman-Charpak Fellowship which is sponsored together with the Department of Science and Technology, Govt. of India, French Institute in India, French Embassy in India, Ministry for Europe and Foreign Affairs, Govt. of France for supporting authors to setup a fruitful collaboration between IIT (ISM), Dhanbad, Dhanbad India, and University of Lille, Lille, France.

References

1. Brady, G.J., Way, A.J., Safron, N.S., Evensen, H.T., Gopalan, P., Arnold, M.S., Quasi-ballistic carbon nanotube array transistors with current density exceeding Si and GaAs. *Sci. Adv.*, 2, e1601240, 2016.
2. Kingston, C., Zepp, R., Andrady, A., Boverhof, D., Fehir, R., Hawkins, D., Roberts, J. *et al.*, Release characteristics of selected carbon nanotube polymer composites. *Carbon*, 68, 337, 2014.
3. Landi, B.J., Ganter, M.J., Cress, C.D., DiLeo, R.A., Raffaelle, R.P., Carbon nanotubes for lithium ion batteries. *Energy Environ. Sci.*, 2, 638, 2009.
4. Dupuis, A.C., The catalyst in the CCVD of carbon nanotubes—A review. *Prog. Mater. Sci.*, 50, 929, 2005.
5. Dresselhaus, M.S., Dressalhaus, G., Eklund, P.C., *Science of fullerenes and carbon nanotubes*, Academic Press, New York, 1996, Berlin: Springer-Verlag, pp. 12–51, 2001.
6. Ming, J.H. and Lin, W.C., Unprecedented re-growth of carbon nanotubes on *in situ* re-activated catalyst. *Nanotechnology*, 20, 0256081, 2009.
7. Rahman, G., Najaf, Z., Mehmood, A., Bilal, S., Shah, A., Mian, S., Ali, G., An overview of the recent progress in the synthesis and applications of carbon nanotubes. *J. Carbon Res.*, 5, 3, 2019.
8. Shaikjee, A. and Coville, N.J., The synthesis, properties and uses of carbon materials with helical morphology. *J. Adv. Res.*, 3, 195, 2012.
9. Volder, D., Michael, F.L., Tawfick, S.H., Baughman, R.H., Hart, A.J., Carbon nanotubes: Present and future commercial applications. *Science*, 339, 535, 2013.
10. Han, Z. and Fina, A., Thermal conductivity of carbon nanotubes and their polymer nanocomposites: A review. *Prog. Polym. Sci.*, 36, 91, 2011.
11. Bao, Q. and Pan, C., Electric field induced growth of well aligned carbon nanotubes from ethanol flames. *Nanotechnology*, 17, 1016, 2006.
12. Chen, L., Bai, J., Wang, C., Pan, Y., Scheer, M., You, X., One-step solid-state thermolysis of a metal–organic framework: A simple and facile route to large-scale of multiwalled carbon nanotubes. *Chem. Commun.*, 3, 1581, 2008.
13. Lee, T., Min, S.H., Gu, M., Jung, Y.K., Lee, W., Lee, J.U., Seong, D.G., Kim, B.S., Layer-by-layer assembly for graphene-based multilayer nanocomposites: Synthesis and applications. *Chem. Mater.*, 27, 3785, 2015.
14. Pan, Y., Sahoo, N.G., Li, L., The application of graphene oxide in drug delivery. *Expert Opin. Drug Deliv.*, 9, 1365, 2012.

15. Lee, W.J., Maiti, U.N., Lee, J.M., Lim, J., Han, T.H., Kim, S.O., Nitrogen-doped carbon nanotubes and graphene composite structures for energy and catalytic applications. *Chem. Commun.*, 50, 6818, 2014.
16. Brownson, D.A.C. and Banks, C.E., Graphene electrochemistry: An overview of potential applications. *Analyst*, 135, 2768, 2010.
17. Choi, W., Lahiri, I., Seelaboyina, R., Kang, Y.S., Synthesis of graphene and its applications: A review. *Crit. Rev. Solid State Mater. Sci.*, 35, 52–71, 2010.
18. Terrones, M., Méndez, A.R.B., Delgado, J.C., Urias, F.L., Cantú, Y.I.V., Macías, F.J.R., Elías, A.L. *et al.*, Graphene and graphite nanoribbons: Morphology, properties, synthesis, defects and applications. *Nano Today*, 5, 351, 2010.
19. Eigler, S., Heim, M.E., Grimm, S., Hofmann, P., Kroener, W., Geworski, A., Dotzer, C. *et al.*, Wet chemical synthesis of graphene. *Adv. Mater.*, 25, 3583, 2013.
20. Xu, B., Yue, S., Sui, Z., Zhang, X., Hou, S., Cao, G., Yang, Y., What is the choice for supercapacitors: Graphene or graphene oxide? *Energy Environ. Sci.*, 4, 2826, 2011.
21. Zhu, Y., Murali, S., Stoller, M.D., Ganesh, K.J., Cai, W., Ferreira, P.J., Pirkle, A. *et al.*, Carbon-based supercapacitors produced by activation of graphene. *Science*, 332, 1537, 2011.
22. Singh, S.K., Dhavale, V.M., Boukherroub, R., Kurungot, S., Szunerits, S., N-doped porous reduced graphene oxide as an efficient electrode material for high performance flexible solid-state supercapacitor. *Appl. Mater. Today*, 8, 141, 2017.
23. Ferrero, G.A., Fuertes, A.B., Sevilla, M., N-doped microporous carbon microspheres for high volumetric performance supercapacitors. *Electrochim. Acta*, 168, 320, 2015.
24. Kao, L.-H. and Hsu, T.-C., Silica template synthesis of ordered mesoporous carbon thick films with 35-nm pore size from mesophase pitch solution. *Mater. Lett.*, 62, 695, 2008.
25. Xiong, W., Gao, Y., Wu, X., Hu, X., Lan, D., Chen, Y., Pu, X., Zeng, Y., Su, J., Zhu, Z., Composite of macroporous carbon with honeycomb-like structure from mollusc shell and NiCo2O4 nanowires for high-performance supercapacitor. *ACS Appl. Mater. Interfaces*, 6, 19416, 2014.
26. Bingfu, L., Xiao, Y., Liu, Y., Large-scale synthesis of porous carbon *via* one-step CuCl 2 activation of rape pollen for high-performance supercapacitors. *J. Mater. Chem. A*, 6, 12046, 2018.
27. Luo, H.M., Chen, H., Chen, Y.Z., Li, P., Zhang, J.Q., Zhao, X., Simple synthesis of porous carbon materials for high-performance supercapacitors. *J. Appl. Electrochem.*, 46, 703, 2016.
28. Christina, S., Otto, P., Leistenschneider, D., Grätz, S., Eßbach, C., Borchardt, L., Upcycling of polyurethane waste by mechanochemistry: Synthesis of N-doped porous carbon materials for supercapacitor applications. *Beilstein J. Nanotechnol.*, 10, 1618, 2019.
29. Yu, P., Liang, Y., Dong, H., Hu, H., Liu, S., Peng, L., Zheng, M., Xiao, Y., Liu, Y., Rational synthesis of highly porous carbon from waste bagasse for

advanced supercapacitor application. *ACS Sustainable Chem. Eng.*, 6, 15325, 2018.
30. Lee, J., Kim, J., Hyeon, T., Recent progress in the synthesis of porous carbon materials. *Adv. Mater.*, 18, 2073, 2006.
31. Yang, Z., Xia, Y., Mokaya, R., Enhanced hydrogen storage capacity of high surface area zeolite-like carbon materials. *J. Am. Chem. Soc.*, 129, 1673, 2007.
32. Hou, P.X., Orikasa, H., Yamazaki, T., Matsuoka, K., Tomita, A., Setoyama, N. *et al.*, Synthesis of nitrogen-containing microporous carbon with a highly ordered structure and effect of nitrogen doping on H_2O adsorption. *Chem. Mater.*, 17, 5187, 2005.
33. Ania, C.O., Khomenko, V., Ero, E.R., Parra, J.B., Béguin, F., The large electrochemical capacitance of microporous doped carbon obtained by using a zeolite template. *Adv. Funct. Mater.*, 17, 1828, 2007.
34. Vinu, A., Ariga, K., Mori, T., Nakanishi, T., Hishita, S., Golberg, D. *et al.*, Preparation and characterization of well-ordered hexagonal mesoporous carbon nitride. *Adv. Mater.*, 17, 1648, 2005.
35. Joo, S.H., Seong, J.C., Oh, I., Kwak, J., Liu, Z., Terasaki, O., Ryoo, R., Ordered nanoporous arrays of carbon supporting high dispersions of platinum nanoparticles. *Nature*, 412, 169, 2001.
36. Acharya, M. and Foley, H.C., Spray-coating of nanoporous carbon membranes for air separation. *J. Membr Sci.*, 161, 1, 1999.
37. Wang, H., Gao, Q., Hu, J., Chen, Z., High performance of nanoporous carbon in cryogenic hydrogen storage and electrochemical capacitance. *Carbon*, 47, 2259, 2009.
38. Xia, K., Gao, Q., Wu, C., Song, S., Ruan, M., Activation, characterization and hydrogen storage properties of the mesoporous carbon CMK-3. *Carbon*, 45, 1989, 2007.
39. Jiang, J., Gao, Q., Xia, K., Hu, J., Enhanced electrical capacitance of porous carbons by nitrogen enrichment and control of the pore structure. *Microporous Mesoporous Mater.*, 118, 28, 2009.
40. Hu, J., Wang, H., Gao, Q., Guo, H., Porous carbons prepared by using metal–organic framework as the precursor for supercapacitors. *Carbon*, 48, 3599, 2010.
41. Liu, B., Shioyama, H., Akita, T., Xu, Q., Metal-organic framework as a template for porous carbon synthesis. *J. Am. Chem. Soc.*, 130, 5390, 2008.
42. Liu, B., Shioyama, H., Jiang, H., Zhang, X., Xu, Q., Metal-organic framework (MOF) as a template for syntheses of nanoporous carbons as electrode materials for supercapacitor. *Carbon*, 48, 456, 2010.
43. Gonzalez, J., Devi, R.N., Tunstall, D.P., Cox, P.A., Wright, P.A., Deuterium NMR studies of framework and guest mobility in the metal-organic framework compound MOF-5, Zn_4O $(O_2CC_6H_4CO_2)_3$. *Microporous Mesoporous Mater.*, 84, 97, 2005.

44. Hafizovic, J., Bjørgen, M., Olsbye, U., Dietzel, P.D.C., Bordiga, S., Prestipino, C. *et al.*, The inconsistency in adsorption properties and powder XRD data of MOF-5 is rationalized by framework interpenetration and the presence of organic and inorganic species in the nanocavities. *J. Am. Chem. Soc.*, 129, 3612, 2007.
45. Ping, W., Gong, P., Sun, J., Wang, J., Yang, S., Design and synthesis of Ni-MOF/CNT composites and rGO/carbon nitride composites for an asymmetric supercapacitor with high energy and power density. *J. Mater. Chem. A*, 3, 13874–13883, 2015.
46. Jia, Y., Mu, C., Yan, B., Qin, X., Shen, C., Xue, H., Pang, H., Nanoparticle/MOF composites: Preparations and applications. *Mater. Horiz.*, 4, 557–569, 2017.
47. Noro, S. and Nakamura, T., Fluorine-functionalized metal–organic frameworks and porous coordination polymers. *NPG Asia Mater.*, 9, e433, 2017.
48. Yuan, N., Church, T.L., Brandt, E.G., Hedin, N., Zou, X., Bernin, D., Insights into Functionalization of Metal-Organic Frameworks Using *In-Situ* NMR Spectroscopy. *Sci. Rep.*, 8, 17530, 2018.
49. Li, B., Ma, D., Li, Y., Zhang, Y., Li, G., Shi, Z., Feng, S., Zaworotko, M.J., Ma, S., Dual Functionalized Cages in Metal–Organic Frameworks *via* Stepwise Postsynthetic Modification. *Chem. Mater.*, 28, 4781, 2016.
50. Hao, J., Xu, X., Fei, H., Li, L., Yan, B., Functionalization of Metal–Organic Frameworks for Photoactive Materials. *Adv. Mater.*, 30, 1705634, 2018.
51. Cohen, S.M., Postsynthetic methods for the functionalization of metal–organic frameworks. *Chem. Rev.*, 112, 970, 2011.
52. Yaghi, O. and Li, H., Hydrothermal Synthesis of a Metal–Organic Framework Containing Large Rectangular Channels. *J. Am. Chem. Soc.*, 117, 10401, 1995.
53. Sharmin, E. and Zafar, F., Introductory chapter: Metal organic frameworks (MOFs), in: *Metal-organic frameworks*, Rijeka: Intech Open, 2016.
54. Long, J.R. and Yaghi, O.M., The pervasive chemistry of metal–organic frameworks. *Chem. Soc. Rev.*, 38, 1213, 2009.
55. Trickett, C.A., Helal, A., Al-Maythalony, B.A., Yamani, Z.H., Cordova, K.E., Yaghi, O.M., The chemistry of metal–organic frameworks for CO_2 capture, regeneration and conversion. *Nat. Rev. Mater.*, 2, 17045, 2017.
56. Wang, L., Han, Y., Feng, X., Wang, B., Metal–organic frameworks for energy storage: Batteries and supercapacitors. *Coord. Chem. Rev.*, 307, 361, 2016.
57. Furukawa, H., Cordova, K.E., O'Keeffe, M., Yaghi, O.M., The Chemistry and Applications of Metal–Organic Frameworks. *Science*, 341, 974, 2013.
58. Wu, Y., Pang, H., Yao, W., Wang, X., Yu, S., Yu, Z., Wang, X., Synthesis of rod-like metal-organic framework (MOF-5) nanomaterial for efficient

removal of U (VI): Batch experiments and spectroscopy study. *Sci. Bull.*, 63, 831, 2018.
59. He, H., Sun, F., Ma, S., Zhu, G., Reticular synthesis of a series of HKUST-like MOFs with carbon dioxide capture and separation. *Inorg. Chem.*, 55, 9071, 2016.
60. Gordon, J., Kazemian, H., Rohani, S., MIL-53 (Fe), MIL-101, and SBA-15 porous materials: Potential platforms for drug delivery. *Mater. Sci. Eng.: C*, 47, 172, 2015.
61. Zafar, F. and Sharmin, E. (Eds.), *Metal-Organic Frameworks*, Rijeka: BoD–Books on Demand, 2016.
62. Chaikittisilp, W., Ariga, K., Yamauchi, Y., A new family of carbon materials: Synthesis of MOF-derived nanoporous carbons and their promising applications. *J. Mater. Chem. A*, 1, 14, 2013.
63. Liu, B., Shioyama, H., Akita, T., Xu, Q., Metal-organic framework as a template for porous carbon synthesis. *J. Am. Chem. Soc.*, 130, 5390, 2008.
64. Salunkhe, R.R., Tang, J., Kamachi, Y., Nakato, T., Kim, J.H., Yamauchi, Y., Asymmetric supercapacitors using 3D nanoporous carbon and cobalt oxide electrodes synthesized from a single metal–organic framework. *ACS Nano*, 9, 6288, 2015.
65. Dresselhaus, M.S., Dresselhaus, G., Avouris, P. (Eds.), *Carbon nanotubes: synthesis, structure, properties, and applications*, Verlag Berlin Heidelberg: Springer Science & Business Media, 2003.
66. Britt, D., Furukawa, H., Wang, B., Glover, T.G., Yaghi, O.M., Highly efficient separation of carbon dioxide by a metal-organic framework replete with open metal sites. *Proc. Natl. Acad. Sci.*, 106, 20637, 2009.
67. Liu, R., Yu, T., Shi, Z., Wang, Z., The preparation of metal–organic frameworks and their biomedical application. *Int. J. Nanomed.*, 11, 1187, 2016.
68. Lu, W., Wei, Z., Gu, Z.Y., Liu, T.F., Park, J., Park, J., Tian, J. *et al.*, Tuning the structure and function of metal–organic frameworks *via* linker design. *Chem. Soc. Rev.*, 43, 5561, 2014.
69. Williams, K., Metal-Organic Frameworks and MOF-derived Carbon Materials for Fuel Cell Applications. Graduate Theses and Dissertations, South Florida: University of South Florida, 2017.
70. Wang, M., Lin, M., Li, J., Huang, L., Zhuang, Z., Lin, C., Zhou, L., Mai, L., Metal–organic framework derived carbon-confined Ni_2P nanocrystals supported on graphene for an efficient oxygen evolution reaction. *Chem. Commun.*, 53, 8372, 2017.
71. Jeong, Y.C., Seo, J.W., Kim, J.H. *et al.*, Function-regeneration of non porous hydrolyzed-MOF-derived materials. *Nano Res.*, 12, 1921, 2019.
72. Chaikittisilp, W., Ariga, K., Yamauchi, Y., A new family of carbon materials: Synthesis of MOF-derived nanoporous carbons and their promising applications. *J. Mater. Chem. A*, 1, 14, 2013.
73. Li, Q., Pan, H., Higgins, D., Cao, R., Zhang, G., Lv, H., Wu, K., Cho, J., Wu, G., Metal–Organic Framework-Derived Bamboo-like Nitrogen-Doped

Graphene Tubes as an Active Matrix for Hybrid Oxygen-Reduction Electrocatalysts. *Small*, 11, 1443, 2015.

74. Xia, B., Yu, Y.Y., Li, N., Wu, H.B., Lou, X.W.D., Wang, X., A metal–organic framework-derived bifunctional oxygen electrocatalyst. *Nat. Energy*, 1, 15006, 2016.
75. Yan, X., Li, X., Yan, Z., Komarneni, S., Porous carbons prepared by direct carbonization of MOFs for supercapacitors. *Appl. Surf. Sci.*, 308, 306, 2014.
76. Su, P., Xiao, H., Zhao, J., Yao, Y., Shao, Z., Li, C., Yang, Q., Nitrogen-doped carbon nanotubes derived from Zn–Fe-ZIF nanospheres and their application as efficient oxygen reduction electrocatalysts with *in situ* generated iron species. *Chem. Sci.*, 4, 294, 2013.
77. Kukulka, W., Cendrowski, K., Michalkiewicz, B., Mijowska, E., MOF-5 derived carbon as material for CO 2 absorption. *RSC Adv.*, 9, 18527, 2019.
78. Ahsan, M.A., Deemer, E., Fernandez-Delgado, O., Wang, H., Curry, M.L., El-Gendy, A.A., Noveron, J.C., Fe nanoparticles encapsulated in MOF-derived carbon for the reduction of 4-nitrophenol and methyl orange in water. *Catal. Commun.*, 130, 105753, 2019.
79. Zhang, W., Wu, Z.Y., Jiang, H.L., Yu, S.H., Nanowire-directed templating synthesis of metal–organic framework nanofibers and their derived porous doped carbon nanofibers for enhanced electrocatalysis. *J. Am. Chem. Soc.*, 136, 14385, 2014.
80. Li, Q., Pan, H., Higgins, D., Cao, R., Zhang, G., Lv, H., Wu, K., Cho, J., Wu, G., Metal–Organic Framework-Derived Bamboo-like Nitrogen-Doped Graphene Tubes as an Active Matrix for Hybrid Oxygen-Reduction Electrocatalysts. *Small*, 11, 1443, 2015.
81. Sui, Y., Zhang, D., Han, Y., Sun, Z., Qi, J., Wei, F., He, Y., Meng, Q., Effects of Carbonization Temperature on Nature of Nanostructured Electrode Materials Derived from Fe-MOF for Supercapacitors. *Electron. Mater. Lett.*, 14, 548, 2018.
82. Huang, L., Wang, H., Chen, J., Wang, Z., Sun, J., Zhao, D. *et al.*, Synthesis, morphology control, and properties of porous metal-organic coordination polymers. *Microporous Mesoporous Mater.*, 58, 105, 2003.
83. Liu, S., Zhang, H., Zhao, Q., Zhang, X., Liu, R., Ge, X., Wang, G., Zhao, H., Cai, W., Metal-organic framework derived nitrogen-doped porous carbon@ graphene sandwich-like structured composites as bifunctional electrocatalysts for oxygen reduction and evolution reactions. *Carbon*, 106, 74, 2016.
84. Zou, F., Chen, Y.M., Liu, K., Yu, Z., Liang, W., Bhaway, S.M., Gao, M., Zhu, Y., Metal organic frameworks derived hierarchical hollow NiO/Ni/graphene composites for lithium and sodium storage. *ACS Nano*, 10, 377, 2015.
85. Xu, Y., Tu, W., Zhang, B., Yin, S., Huang, Y., Kraft, M., Xu, R., Nickel Nanoparticles Encapsulated in Few-Layer Nitrogen-Doped Graphene Derived from Metal–Organic Frameworks as Efficient Bifunctional Electrocatalysts for Overall Water Splitting. *Adv. Mater.*, 29, 1605957, 2017.

86. Cao, X., Zheng, B., Rui, X., Shi, W., Yan, Q., Zhang, H., Metal oxide-coated three-dimensional graphene prepared by the use of metal–organic frameworks as precursors. *Angew. Chem. Int. Ed.*, 53, 1404, 2014.
87. Cao, X., Zheng, B., Shi, W., Yang, J., Fan, Z., Luo, Z., Rui, X., Chen, B., Yan, Q., Zhang, H., Reduced Graphene Oxide-Wrapped MoO_3 Composites Prepared by Using Metal–Organic Frameworks as Precursor for All-Solid-State Flexible Supercapacitors. *Adv. Mater.*, 27, 4695, 2015.
88. Shao, J., Zhou, H., Zhu, M., Feng, J., Yuan, A., Facile synthesis of metal-organic framework-derived Co_3O_4 with different morphologies coated graphene foam as integrated anodes for lithium-ion batteries. *J. Alloys Compd.*, 768, 1049, 2018.
89. Chen, J.J., Chen, Y.T., Raja, D.S., Kang, Y.H., Tseng, P.C., Lin, C.H., Carbonization and oxidation of metal–organic frameworks based on 1, 4-naphthalene dicarboxylates. *Sci. Technol. Adv. Mater.*, 16, 054203, 2015.
90. Tang, J. and Yamauchi, Y., Carbon materials: MOF morphologies in control. *Nat. Chem.*, 8, 638, 2016.
91. Zhang, L., Su, Z., Jiang, F., Yang, L., Qian, J., Zhou, Y., Li, W., Hong, M., Highly graphitized nitrogen-doped porous carbon nanopolyhedra derived from ZIF-8 nanocrystals as efficient electrocatalysts for oxygen reduction reactions. *Nanoscale*, 6, 6590, 2014.
92. Chaikittisilp, W., Hu, M., Wang, H., Huang, H.S., Fujita, T., Wu, K.C., Chen, L.C., Yamauchi, Y., Ariga, K., Nanoporous carbons through direct carbonization of a zeolitic imidazolate framework for supercapacitor electrodes. *Chem. Commun.*, 48, 7259, 2012.
93. Lim, S., Suh, K., Kim, Y., Yoon, M., Park, H., Dybtsev, D.N., Kim, K., Porous carbon materials with a controllable surface area synthesized from metal-organic frameworks. *Chem. Commun.*, 48, 7447, 2012.
94. Aiyappa, H.B., Pachfule, P., Banerjee, R., Kurungot, S., Porous carbons from nonporous MOFs: Influence of ligand characteristics on intrinsic properties of end carbon. *Cryst. Growth Des.*, 13, 4195, 2013.
95. Xing, C., Liu, Y., Su, Y., Chen, Y., Hao, S., Wu, X., Wang, X., Cao, H., Li, B., Structural Evolution of Co-Based Metal Organic Frameworks in Pyrolysis for Synthesis of Core–Shells on Nanosheets: Co@CoO_x@Carbon-rGO Composites for Enhanced Hydrogen Generation Activity. *ACS Appl. Mater. Interfaces*, 8, 15430, 2016.
96. Yang, L., Zeng, X., Wang, W., Cao, D., Recent Progress in MOF-Derived, Heteroatom-Doped Porous Carbons as Highly Efficient Electrocatalysts for Oxygen Reduction Reaction in Fuel Cells. *Adv. Funct. Mater.*, 28, 1704537, 2018.
97. Abbasi, Z., Shamsaei, E., Leong, S.K., Ladewig, B., Zhang, X., Wang, H., Effect of carbonization temperature on adsorption property of ZIF-8 derived nanoporous carbon for water treatment. *Microporous Mesoporous Mater.*, 236, 28–37, 2016.

98. Zhong, Y., Shi, T., Huang, Y., Cheng, S., Liao, G., Tang, Z., One-step synthesis of porous carbon derived from starch for all-carbon binder-free high-rate supercapacitor. *Electrochim. Acta*, 269, 676, 2018.
99. Chen, J., Chen, Y., Senthil Raja, D., Kang, Y., Tseng, P., Lin, C., Carbonization and oxidation of metal–organic frameworks based on 1, 4-naphthalene dicarboxylates. *Sci. Technol. Adv. Mater.*, 16, 054203, 2015.
100. Pandiaraj, S., Aiyappa, H.B., Banerjee, R.l, Kurungot, S., Post modification of MOF derived carbon *via* gC_3N_4 entrapment for an efficient metal-free oxygen reduction reaction. *Chem. Commun.*, 50, 3363, 2014.
101. Huang, J., Qiao, R., Sumpter, B.G., Meunier, V., Effect of diffuse layer and pore shapes in mesoporous carbon supercapacitors. *J. Mater. Res.*, 25, 1469, 2010.
102. Halper, M.S. and Ellenbogen, J.C., *Supercapacitors: A brief overview*, p. 1, The MITRE Corporation, McLean, Virginia, USA, 2006.
103. Baptista, J.M., Sagu, J., Wijayantha, U., Lobato, K., State-of-the-art materials for high power and high energy supercapacitors: Performance metrics and obstacles for the transition from lab to industrial scale-A critical approach. *Chem. Eng. J.*, 374, 1153, 2019.
104. Li, Z., Xu, K., Pan, Y., Recent development of Supercapacitor Electrode Based on Carbon Materials. *Nanotechnol. Rev.*, 8, 35, 2019.
105. Cendrowski, K., Kukulka, W., Kedzierski, T., Zhang, S., Mijowska, E., Poly (vinylidene fluoride) and carbon derivative structures from eco-friendly MOF-5 for supercapacitor electrode preparation with improved electrochemical performance. *Nanomaterials*, 8, 890, 2018.
106. Li, Z., Zou, K., Zhang, X., Han, T., Yang, Y., Hierarchically flower-like N-doped porous carbon materials derived from an explosive 3-fold interpenetrating diamondoid copper metal–organic framework for a supercapacitor. *Inorg. Chem.*, 55, 6552, 2016.
107. Jiang, W., Pan, J., Liu, X., A novel rod-like porous carbon with ordered hierarchical pore structure prepared from Al-based metal-organic framework without template as greatly enhanced performance for supercapacitor. *J. Power Sources*, 409, 13, 2019.
108. Hulicova-Jurcakova, D., Kodama, M., Shiraishi, S., Hatori, H., Zhu, Z.H., Lu, G.Q., Nitrogenenriched nonporous carbon electrodes with extraordinary supercapacitance. *Adv. Funct. Mater.*, 19, 1800, 2009.
109. Xia, X., Shi, L., Liu, H., Yang, L., He, Y., A facile production of microporous carbon spheres and their electrochemical performance in EDLC. *J. Phys. Chem. Solids*, 73, 385, 2012.
110. Van, K.L., Maleski, K., Mathis, T.S., Breslin, J.P., Gogotsi, Y.G., Processing of Onion-like Carbon for Electrochemical Capacitors. *ECS J. Solid State Sc.*, 6, M3103, 2017.
111. 44. Yang, C., Liu, P., Wang, T., Well-Defined core-shell carbon black/polypyrrole nanocomposites for electrochemical energy storage. *ACS Appl. Mater. Interfaces*, 3, 1109, 2011.

112. Zapata-Benabithe, Z., Carrasco-Marín, F., Moreno-Castilla, C., Preparation, surface characteristics, and electrochemical double-layer capacitance of KOH-activated carbon aerogels and their O-and N-doped derivatives. *J. Power Sources*, 219, 80, 2012.
113. Ko, W., Chen, Y., Lu, K., Lin, K., Porous honeycomb structures formed from interconnected MnO_2 sheets on CNT-coated substrates for flexible all-solid-state supercapacitors. *Sci. Rep.*, 6, 18887, 2016.
114. Majumder, M., Choudhary, R.B., Thakur, A.K., Hemispherical nitrogen-doped carbon spheres integrated with polyindole as high performance electrode material for supercapacitor applications. *Carbon*, 142, 650, 2019.
115. Snook, G.A., Kao, P., Best, A.S., Conducting-polymer-based supercapacitor devices and electrodes. *J. Power Sources*, 196, 1, 2011.
116. Talbi, H., Just, P.E., Dao, L.H., Electropolymerization of aniline on carbonized polyacrylonitrile aerogel electrodes: Applications for supercapacitors. *J. Appl. Electrochem.*, 33, 465, 2003.
117. Li, Z., Xu, K., Pan, Y., Recent development of Supercapacitor Electrode Based on Carbon Materials. *Nanotechnol. Rev.*, 8, 35, 2019.
118. Peng, Z., Liu, X., Meng, H., Li, Z., Li, B., Liu, Z., Liu, S., Design and tailoring of the 3D macroporous hydrous RuO2 hierarchical architectures with a hard-template method for high-performance supercapacitors. *ACS Appl. Mater. Interfaces*, 9, 4577, 2016.
119. Larasati, T.D., Prakoso, T., Rizkiana, J., Devianto, H., Widiatmoko, P., Nurdin, I., Nano Carbon Produced by Advanced Mild Hydrothermal Process of Oil Palm Biomass for Supercapacitor Material, in: *IOP Conference series: Materials Science and Engineering*, vol. 543, p. 012031, 2019.
120. Díaz, R., Orcajo, M.G., Botas, J.A., Calleja, G., Palma, J., Co_8-MOF-5 as electrode for supercapacitors. *Mater. Lett.*, 68, 126, 2012.
121. Xu, G., Nie, P., Dou, H., Ding, B., Li, L., Zhang, X., Exploring metal organic frameworks for energy storage in batteries and supercapacitors. *Mater. Today*, 20, 191, 2017.
122. Lee, D., Yoon, Y.S.J., Shrestha, N.K., Lee, S.H., Ahn, H., Han, S.H., Unusual energy storage and charge retention in Co-based metal–organic-frameworks. *Microporous Mesoporous Mater.*, 153, 163, 2012.
123. Wang, L., Feng, X., Ren, L., Piao, Q., Zhong, J., Wang, Y., Li, H., Chen, Y., Wang, B., Flexible solid-state supercapacitor based on a metal–organic framework interwoven by electrochemically-deposited PANI. *J. Am. Chem. Soc.*, 137, 4920, 2015.
124. Khan, I.A., Badshah, A., Nadeem, M.A., Haider, N., Nadeem, M.A., A copper based metal-organic framework as single source for the synthesis of electrode materials for high-performance supercapacitors and glucose sensing applications. *Int. J. Hydrogen Energy*, 39, 19609, 2014.
125. Lucier, B.E.G., Chen, S., Huang, Y., Characterization of metal–organic frameworks: Unlocking the potential of solid-state NMR. *Acc. Chem. Res.*, 51, 319, 2017.

6

Application of Two-Dimensional Monoelements-Based Material in Field-Effect Transistor for Sensing and Biosensing

Tejaswini Sahoo, Jnana Ranjan Sahu, Jagannath Panda, Neeraj Kumari and Rojalin Sahu*

School of Applied Sciences, Kalinga Institute of Industrial Technology, Deemed to be University, Bhubaneswar, India

Abstract

The development of new two-dimensional (2D) monoelements-based material, also known as Xenes (like tellurene, bismuthene, antimonene, arsenene, phosphorene, stanene, germanene grapheme, silicone, and borophene) has attracted many researchers due to their wide range of applications in diverse sector along with promoting novel innovations in the field of science. These 2D monoelements due to their impressive physical, chemical, electronic, and optical properties have been identified as potential agents for a variety of applications such as theranostics, therapeutic delivery, bioimaging, gas sensors, biosensors, field-effect transistors, semiconductors, etc. This chapter covers basic idea of 2D monoelements, classifications, and application in field-effect transistors for sensing and biosensing. Finally, various challenges, future developments and research progress are also focused.

Keywords: Two-dimensional, monoelements, field-effect transistor, sensor, biosensor

**Corresponding author*: rsahufch@kiit.ac.in

Inamuddin, Rajender Boddula, Mohd Imran Ahamed and Abdullah M. Asiri (eds.)
Monoelements: Properties and Applications, (119–148) © 2020 Scrivener Publishing LLC

6.1 Introduction

The two-dimensional (2D) material comprises of thin sheets of atom have lured comprehensive consideration because of their peerless and unusual properties [1–3]. Supported by victory of graphene, a superabundance of 2D material monoelements have been developed to be used in environment, energy, catalyst, field-effect transistor (FET), sensors, biomedicines, biosensors, etc. Few examples of sheet-like structure which have weak bonding among layers and have in-plane strong covalent bond are, transition metal dichalcogenides (TMD), boron nitride (hexagonal), carbo nitrides, and nitrides (MXenes), and their derivative. After exfoliation, all the atoms get exposed on the surface (ultra thin form), as a result the surface area of 2D materials gets enlarged, thereby reactivating its chemical and physical properties affecting the 2D wave function from quantum confinement effects. These ultrathin 2D materials exhibit special catalysis, electronic, magnetic properties better than those of bulk structured materials and promoting the above-mentioned applications. Recently, developed phosphorene and borophene were reported to exhibit extraordinary role in the field of nanotechnology [4, 5], thereby globally addressing difficulties related to electronics, medical, environmental, etc. These 2D monoelements (Xenes) have been reported as potential candidate in the field of biological theranostics for diagnosis such as PTT (photothermal therapy) [6], PDT (photodynamic therapy) [7], FI (fluroscence imaging) [8], PLTI (photo luminescence lifetime imaging) [9], CT (computed tomography) [10], PTI (photothermal imaging) [6], PAI (photoacoustic imaging) [11]. The unique surface characteristics of these 2D monoelements (Xenes) enhance their bio-sensing applications such as DNA sensors, cellular growth factors, hydrogen peroxides, etc. [12]. The biosensing activity of these materials can be enhanced by improving their electrical properties and chemical fictionalizations. TMD are associated with low carrier mobility which creates hurdle in the biosensing activity. Even though graphenes and Mxenes exhibit high carrier mobility, the zero band gap and limited band gap of MXenes and graphene, respectively, restrict their performance in the field of sensing, bioimaging, and FET. Hexagonal boron nitride acts as good conductor due to their excellent electrical resistance but its insulating properties limit its biomedical application. In contrast to these 2D materials, 2D Xenes like phosphorene exhibited potential to join the space between insulator, zero band gap, limited band gap and large band gap [5]. By, manipulating the interactions between Xenes and electromagnetic wave (UV to near infrared region) contributes in large extinction coefficient of

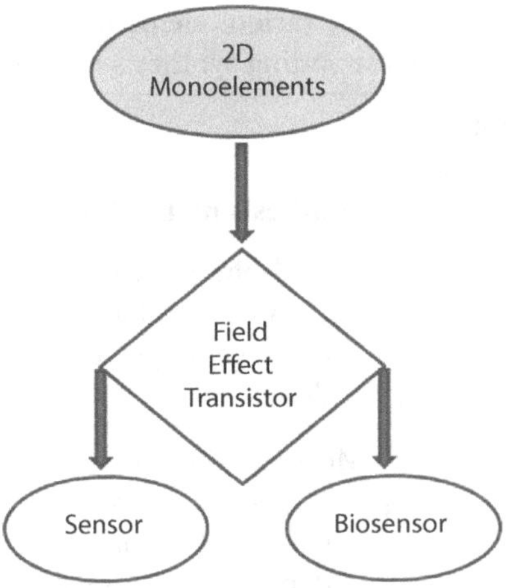

Figure 6.1 Application of 2D monoelements in field-effect transistor for sensing and biosensing.

Xenes and photoredox capability, thus making it a promising candidate in the field of biosensing, bioimaging, optical therapy, etc. [6, 7, 12, 13].

This chapter summarizes various categories of 2D monoelements, application in FET for sensing and biosensing applications as illustrated in Figure 6.1. Finally, various challenges and future insights are also covered.

6.1.1 Classification of 2D Monoelement (Xenes) in the Periodic Table

A concise detail of different types of novel 2D monoelements (Xenes) in the periodic table like in group III (gallenene and borophene), group IV (stanene, germanene, silicene), group V (bismuthene, phosphorene, arsenene, antimonene), and groupVI (tellurene and selenene), and their respective methods of synthesis and various characteristics are discussed in this section. Different elements in the periodic table which can form xenes and their corresponding synthesis methods are summarized in Table 6.1. Various synthesis methods of xenes are systematically categorized for various groups of the periodic table as well as their applications are shown in Table 6.2.

6.1.2 Group III

In this family, the 2D form of monoelements Boron (borophene) and gallium (gallenene) have been reported to be synthesized successfully.

Table 6.1 Below table summarizes various elements in the periodic table which can form Xenes and their corresponding synthesis methods [9].

Serial number	Elements (Atomic number)	Synthesis method of Xenes	Reference
1	B (5)	Liquid-phase exfoliation, Etching, Mechanical cleavage, Physical vapor deposition	[4, 13]
2	C (6)	Mechanical cleavage, Liquid-phase exfoliation, Chemical vapor deposition, Physical vapor deposition	
3	N (7)	No Xenes till date	
4	O (8)	No Xenes till date	
5	Al (13)	No Xenes till date	
6	Si (14)	Physical vapor deposition, Chemical exfoliation	[20, 21]
7	P (15)	Mechanical cleavage, Liquid-phase exfoliation, Etching, Chemical vapor deposition, Physical vapor deposition, Wet-chemistry	[5, 8, 22]
8	S (16)	No Xenes till date	
9	Ga (31)	Solid-melt exfoliation	[23, 24]
10	Ge (32)	Physical vapor deposition, Mechanical cleavage	[2, 25]
11	As (33)	Mechanical cleavage, Liquid-phase exfoliation	[16, 26]
12	Se (34)	Physical vapor deposition	[27]
13	In (49)	No Xenes till date	
14	Sn (50)	Physical vapor deposition	[15]

(Continued)

Table 6.1 Below table summarizes various elements in the periodic table which can form Xenes and their corresponding synthesis methods [9]. (*Continued*)

Serial number	Elements (Atomic number)	Synthesis method of Xenes	Reference
15	Sb (51)	Physical vapor deposition, Mechanical cleavage, Liquid-phase exfoliation	[6, 11, 16, 29]
16	Te (52)	Liquid-phase exfoliation, Physical vapor deposition, Wet-chemistry	[28]
17	Ti (81)	No Xenes till date	
18	Pb (82)	No Xenes till date	
19	Bi (83)	Physical vapor deposition, Mechanical cleavage, Liquid-phase exfoliation	[2, 16, 30]
20	Po (84)	No Xenes till date	

6.1.2.1 Borophene

The trivalent nature of boron is responsible for its chemically complex characteristics. The octet rule is not implemented in case of boron resulting in electron poor bonding configuration. Borophene is categorized as a class of 2D boron sheets due to many unique characteristics. Borophene structures are characterized by similar chemically structured polymorphism and anisotropy, as a result of which the flexibility of its properties are intensified [4]. Borophene is reported to be metallic and also lightest 2D metal. It exhibited optical clarity and mechanical adherence simultaneously. Borophene metallic property differs from phosphorene (semiconductor) and silicone (semimetals). Borophene exhibited high temperature conductivity [9].

6.1.2.2 Gallenene

At room temperature, it exists as liquid. It has a low temperature phase diagram. The chemical stability of gallenene demonstrated phase change memories and low power consumption optically [9].

Table 6.2 Different types of two-dimensional monoelements (Xenes) and their various types of applications, synthesis methods are shown below.

Element (2D form)	Group in periodic table	Morphology	Method of synthesis	Application in the bio medical field	Reference
B (Borophene)	III	Nanosheets	etching, liquid-phase exfoliation, mechanical cleavage, PVD	PIT, FI, drug delivery, PAI	[4, 13]
Ga (Gallenene)	III	Nanosheets	solid-melt exfoliation and PVD	–	[23, 24]
Si (Silicene)	IV	Nanosheets	chemical exfoliation and PVD	Biosensors	[20, 21]
Ge (Germanene)	IV	Nanosheets	PVD and Mechanical cleavage	–	[2, 57, 58]
Sn (Stanene)	IV	Nanosheets	PVD	–	[15]
P (Phosphorene)	V	quantum dots, Nano-particles and Nanosheets,	etching, wet-chemistry, PVD, Liquid-phase exfoliation, CVD, and Mechanical cleavage	neuroprotectivenanomedicine, bone regeneration Biosensors, PTI, PAI, drug deliver, PTT, PDT, FI, PLTI, antimicrobials, etc	[5, 9, 8, 59, 22, 60–63]

(Continued)

Table 6.2 Different types of two-dimensional monoelements (Xenes) and their various types of applications, synthesis methods are shown below. (*Continued*)

Element (2D form)	Group in periodic table	Morphology	Method of synthesis	Application in the bio medical field	Reference
As (Arsenene)	V	Nanosheets	Liquid-phase exfoliation and Mechanical cleavage	Biosensors	[16, 64, 65]
Sb (Antimonene)	V	quantum dots and Nanosheets	Liquid-phase exfoliation PVD, and Mechanical cleavage	drug delivery, PTT, Biosensors, PTI, PAI	[6, 11, 16, 29]
Bi (Bismuthene)	V	Liquid phase exfoliation, Mechanical cleavage, PVD and	nanodots, nanoparticles, and Nanosheets	PAI, PTI, PTT, CT, RT	[2, 16, 59, 30, 65]
Se (Selenene)	VI	Nanosheets	PVD	–	[19]
Te (Tellurene)	VI	nanodots and Nanosheets	PVD, liquid-phase Wet-chemistry	PAI, PTT, PDT, PTI	[7, 66–68]

6.1.3 Group IV

6.1.3.1 Silicene

Silicon differs from carbon. Its hybridization is Sp3 and structure is tetrahedral in the bulk structure. The monolayer of silicene exhibited honeycomb like crystal structure formed by stable combination of non-planar buckled configuration. Like graphene, it exhibited intensified spin orbit coupling and also low symmetry as well as honeycomb structure. The flexible gate band gaps, non-trivial electronic states enable silicene to be used in transistors and photosensors [14].

6.1.3.2 Germanene

Analogous to silicene, germanene are also reported of having similar honeycomb structure. It is also called as brother of silicene. Germanium is a heavy metal and its high atomic number supports spin orbit coupling. Hence, germanene experiences better insulator characteristics than silicene [9].

6.1.3.3 Stanene

Stannum (tin) is located very deep in group IV in the periodic table. Its structure is also identified as honeycomb structure. In comparison to carbon, silicon, and germanium, tin is the heaviest of all, and hence, it has strongest spin orbit coupling. Stanene is identified as promising candidate for insulator because of its unique electronic structure. It is also recognized as cousin of graphene [15].

6.1.4 Group V

6.1.4.1 Phosphorene

The list of allotropes of phosphorus consists of red phosphorus, violet phosphorus, white phosphorus, blue phosphorus, black phosphorus, gaseous phosphorus. White phosphorus is also considered as yellow phosphorus and is not a suitable candidate for biomedical application due to its toxicity. Its structure is tetrahedral P4 in which every phosphorus atom is linked to remaining three phosphorus atoms *via* single bonding. Violet phosphorus is produced *via* annealing of red phosphorus by day long heating and it has monoclinic structure. Green phosphorus or diphosphorus exists in gaseous form and is produced by thermolysis process. Another stable form is blue phosphorus which has hexagonal structure. Among all the highest stability, thermodynamically allotropic form is black phosphorus. It consists of orthorhombic structures with layers where every phosphorus atom exhibit five valence electrons in each atomic layer and undergo hybridization to form covalent bonds with nearby phosphorus atoms [9].

6.1.4.2 *Arsenene*

The general allotropes of arsenene include black, yellow, and grey arsenic. Among these three, grey arsenic is the most stable and unique due to its rhombohedral structure. Gray arsenic is a semi-metal material. Upon heating arsenic layered orthorhombic structure can be obtained alike black phosphorus. Methods like liquid phase exfoliation and mechanical cleavage can be used to produce arsenene [9].

6.1.4.3 *Antimonene*

The known allotropes of Sb are black, gray, and explosive Sb. Similar to gray arsenic, graySb has also rhombohedral structure and it is most stable in comparison to other allotropic form. It has also semi-metal characteristics. On rigorous cooling vapors of Sb, black Sb allotropic form can be synthesized. In atmospheric environment, black Sb is highly active chemically. In the presence of vacuum, it can be converted to stable form of crystalline gray Sb. Also, in presence of mechanical or heating stress the explosive Sb can be forcefully transformed to gray Sb [16]. Exfoliation of Sb into antimonene leads to development of semiconductor behavior with band gap of 2.28 eV.

6.1.4.4 *Bismuthene*

Bi has only form which is stable and rhombohedral in shape. It is metallic and heaviest than other elements in group V. The large atomic mass and strong spin orbit coupling facilitates bismuth to be identified as component of non-trivial system. Bismuthene is the 2D form of bismuth and it has the ability to act as insulator. The electrical transport properties of bismuthene can be measured by high resolution diffraction microscopy and scanning tunnelling microscopy [17].

6.1.5 Group VI

6.1.5.1 *Selenene*

Se structures are identified to be composed of zero-dimensional atomic rings or one-dimensional helical chains of atom [18]. The distance between rings in Se-Se bond is about 2.4 Angstrom. In this structure, Se atoms are decorated in square lattice form with strong covalent bonds joining each of the Se atoms. Selenium is reported anti-angiogenic agent which is applied in the field of cancer chemotherapy. Also, it has been reported that selenium-based materials are effective agents for cancer radio therapy where it is used radio sensitizers. Two-dimensional Se, i.e., selenene, can be produced using PVD methods [19].

6.1.5.2 Tellurene

Each tellurium atom is bonded to nearby tellurium atoms in the same chain. Bulk tellurium has very high hole mobilities due to spin orbit coupling. It is a semiconductor and capable of generating vacancies and active electrons. Oxidative stress can be also induced when they react with ambient surrounding. This special nature of tellurene helps tellurene to be applied in bio catalysis [9].

6.2 Field-Effect Transistor

This electronic device has the property of controlling flow of current using an electric field. It comprises of three parts, i.e., source, gate, and drain. The current flow in FET is controlled by applying voltage to gate, as a result of which conductivity in between source and drain is altered. During the operative process, FET does not carry both electrons and holes as charge carriers. It is a uni-charge carrier transistor system, i.e., either electron or holes are used for carrying charge. There are many types of FETs. It shows high resistance at low frequency. Most extensively used FET is metal oxide semiconductor FET (MOSFET). FET has an active pathway along which holes or electrons (charge carriers) are carried between sources to drain. Ohmic contacts in the semiconductor connect the source terminal and drain terminal conductor. The three different terminals of FET are source, from where the charge carriers penetrate into the channel, drain terminal from where the charge carriers depart, and gate terminal which regulates the conductivity of the channel.

Among the modern electric products, FETs are most extensively used as semiconductor. An appropriate band gap is needed for fast switching and proper off-on ratio of current. In comparison to three-dimensional electronic materials, 2D semiconductors have better properties like pure surface without drooping bonds, absence of phonons coupling, and substantial band gaps (1 to 2 eV). Schematic diagram of field effect transistor is illustrated in Figure 6.2.

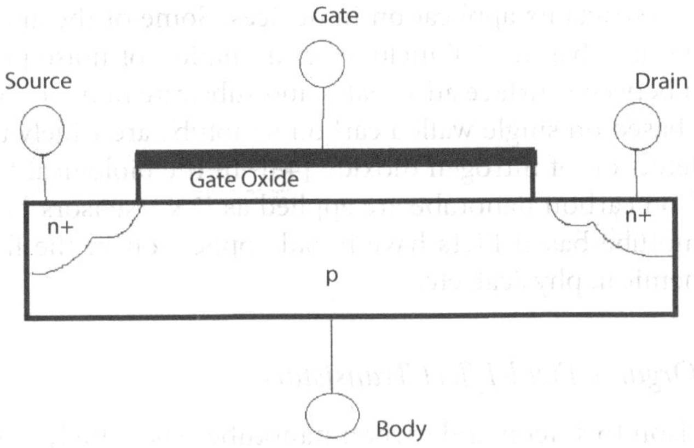

Figure 6.2 Schematic figure of field-effect transistor.

6.2.1 Different Types of Recently Developed Field-Effect Transistors

6.2.1.1 Field-Effect Transistors Based on Silicon

The traditional-based silicon-based FETs have high ratio of on/off [31]. These silicon-based nanowires have alluring properties like excellent charge carrier mobility and efficient use in the sector of biosensing, nanosensors, etc. [32–41]. The strength and sensitivity of silicon nanowire helps in wide applications in biological fields for detecting steroids [42]. Silicon-based FETs are also effective in physical field for sensing such as detection of plasmonic terahertz [43]. But these silicon-based FET are also related with few constraints such as reproduction, stability, error endurance during the manufacture of devices with high performance which creates hurdles in the production process, hence these silicon-based FET are not applied on large scale in devices. Detailed investigation is needed to conquer these hurdles.

6.2.1.2 Field-Effect Transistors Based on Carbon Nanotube

These carbon nano tubes possess unique electrical properties and extensively applied in field of nanoelectronics. The first carbon nanotube-based FET was manufactured in the year 1998 [44]. After that, the silicon-based FET were preferably substituted by carbon nanotube FET due to their better performance in on/off ratio of current, cut-off frequency, mobility, etc. [45–47]. Gate hysteresis is one issue associated with carbon nanotube

FET which restricts its application in devices. Some of the application of carbon nanotube-based FET includes examination of noise produced by interaction between surface adsorbates and substrate in nano FET biosensors. FETs based on single walled carbon nanotube are widely used in the selective detection of nitrogen dioxide [48]. In the biological fields, these FET based on carbon nanotube are applied as DNA sensors [49]. Overall carbon nanotube-based FETs have broad application in the field of biosensing, chemical, physical, etc.

6.2.1.3 Organic Field-Effect Transistors

In comparison to silicon and carbon nanotube-based FET, organic FETs are better due to superior properties in terms of cost effectiveness, flexibility, light weight, etc. There are four different categories, i.e., top gate/bottom contact, topgate/top contact, bottom gate/top contact, and bottom gate/bottom contact [50]. Organic semiconductors in FET are used as active candidate in processing of plastic circuits [51]. Organic FET with inkjet printed n-channel show 0.05–0.06 $cm^2V^{-1}s^{-1}$ mobility and approximate on/off ratio of 10.6 [52]. Even though organic FET is very potential, its low charge carrying mobility obstructs its development up to some extent.

6.2.1.4 Field-Effect Transistors Based on Graphene

Graphene has special band structure and high carrier mobility and this property enables it to be successfully applied in FET. The limitations related to silicon, carbon nanotube, organic materials, and other such FET were rectified by graphene-based FET- as the channel material. These graphene FET in order to realize the downsizing of incorporated circuits follows Moore's law [53].

6.3 Application of 2D Monoelements in Field-Effect Transistor for Sensing and Biosensing

6.3.1 Biosensor

The biomarkers present in blood samples are in very limited concentration. Hence, in order to detect these, highly specified and very sensitive sensors in the field of biomedical are required. In this part, 2D monoelement (Xenes)–based FET for sensing and biosensing, gas sensors, calorimetric, electrochemical, and fluorescent biosensor are explained. The extraordinary physical and electrical properties of Xenes enable it to perform as a promising agent for manufacture and designing of biosensors in order to detect biomolecules such as DNA, enzymes, protein, hormones, glucose, etc. Germanene and silicene has a specific role in conductors in FET biosensor because they are capable of transporting electron like metal conducting devices. These biosensors are designed in a way to change binding information in bio-macromolecules at the interface of liquid/solid into highly selective and sensitive detectable electronic results [54]. Additionally, an electrochemical biosensor to recognize specific single-strand DNA, based on silicon biosensing device was developed. Pnictogens, Group VA elements like 2D bismuthene, arsenene, antimonene, exhibit high stability due to their heavier weight which promotes their performance as biosensor, such as sensing of chemical substances like phenol [55]. Antimonene were found efficient of all in detecting phenol due to their unique property of low oxidation to bulk ratio.

Phosphorene has also developed recently as biosensor in FET devices in which phosphorene has shown reduction in current fluctuation with very high mobility of hole. Phosphorene decorated with gold nanoparticles demonstrated very good reduction catalytic activity of 4-nitrophenol which is yellow in color to colorless 4-aminophenol. This color change acts like biomarker indicating carcino embryonic antigen present in colon and breast cancer patients with high selectivity and sensitivity. Further, phosphorene-based biosensor also detected hydrogen peroxide which is an important part in biomedicine [56]. It can also detect myoglobin which is an important biomarker for cardiovascular diseases. But, the instability of phosphorene in air causes obstacles in its performance even though it exhibit smaller band gap as well as high mobility.

Fast and specific differentiation of different types of odorous substances is very important in designing highly developed sensing devices for diagnosis of diseases, monitoring of environment and safety in food. A highly sensitive FET-based olfactory system called B-nose or bioelectronics nose in which the olfactory receptor is conjugated with graphene bilayer (plasma treated). The oxygen and ammonia controlled plasma treatment brought stability to n- and p-type property of bilayer graphene. This on combination with olfactory receptor of human formed a FET-type system which is liquid ion gated [69].

These bioelectronics noses are capable of recognizing an odorant substance due to their transparent and flexible property of sensing. These B-noses exhibit stability for long duration and durability. Manufacture of such devices which have the property of differentiating odorant sensitively needs to be designed properly in order to use it in various sectors such as food, medicine, environment, etc. [70–72].

Graphene is sheet of carbon of Sp2 hybridization with 2D structure [73]. The unique properties of graphene like electrical, mechanical, and thermal make its application possible in a wide area [74]. The devices in which graphene films are used were reported to demonstrate high mobility in the presence of room temperature which leads its application at nanoscale as biological and chemical sensors, storing energy, electronic devices, etc. [75–83].

Particularly, biosensor based on graphene oxide in reduced form is identified to exhibit simplicity in structure and label-free detection [84–89]. The extraordinary characteristic of graphene with summation of electronic and physical properties enables it to become a promising agent in the field of bioelectronics and biosensing. In this context, FET developed of graphene solution gate which are capable to be operated even in aqueous surrounding are very much attractive. FET based on graphene solution show high sensitivity and trans-conductance, collectively produce very low gate noise [90].

The electronic devices based on silicon for applications as sensors are affected by high noise and this obstacle hinders the sensitivity of the device [91]. FETs based on carbon nanowire [92, 93] and nanotube [94, 95] are reported to be as potential agent for biological and chemical species detection electrically. Even though they possess such extraordinary performance but due to certain issues such as reproducibility, hindrance in the process of integration of carbon nanotube on large scale limits its performance. Graphene is reported to be free from all the associated disadvantages related to silicon and carbon nanotube for the application in the field of bioelectronics and biosensor [96, 97]. In comparison to silicon, graphene demonstrated higher stability chemically and inert in biological field. Additionally, graphene electric characteristics enable graphene to over shine the sensing capacity of silicon-based devices.

Graphene is an exemplary 2D system which demonstrates excellent mobility of both electrons and hole at room temperature and also bears electrical sensitivity to adsorption of charge [96–101]. FET based on graphene was identified to possess superior sensitivity in sensing gas. It is reported that graphene-based FET operates mostly in vacuum environment. It was also recently identified that graphene FET based on electrolyte gate act as detectors of protein adsorption and pH detection [102]. Another such biosensor was reported as graphene-based FET which is aptamer modified, used for electrical detection of immunoglobulin E protein selectively [103].

Graphene-based FETs due to their extraordinary properties were preferred for various types of sensing application such as DNA, gas, glucose, pH, sensing, etc.

6.3.1.1 DNA Sensors

Mohanty and Berry in the year 2008 were first to report their investigation result on biomolecules and graphene [104]. It was reported that single DNA strand was connected on graphene sheet surface. Functionalization of complementary DNA through fluorescence and hybridization with probe DNA affected electric field and current. As reported by Dong *et al.*, use of graphene-based FET can support sensitive and label-free detection of DNA *via* direct point response [105]. Graphene sheet grown *via* chemical vapor deposition, with probe DNA which are pre-grounded on graphene surface film during solution process were used to fabricate graphene-based FET. Left shift presented by direct point on post hybridization with targeted DNA revealed n-doping as a result of which electron were transferred to graphene-based FET from DNA. The limit of detection was reported to be 0.01 nm. Further, on modifying the surface of graphene with gold nanoparticles, it was found that there was extension in the upper detection limit (500 nm). It was believed that detection of DNA electrically was affected majorly by n-doping. Further, in the year 2011, another work was reported by Dong regarding reduced graphene oxide and gold nanoparticles [106]. It was reported that fabrication of transistor with mixture of graphene and gold nanoparticles resulted in better conductance and mobility of hole in comparison to naked sheets of graphene. Hence, it was concluded that transistors incorporated with both graphene and nanocomposites of gold displayed better sensitivity and stability during detection of DNA. Platinum nanoparticles can be also used for coating in place of gold for detecting DNA [107].

6.3.1.2 Protein Sensors

Along with DNA detection, other biological components detection was also reported recently. Aromatic ring present in amino acids make the protein detection, which are binded on the surface of graphene *via* pi-pi bond, possible through doping effects. Bovine serum albumin was reported to be detected by Ohno *et al.* using electrolyte gated graphene-based FET [108]. The limit of detection was very less, i.e., 0.3 nm. For detection of target protein, it was necessary to functionalize the surface of graphene by receptor. Further, Ohno *et al.* again reported a biosensor which was based on aptamer modified graphene FET, used for detection of immunoglobulin E protein selectively [109]. In the following year, Mao *et al.* reported another device based on reduced graphene oxide sheets for detection of specific immunoglobulin G.

The recognizable alteration electrically around the instrument was negative charge of immunoglobulin G protein with detection limit of 2 ng ml^{-1}. Another protein FET sensors were designed by using gold nano antibody conjugates incorporated on vertically oriented sheets of graphene [110]. Chemical vapor deposition which was plasma enhanced were used for assembling the vertically oriented sheets of graphene. This special orientation and structure of vertically oriented sheets of graphene enabled it to behave as a suitable material for sensing. It eases the accessing of molecules under analysis on the entire surface of FET. These sensors were reported to be superior for sensing immunoglobulin G protein with excellent sensitivity and high selectivity. The above results indicated suitable route to improve the graphene devices incorporated with gold nanoparticle for laboratory diagnosis through detection of number of proteins. Recently, it was reported that graphene-based FET were used for diagnosis of prostate cancer with enhanced detection sensitivity [111]. Fibronectin is another protein which are reported to be detected using reduced graphene oxide-based FET with very low limit of detection, i.e., 0.5 nm.

6.3.1.3 Glucose Sensor

Other biological application of graphene-based biosensor was reported by Huang *et al.* for glucose and glutamate detection sensitively *via* chemical vapor deposition grown graphene [112]. The surface of graphene-based FET was functionalized *via* a 1-pyrenebutanoic acid succinimidyl ester (common linker molecule) with glutamate dehydrogenase and glucose oxidase. Due to this modification, there was change in the current of graphene-based FET, thereby detecting glutamate and glucose due to interaction between glucose oxidase and glutamate dehydrogenase with glucose and glutamate, respectively. It was also reported that in comparison to single walled nanotube FET, graphene-based FETs were superior as they gave detection result efficiently even in the lower concentration of approximate 0.01 mm glucose and 5 μm. Kavitha reported graphene films coated with ZnO nanoparticles exhibited better sensing property in detecting glucose [113].

6.3.1.4 Living Cell and Bacteria Sensors

In comparison to other chemical substances, living cells detection is not easy due to the complex reaction of living cell membrane with graphene.

Berry reported a bio-device which detected single bacteria *via* current response observed after the bacteria did absorption and also changed graphene chemically [104]. Much investigation is done in order to explore such devices which detect living cells and bacteria. Huang and his co-workers reported specified biosensor (E. coli antibody coated graphene FET) to detect E.coli with very high selectivity and sensitivity with detection limit of 10 cfu ml^{-1} [114]. Additionally, graphene-based FETs are also pH value sensitive, as a result of which current change reveals the metabolism of living cells. Modification of pH occurs by the release of pyruvic, lactic, citric acids (organic metabolic acids) which further affects the current change. Ang *et al.* in the year 2011 reported graphene-based transitors which detected red blood cells infected by malaria [115]. The most important part of the work was that these graphene base transistor sensors were functionalized with microfluids and resulted in single cell flow sensing. Parasitized erythrocytes (schizont and trophozoite) *via* plasmodium falciparum differing in structures can be differentiated *via* change in conductance. This has vital application in laboratory diagnosis in coming future.

6.3.2 Sensor

Researches on chemical sensing of gaseous substance like nitrogen dioxide, ammonia, water, and metallic ions are reported recently performed by graphene FET. The different selectivity and sensitivity of graphene FET for varieties of ions and molecules are studied and investigated in details by scientists.

6.3.2.1 Gas Sensor

The concept behind the gas sensor based on graphene FET is that gaseous molecule either behaves as donor or acceptor or gets absorbed on the surface of graphene which finally modifies the surrounding and current is transported that can be analysed [116]. In this context, the very first sensor based on graphene in order to detect gas molecules was reported by Novoselov and his team [117]. Pure layer of graphene undergoes mechanical exfoliation and has a vital role as sensing device in the detection of nitrogen dioxide. The detection of nitrogen can be performed by studying the alteration in the drain current and gate voltage. At room temperature, most sensitive detection of gaseous molecule takes place which shows that during this, the device experiences high conductance and very low noise of graphene. Graphene-based FET designed with substrate (SiC-6H) were found to be better in their performance in detecting nitrogen dioxide at lower concentration (10 parts per billion) [118] with in comparison to traditional substrate like carbon nanotube-based FET which has the sensing limit below 100 parts per billion [119]. The interrelation between humidity, temperature and nitrogen dioxide sensitive detection with graphene-based FET which are bottom gated incorporated with single atomic layer of graphene [120]. Sensor based on new reduced graphene oxide configuration for detection of nitrogen dioxide demonstrated better result in comparison to other sensor, i.e., 5.8 parts per billion [121]. Another such work was reported by Ren *et al.* for detecting sulphur dioxide at various temperatures [122]. Here, sulphur dioxide behaved as acceptor causes induction of p-doping to graphene, thereby shifting Dirac point in a positive direction at room temperature with 0.0678 V/ppm and with 1.09 V/ppm at 100°C. The sulphur dioxide concentration of 50 parts per million when investigated, a 100% decrease in resistance was observed at 100°C and to 60% at room temperature. It was finally concluded that the sensitivity of sulphur dioxide sensor increased with the increase in temperature. In order to increase the specificity and sensitiveness of sensors based on the graphene for detection of gaseous molecule, it is necessary to coat the surface of graphene with such elements which can react with specific targeted species by giving required signal. Coating palladium on the surface of graphene FET for detection of hydrogen resulted in high sensitivity [123]. In the year 2012, it was investigated that the process of nitrogen dioxide sensing can be enhanced by ozone treatment [124]. After the ozone treatment, it was found that the performance of the sensor for nitrogen dioxide detection was improved exhibiting twice increment in the percentage of response towards nitrogen dioxide with 200 parts per million concentration and overall eight times faster reaction in comparison to bare graphene sheets. The performance of sensor and its selectivity can also be enhanced by modifying the structure of the graphene like nano ribbon or nano mesh. It was reported that device with nano mesh configuration demonstrated better selectivity and sensitivity in detection of nitrogen dioxide and ammonia at room temperature with detection limit of 15 and 160 parts per billion, respectively.

6.3.2.2 pH Sensor

In comparison to traditional carbon material, graphene was identified as excellent and promising candidate for sensing due to high carrier mobility and large surface area. The very first pH sensing device was designed by using single, double, and triple layer sheets of graphene developed on SiC with the help of silver/silver chloride gate electrode [125]. These graphene transistors layered differently resulted in same pH sensitivity of approximate 99mV/pH. New graphene FET suspended on gold electrode resulted in 20 mV/pH sensitivity was reported [126]. Enhanced quality of solution gated FET with graphene which undergoes mechanical exfoliation was designed and reported by Ohno *et al.* [127]. On increasing pH, direct points shifted towards positive end in the graphene devices. On immersing pH, sensor based on graphene in electrolyte demonstrated 30 times higher transconductance in comparison to vacuum. It was reported that pH value in the aqueous environment was resulted in between 4 and 10, as a result of which, the device resistance decreased on increment of pH value. Sohn reported solution gated reduced graphene oxide FET for pH sensor [128]. There are a large number of functional groups present on the surface of reduced graphene oxide channel which resulted in site binding theory which was the reason behind sensing mechanism of hydrogen ion. Modification of structure cans enhance the selectivity and activity of the sensors. Park reported that integration of graphene with graphite monolithically resulted in graphene-based FET arrays which can give simultaneous outcome of both reproducibility and detection of pH [129]. This was identified as effective way to enhance the sensitivity and activity of pH sensor.

6.3.2.3 *Metal Ion and Other Chemical Sensors*

Undoubtedly, few heavy metal ions like cadmium (Cd^{2+}), lead (Pb^{2+}), and mercury (Hg^{2+}) are responsible for human health hazard even if it is present in minute amount. It is very much necessary to detect these toxic substances and for which development sensing devices are very much necessary. It was reported that new sensitive device integrated with graphene nanosheets and special properties of Nafion was designed in order to detect lead (Pb^{2+}) ion and cadmium (Cd^{2+}) ion [130]. Wen *et al.* reported his work recently regarding the device made of graphene coated with gold nanoparticles and DNA [131]. Further, Gong and his workers also reported another such graphene nanosheet-based device coated gold nanoparticles used for detecting mercury (Hg^{2+}) selectively [132]. In the year 2011, Zhang *et al.* reported a device based on graphene FET fabricated by self-assembly of 1-octadecanethiol on dielectric silicon oxide amorphous substrate for sensitive detection of mercury (Hg^{2+}) [133]. The charge carrier are generally affected by roughness on the surface of graphene, hence in order to solve this problem, alkane thiol fabrication on the surface of graphene is very essential. The detection limit of functional graphene-based FET for mercury (Hg^{2+}) can successfully reach to 10 parts per million. Sudibya reported reduced graphene oxide FET sensor coated with protein [134]. One such protein named calmodulin were used to functionalize reduced graphene oxide FET for detecting magnesium (Mg^{2+}) and calcium (Ca^{2+}) with detection limit for calcium ion 1 μm. Further, reduced graphene oxide sensors altered with type II metalloprotein demonstrated better performance in mercury ion detection with 1-nm detection limit. Maehashi *et al.* designed graphene-based FET which was electrolyte gated that was altered with the-valinomycin to design potassium ion (K^+) sensor [135]. Fu *et al.* reported that graphene-based FET coated with noncovalent substance phenol which resulted in preservation of mobility and detection of hydrogen and potassium ion [136]. It was recently reported that integration of microfluidic system with graphene-based FET gave better outcome. Graphene-based FET which is solution gated are reported to be promising agent in all type of applications in sensing.

6.4 Conclusions and Perspectives

In this chapter, it is decoratively classified and explained the recent works of 2D monoelements reflecting their classification and application. Many investigations have been reported regarding the credential of 2D monoelements in their application as sensor in many chemical and biomedical fields. Even though these Xenes are connected with number of benefits in terms of degradation and metabolic activity, few Xenes (antimonene) were reported to accumulate in livers in comparison to other organ while other materials get released from body within a week. Hence, detail investigation is needed to study the side effects related to these materials and to confirm their easy excretion from the body without affecting any organ [9].

The exceptional characteristics of graphene are like charge carrier mobility, electron transfer at high rate structural benefits, etc. Collectively, all these properties enabled this graphene-based devices to be applied efficiently in various type of sensing and biosensing activity in comparison to traditional sensors. Instead of so many advantages, it is a topic of concern to produce reproducible, consistent, and large surface area-based sensors. Hence, it is necessary to remove these and clean the device to enhance the sensor performance. It is believed these graphene-based transistors have a bright future if its loopholes are filled by efforts of researchers and scientists. It is expected that the young readers should research and investigate attractive facts related to these concept. In this chapter, it is strongly indicated that graphene-based FET can create a new standard in the field of sensing and biosensing and bioelectronics [53].

References

1. Reina, G., González-Domínguez, J.M., Criado, A., Vázquez, E., Bianco, A., Prato, M., Promises, facts and challenges for graphene in biomedical applications. *Chem. Soc. Rev.*, 46, 15, 4400–4416, 2017.
2. Kong, X., Liu, Q., Zhang, C., Peng, Z., Chen, Q., Elemental two-dimensional nanosheets beyond graphene. *Chem. Soc. Rev.*, 46, 8, 2127–2157, 2017.
3. Molle, A., Goldberger, J., Houssa, M., Xu, Y., Zhang, S.-C., Akinwande, D., Buckled two-dimensional Xene sheets. *Nat. Mater.*, 16, 2, 163, 2017.
4. Mannix, A.J., Zhang, Z., Guisinger, N.P., Yakobson, B.I., Hersam, M.C., Borophene as a prototype for synthetic 2D materials development. *Nat. Nanotechnol.*, 13, 6, 444, 2018.
5. Qiu, M., Ren, W.X., Jeong, T., Won, M., Park, G.Y., Sang, D.K., Liu, L.-P., Zhang, H., Kim, J.S., Omnipotent phosphorene: A next-generation, two-dimensional nanoplatform for multidisciplinary biomedical applications. *Chem. Soc. Rev.*, 47, 15, 5588–5601, 2018.
6. Tao, W., Ji, X., Xu, X., Islam, M.A., Li, Z., Chen, S., Saw, P.E. *et al.*, Antimonene Quantum Dots: Synthesis and Application as Near-Infrared Photothermal Agents for Effective Cancer Therapy. *Angew. Chem. Int. Ed.*, 56, 39, 11896–11900, 2017.
7. Lin, Y., Wu, Y., Wang, R., Tao, G., Luo, P.-F., Lin, X., Huang, G., Li, J., Yang, H.-H., Two-dimensional tellurium nanosheets for photoacoustic imaging-guided photodynamic therapy. *Chem. Commun.*, 54, 62, 8579–8582, 2018.
8. Tao, W., Zhu, X., Yu, X., Zeng, X., Xiao, Q., Zhang, X., Ji, X. *et al.*, Black phosphorus nanosheets as a robust delivery platform for cancer theranostics. *Adv. Mater.*, 29, 1, 1603276, 2017.
9. Tao, W., Kong, N., Ji, X., Zhang, Y., Sharma, A., Ouyang, J., Qi, B. *et al.*, Emerging two-dimensional monoelemental materials (Xenes) for biomedical applications. *Chem. Soc. Rev.*, 48, 11, 2891–2912, 2019.
10. Lei, P., An, R., Zhang, P., Yao, S., Song, S., Dong, L., Xu, X., Du, K., Feng, J., Zhang, H., Ultrafast synthesis of ultrasmall poly (Vinylpyrrolidone)-protected Bismuth Nanodots as a multifunctional theranostic agent for *in vivo* dual modal CT/Photothermal-imaging-guided photothermal therapy. *Adv. Funct. Mater.*, 27, 35, 1702018, 2017.
11. Tao, W., Ji, X., Zhu, X., Li, L., Wang, J., Zhang, Y., Saw, P.E. *et al.*, Two-Dimensional Antimonene-Based Photonic Nanomedicine for Cancer Theranostics. *Adv. Mater.*, 30, 38, 1802061, 2018.

12. Xue, T., Liang, W., Li, Y., Sun, Y., Xiang, Y., Zhang, Y., Dai, Z. *et al.*, Ultrasensitive detection of miRNA with an antimonene-based surface plasmon resonance sensor. *Nat. Commun.*, 10, 1, 28, 2019.
13. Ji, X., Kong, N., Wang, J., Li, W., Xiao, Y., Gan, S.T., Zhang, Y. *et al.*, A Novel Top-Down Synthesis of Ultrathin 2D Boron Nanosheets for Multimodal Imaging-Guided Cancer Therapy. *Adv. Mater.*, 30, 36, 1803031, 2018.
14. Mannix, A.J., Kiraly, B., Hersam, M.C., Guisinger, N.P., Synthesis and chemistry of elemental 2D materials. *Nat. Rev. Chem.*, 1, 2, 0014, 2017.
15. Zhu, F.-f., Chen, W.-j., Xu, Y., Gao, C.-l., Guan, D.-d., Liu, C.-h., Qian, D., Zhang, S.-C., Jia, J.-f., Epitaxial growth of two-dimensional stanene. *Nat. Mater.*, 14, 10, 1020, 2015.
16. Zhang, S., Guo, S., Chen, Z., Wang, Y., Gao, H., Gómez-Herrero, J., Ares, P., Zamora, F., Zhu, Z., Zeng, H., Recent progress in 2D group-VA semiconductors: From theory to experiment. *Chem. Soc. Rev.*, 47, 3, 982–1021, 2018.
17. Chen, S.-C., Wu, J.-Y., Lin, M.-F., Feature-rich magneto-electronic properties of bismuthene. *New J. Phys.*, 20, 6, 062001, 2018.
18. Xian, L., Pérez Paz, A., Bianco, E., Ajayan, P.M., Rubio, A., Square selenene and tellurene: Novel group VI elemental 2D materials with nontrivial topological properties. *2D Mater.*, 4, 4, 041003, 2017.
19. Qin, J., Qiu, G., Jian, J., Zhou, H., Yang, L., Charnas, A., Zemlyanov, D.Y. *et al.*, Controlled growth of a large-size 2D selenium nanosheet and its electronic and optoelectronic applications. *ACS Nano*, 11, 10, 10222–10229, 2017.
20. Tao, L., Cinquanta, E., Chiappe, D., Grazianetti, C., Fanciulli, M., Dubey, M., Molle, A., Akinwande, D., Silicene field-effect transistors operating at room temperature. *Nat. Nanotechnol.*, 10, 3, 227, 2015.
21. Tchalala, M.R., Ali, M.A., Enriquez, H., Kara, A., Lachgar, A., Yagoubi, S., Foy, E. *et al.*, Silicon sheets by redox assisted chemical exfoliation. *J. Phys.: Condens. Matter*, 25, 44, 442001, 2013.
22. Qiu, M., Wang, D., Liang, W., Liu, L., Zhang, Y., Chen, X., Sang, D.K. *et al.*, Novel concept of the smart NIR-light–controlled drug release of black phosphorus nanostructure for cancer therapy. *Proc. Natl. Acad. Sci.*, 115, 3, 501–506, 2018.
23. Kochat, V., Samanta, A., Zhang, Y., Bhowmick, S., Manimunda, P., Asif, S., Stender, A.S. *et al.*, Atomically thin gallium layers from solid-melt exfoliation. *Sci. Adv.*, 4, 3, e1701373, 2018.
24. Xing, Y., Zhang, H.-M., Fu, H.-L., Liu, H., Sun, Y., Peng, J.-P., Wang, F. *et al.*, Quantum Griffiths singularity of superconductor-metal transition in Ga thin films. *Science*, 350, 6260, 542–545, 2015.
25. Bianco, E., Butler, S., Jiang, S., Restrepo, O.D., Windl, W., Goldberger, J.E., Stability and exfoliation of germanane: A germanium graphane analogue. *ACS Nano*, 7, 5, 4414–4421, 2013.

26. Gusmão, R., Sofer, Z., Bouša, D., Pumera, M., Pnictogen (As, Sb, Bi) nanosheets for electrochemical applications are produced by shear exfoliation using kitchen blenders. *Angew. Chem.*, 129, 46, 14609–14614, 2017.
27. Xing, C., Xie, Z., Liang, Z., Liang, W., Fan, T., Ponraj, J.S., Dhanabalan, S.C., Fan, D., Zhang, H., 2D Nonlayered Selenium Nanosheets: Facile Synthesis, Photoluminescence, and Ultrafast Photonics. *Adv. Opt. Mater.*, 5, 24, 1700884, 2017.
28. Wang, Q., Safdar, M., Xu, K., Mirza, M., Wang, Z., He, J., Van der Waals epitaxy and photoresponse of hexagonal tellurium nanoplates on flexible mica sheets. *ACS Nano*, 8, 7, 7497–7505, 2014.
29. Ji, J., Song, X., Liu, J., Yan, Z., Huo, C., Zhang, S., Su, M. *et al.*, Two-dimensional antimonene single crystals grown by van der Waals epitaxy. *Nat. Commun.*, 7, 13352, 2016.
30. Yang, Q.-Q., Liu, R.-T., Huang, C., Huang, Y.-F., Gao, L.-F., Sun, B., Huang, Z.-P. *et al.*, 2D bismuthene fabricated *via* acid-intercalated exfoliation showing strong nonlinear near-infrared responses for mode-locking lasers. *Nanoscale*, 10, 45, 21106–21115, 2018.
31. Yu, C., Liu, H., Ni, W., Gao, N., Zhao, J., Zhang, H., Theoretical study of the source-drain current and gate leakage current to understand the graphene field-effect transistors. *Phys. Chem. Chem. Phys.*, 13, 8, 3461–3467, 2011.
32. Ng, T.N., Wong, W.S., Lujan, R.A., Street, R.A., Characterization of charge collection in photodiodes under mechanical strain: Comparison between organic bulk heterojunction and amorphous silicon. *Adv. Mater.*, 21, 18, 1855–1859, 2009.
33. Wu, X., Kulkarni, J.S., Collins, G., Petkov, N., Almecija, D., Boland, J.J., Erts, D., Holmes, J.D., Synthesis and electrical and mechanical properties of silicon and germanium nanowires. *Chem. Mater.*, 20, 19, 5954–5967, 2008.
34. Barth, S., Hernandez-Ramirez, F., Holmes, J.D., Romano-Rodriguez, A., Synthesis and applications of one-dimensional semiconductors. *Prog. Mater. Sci.*, 55, 6, 563–627, 2010.
35. Tisch, U. and Haick, H., Nanomaterials for cross-reactive sensor arrays. *MRS Bull.*, 35, 10, 797–803, 2010.
36. Bashouti, M.Y., Tung, R.T., Haick, H., Tuning the electrical properties of Si nanowire field-effect transistors by molecular engineering. *Small*, 5, 23, 2761–2769, 2009.
37. Paska, Y., Stelzner, T., Assad, O., Tisch, U., Christiansen, S., Haick, H., Molecular gating of silicon nanowire field-effect transistors with nonpolar analytes. *ACS Nano*, 6, 1, 335–345, 2011.
38. Assad, O., Leshansky, A.M., Wang, B., Stelzner, T., Christiansen, S., Haick, H., Spray-coating route for highly aligned and large-scale arrays of nanowires. *ACS Nano*, 6, 6, 4702–4712, 2012.
39. Wong, W.S., Raychaudhuri, S., Lujan, R., Sambandan, S., Street, R.A., Hybrid Si nanowire/amorphous silicon FETs for large-area image sensor arrays. *Nano Lett.*, 11, 6, 2214–2218, 2011.

40. Street, R.A., Thin-film transistors. *Adv. Mater.*, 21, 20, 2007–2022, 2009.
41. Pui, T.-S., Agarwal, A., Ye, F., Balasubramanian, N., Chen, P., CMOS-compatible nanowire sensor arrays for detection of cellular bioelectricity. *Small*, 5, 2, 208–212, 2009.
42. Chang, K.S., Chen, C.C., Sheu, J.T., Li, Y.-K., Detection of an uncharged steroid with a silicon nanowire field-effect transistor. *Sens. Actuators B: Chem.*, 138, 1, 148–153, 2009.
43. Hwang, H.C., Park, K., Park, W.-K., Han, S.-T., Kim, K.R., Design and characterization of plasmonic terahertz wave detectors based on silicon field-effect transistors. *Jpn. J. Appl. Phys.*, 51, 6S, 06FE17, 2012.
44. Tans, S.J., RM Verschueren, A., Dekker, C., Room-temperature transistor based on a single carbon nanotube. *Nature*, 393, 6680, 49, 1998.
45. Avouris, P., Carbon nanotube electronics and photonics. *Phys. Today*, 62, 1, 34–40, 2009.
46. Rosenblatt, S., Lin, H., Sazonova, V., Tiwari, S., McEuen, P.L., Mixing at 50 GHz using a single-walled carbon nanotube transistor. *Appl. Phys. Lett.*, 87, 15, 153111, 2005.
47. Le Louarn, A., Kapche, F., Bethoux, J.-M., Happy, H., Dambrine, G., Derycke, V., Chenevier, P., Izard, N., Goffman, M.F., Bourgoin, J.-P., Intrinsic current gain cutoff frequency of 30 GHz with carbon nanotube transistors. *Appl. Phys. Lett.*, 90, 23, 233108, 2007.
48. Fu, D., Lim, H., Shi, Y., Dong, X., Mhaisalkar, S.G., Chen, Y., Moochhala, S., Li, L.-J., Differentiation of gas molecules using flexible and all-carbon nanotube devices. *J. Phys. Chem. C*, 112, 3, 650–653, 2008.
49. Dong, X., Fu, D., Xu, Y., Wei, J., Shi, Y., Chen, P., Li, L.-J., Label-free electronic detection of DNA using simple double-walled carbon nanotube resistors. *J. Phys. Chem. C*, 112, 26, 9891–9895, 2008.
50. Wang, C., Dong, H., Hu, W., Liu, Y., Zhu, D., Semiconducting π-conjugated systems in field-effect transistors: A material odyssey of organic electronics. *Chem. Rev.*, 112, 4, 2208–2267, 2011.
51. Garnier, F., Hajlaoui, R., Yassar, A., Srivastava, P., All-polymer field-effect transistor realized by printing techniques. *Science*, 265, 5179, 1684–1686, 1994.
52. Baeg, K.-J., Khim, D., Kim, J.-H., Kang, M., You, I.-K., Kim, D.-Y., Noh, Y.-Y., Improved performance uniformity of inkjet printed n-channel organic field-effect transistors and complementary inverters. *Org. Electron.*, 12, 4, 634–640, 2011.
53. Zhan, B., Li, C., Yang, J., Jenkins, G., Huang, W., Dong, X., Graphene Field-Effect Transistor and Its Application for Electronic Sensing. *Small*, 10, 20, 4042–4065, 2014.
54. Kaneko, S., Tsuchiya, H., Kamakura, Y., Mori, N., Ogawa, M., Theoretical performance estimation of silicene, germanene, and graphene nanoribbon field-effect transistors under ballistic transport. *Appl. Phys. Express*, 7, 3, 035102, 2014.

55. Mayorga-Martinez, C.C., Gusmão, R., Sofer, Z., Pumera, M., Pnictogen-Based Enzymatic Phenol Biosensors: Phosphorene, Arsenene, Antimonene, and Bismuthene. *Angew. Chem. Int. Ed.*, 58, 1, 134–138, 2019.
56. Choi, J.R., Yong, K.W., Choi, J.Y., Nilghaz, A., Lin, Y., Xu, J., Lu, X., Black phosphorus and its biomedical applications. *Theranostics*, 8, 4, 1005, 2018.
57. Molle, A., Goldberger, J., Houssa, M., Xu, Y., Zhang, S.-C., Akinwande, D., Buckled two-dimensional Xene sheets. *Nat. Mater.*, 16, 2, 163, 2017.
58. Jiang, S., Butler, S., Bianco, E., Restrepo, O.D., Windl, W., Goldberger, J.E., Improving the stability and optical properties of germanane *via* one-step covalent methyl-termination. *Nat. Commun.*, 5, 3389, 2014.
59. Pumera, M. and Sofer, Z., 2D monoelemental arsenene, antimonene, and bismuthene: Beyond black phosphorus. *Adv. Mater.*, 29, 21, 1605299, 2017.
60. Zeng, X., Luo, M., Liu, G., Wang, X., Tao, W., Lin, Y., Ji, X., Nie, L., Mei, L., Polydopamine-modified black phosphorous nanocapsule with enhanced stability and photothermal performance for tumor multimodal treatments. *Adv. Sci.*, 5, 10, 1800510, 2018.
61. Chen, W., Ouyang, J., Liu, H., Chen, M., Zeng, K., Sheng, J., Liu, Z. *et al.*, Black Phosphorus Nanosheet-Based Drug Delivery System for Synergistic Photodynamic/Photothermal/Chemotherapy of Cancer. *Adv. Mater.*, 29, 5, 1603864, 2017.
62. Wang, H., Yang, X., Shao, W., Chen, S., Xie, J., Zhang, X., Wang, J., Xie, Y., Ultrathin black phosphorus nanosheets for efficient singlet oxygen generation. *J. Am. Chem. Soc.*, 137, 35, 11376–11382, 2015.
63. Ouyang, J., Liu, R.-Y., Chen, W., Liu, Z., Xu, Q., Zeng, K., Deng, L., Shen, L., Liu, Y.-N., A black phosphorus based synergistic antibacterial platform against drug resistant bacteria. *J. Mater. Chem. B*, 6, 39, 6302–6310, 2018.
64. Gu, J., Du, Z., Zhang, C., Ma, J., Li, B., Yang, S., Liquid-Phase Exfoliated Metallic Antimony Nanosheets toward High Volumetric Sodium Storage. *Adv. Energy Mater.*, 7, 17, 1700447, 2017.
65. Meng, X., Wang, X., Cheng, Z., Tian, N., Lang, M.C., Yan, W., Liu, D., Zhang, Y., Wang, P., Photoluminescence lifetime of black phosphorus nanoparticles and their applications in live cell imaging. *ACS Appl. Mater. Interfaces*, 10, 37, 31136–31145, 2018.
66. Wang, Y., Qiu, G., Wang, R., Huang, S., Wang, Q., Liu, Y., Du, Y. *et al.*, Field-effect transistors made from solution-grown two-dimensional tellurene. *Nat. Electron.*, 1, 4, 228, 2018.
67. Qiao, J., Pan, Y., Yang, F., Wang, C., Chai, Y., Ji, W., Few layer Tellurium: One-dimensional-like layered elementary semiconductor with striking physical properties. *Sci. Bull.*, 63, 3, 159–168, 2018.
68. Yang, T., Ke, H., Wang, Q., Tang, Y., Deng, Y., Yang, H., Yang, X. *et al.*, Bifunctional tellurium nanodots for photo-induced synergistic cancer therapy. *ACS Nano*, 11, 10, 10012–10024, 2017.

69. Park, S.J., Kwon, O.S., Lee, S.H., Song, H.S., Park, T.H., Jang, J., Ultrasensitive flexible graphene based field-effect transistor (FET)-type bioelectronic nose. *Nano Lett.*, 12, 10, 5082–5090, 2012.
70. Lee, S.H. and Park, T.H., Recent advances in the development of bioelectronic nose. *Biotechnol. Bioprocess Eng.*, 15, 1, 22–29, 2010.
71. Göpel, W., Ziegler, Ch, Breer, H., Schild, D., Apfelbach, R., Joerges, J., Malaka, R., Bioelectronic noses: A status report part I. *Biosens. Bioelectron.*, 13, 3-4, 479–493, 1998.
72. Liu, Q., Cai, H., Xu, Y., Li, Y., Li, R., Wang, P., Olfactory cell-based biosensor: A first step towards a neurochip of bioelectronic nose. *Biosens. Bioelectron.*, 22, 2, 318–322, 2006.
73. Novoselov, K.S., Geim, A.K., Morozov, SVb, Jiang, D., Katsnelson, M.I., Grigorieva, IVa, Dubonos, SVb, Firsov, A.A., Two-dimensional gas of massless Dirac fermions in graphene. *Nature*, 438, 7065, 197, 2005.
74. Shafraniuk, S., *Thermoelectricity and heat transport in graphene and other 2D nanomaterials*, Elsevier, 2017.
75. Wei, D., Liu, Y., Wang, Y., Zhang, H., Huang, L., Yu, G., Synthesis of N-doped graphene by chemical vapor deposition and its electrical properties. *Nano Lett.*, 9, 5, 1752–1758, 2009.
76. Jin, H.J., Lee, S.H., Kim, T.H., Park, J., Song, H.S., Park, T.H., Hong, S., Nanovesicle-based bioelectronic nose platform mimicking human olfactory signal transduction. *Biosens. Bioelectron.*, 35, 1, 335–341, 2012.
77. Park, H., Chang, S., Jean, J., Cheng, J.J., Araujo, P.T., Wang, M., Bawendi, M.G. *et al.*, Graphene cathode-based ZnO nanowire hybrid solar cells. *Nano Lett.*, 13, 1, 233–239, 2012.
78. Wang, X., Zhi, L., Müllen, K., Transparent, conductive graphene electrodes for dye-sensitized solar cells. *Nano Lett.*, 8, 1, 323–327, 2008.
79. Liu, X., Aizen, R., Freeman, R., Yehezkeli, O., Willner, I., Multiplexed aptasensors and amplified DNA sensors using functionalized graphene oxide: Application for logic gate operations. *ACS Nano*, 6, 4, 3553–3563, 2012.
80. Yu, G., Hu, L., Vosgueritchian, M., Wang, H., Xie, X., McDonough, J.R., Cui, X., Cui, Y., Bao, Z., Solution-processed graphene/MnO2 nanostructured textiles for high-performance electrochemical capacitors. *Nano Lett.*, 11, 7, 2905–2911, 2011.
81. Wang, H., Yang, Y., Liang, Y., Robinson, J.T., Li, Y., Jackson, A., Cui, Y., Dai, H., Graphene-wrapped sulfur particles as a rechargeable lithium–sulfur battery cathode material with high capacity and cycling stability. *Nano Lett.*, 11, 7, 2644–2647, 2011.
82. Liang, Y.T., Vijayan, B.K., Gray, K.A., Hersam, M.C., Minimizing graphene defects enhances titania nanocomposite-based photocatalytic reduction of CO2 for improved solar fuel production. *Nano Lett.*, 11, 7, 2865–2870, 2011.
83. He, R.X., Lin, P., Liu, Z.K., Zhu, H.W., Zhao, X.Z., LW Chan, H., Yan, F., Solution-gated graphene field effect transistors integrated in microfluidic systems and used for flow velocity detection. *Nano Lett.*, 12, 3, 1404–1409, 2012.

84. Li, W., Geng, X., Guo, Y., Rong, J., Gong, Y., Wu, L., Zhang, X. *et al.*, Reduced graphene oxide electrically contacted graphene sensor for highly sensitive nitric oxide detection. *ACS Nano*, 5, 9, 6955–6961, 2011.
85. Lu, G., Park, S., Yu, K., Ruoff, R.S., Ocola, L.E., Rosenmann, D., Chen, J., Toward practical gas sensing with highly reduced graphene oxide: A new signal processing method to circumvent run-to-run and device-to-device variations. *ACS Nano*, 5, 2, 1154–1164, 2011.
86. Kwon, O.S., Park, S.J., Hong, J.-Y., Han, A.-R., Lee, J.S., Lee, J.S., Oh, J.H., Jang, J., Flexible FET-type VEGF aptasensor based on nitrogen-doped graphene converted from conducting polymer. *ACS Nano*, 6, 2, 1486–1493, 2012.
87. Dan, Y., Lu, Y., Kybert, N.J., Luo, Z., Charlie Johnson, A.T., Intrinsic response of graphene vapor sensors. *Nano Lett.*, 9, 4, 1472–1475, 2009.
88. He, Q., Sudibya, H.G., Yin, Z., Wu, S., Li, H., Boey, F., Huang, W., Chen, P., Zhang, H., Centimeter-long and large-scale micropatterns of reduced graphene oxide films: Fabrication and sensing applications. *ACS Nano*, 4, 6, 3201–3208, 2010.
89. Sudibya, H.G., He, Q., Zhang, H., Chen, P., Electrical detection of metal ions using field-effect transistors based on micropatterned reduced graphene oxide films. *ACS Nano*, 5, 3, 1990–1994, 2011.
90. Dankerl, M., Hauf, M.V., Lippert, A., Hess, L.H., Birner, S., Sharp, I.D., Mahmood, A. *et al.*, Graphene Solution-Gated Field-Effect Transistor Array for Sensing Applications. *Adv. Funct. Mater.*, 20, 18, 3117–3124, 2010.
91. Voelker, M. and Fromherz, P., Signal transmission from individual mammalian nerve cell to field-effect transistor. *Small*, 1, 2, 206–210, 2005.
92. Cui, Y., Wei, Q., Park, H., Lieber, C.M., Nanowire nanosensors for highly sensitive and selective detection of biological and chemical species. *Science*, 293, 5533, 1289–1292, 2001.
93. Misra, N., Martinez, J.A., Huang, S.-C.J., Wang, Y., Stroeve, P., Grigoropoulos, C.P., Noy, A., Bioelectronic silicon nanowire devices using functional membrane proteins. *Proc. Natl. Acad. Sci.*, 106, 33, 13780–13784, 2009.
94. Kong, J., Franklin, N.R., Zhou, C., Chapline, M.G., Peng, S., Cho, K., Dai, H., Nanotube molecular wires as chemical sensors. *Science*, 287, 5453, 622–625, 2000.
95. Chen, R.J., Bangsaruntip, S., Drouvalakis, K.A., Kam, N.W.S., Shim, M., Li, Y., Kim, W., Utz, P.J., Dai, H., Noncovalent functionalization of carbon nanotubes for highly specific electronic biosensors. *Proc. Natl. Acad. Sci.*, 100, 9, 4984–4989, 2003.
96. Neto, A.H.C., Guinea, F., MR Peres, N., Novoselov, K.S., Geim, A.K., The electronic properties of graphene. *Rev. Mod. Phys.*, 81, 1, 109, 2009.
97. Geim, A.K. and Novoselov, K.S., The rise of graphene. *Nat. Mater.*, 6, 183–191, 2007, https://doi.org/10.1038/nmat1849
98. Hatakeyama, R., Overview of the Nanocarbon Network. *Science*, 306, 666, 2004.

99. Bolotin, K.I., Sikes, K.J., Hone, J., Stormer, H.L., Kim, Ph, Temperature-dependent transport in suspended graphene. *Phys. Rev. Lett.*, 101, 9, 096802, 2008.
100. Farmer, D.B., Chiu, H.-Y., Lin, Y.-M., Jenkins, K.A., Xia, F., Avouris, P., Utilization of a buffered dielectric to achieve high field-effect carrier mobility in graphene transistors. *Nano Lett.*, 9, 12, 4474–4478, 2009.
101. Hwang, E.H., Adam, S., Das Sarma, S., Transport in chemically doped graphene in the presence of adsorbed molecules. *Phys. Rev. B*, 76, 19, 195421, 2007.
102. Ohno, Y., Maehashi, K., Yamashiro, Y., Matsumoto, K., Electrolyte-gated graphene field-effect transistors for detecting pH and protein adsorption. *Nano Lett.*, 9, 9, 3318–3322, 2009.
103. Ohno, Y., Maehashi, K., Matsumoto, K., Label-free biosensors based on aptamer-modified graphene field-effect transistors. *J. Am. Chem. Soc.*, 132, 51, 18012–18013, 2010.
104. Mohanty, N. and Berry, V., Graphene-based single-bacterium resolution bio-device and DNA transistor: Interfacing graphene derivatives with nanoscale and microscale biocomponents. *Nano Lett.*, 8, 12, 4469–4476, 2008.
105. Dong, X., Shi, Y., Huang, W., Chen, P., Li, L.-J., Electrical detection of DNA hybridization with single-base specificity using transistors based on CVD-grown graphene sheets. *Adv. Mater.*, 22, 14, 1649–1653, 2010.
106. Dong, X., Huang, W., Chen, P., *In situ* synthesis of reduced graphene oxide and gold nanocomposites for nanoelectronics and biosensing. *Nanoscale Res. Lett.*, 6, 1, 60, 2011.
107. Yin, Z., He, Q., Huang, X., Zhang, J., Wu, S., Chen, P., Lu, G., Zhang, Q., Yan, Q., Zhang, H., Real-time DNA detection using Pt nanoparticle-decorated reduced graphene oxide field-effect transistors. *Nanoscale*, 4, 1, 293–297, 2012.
108. C.-S. Ah, A. Kim, C.-W. Park, C.-g. Ahn, J.-H. Yang, I.-b. Baek, T. Kim *et al.*, Apparatus and method for detecting biomolecules, U.S. Patent 8,529,750, issued September 10, 2013.
109. Mao, S., Lu, G., Yu, K., Bo, Z., Chen, J., Specific protein detection using thermally reduced graphene oxide sheet decorated with gold nanoparticle-antibody conjugates. *Adv. Mater.*, 22, 32, 3521–3526, 2010.
110. Mao, S., Yu, K., Chang, J., Steeber, D.A., Ocola, L.E., Chen, J., Direct growth of vertically-oriented graphene for field-effect transistor biosensor. *Sci. Rep.*, 3, 1696, 2013.
111. Mao, K., Wu, D., Li, Y., Ma, H., Ni, Z., Yu, H., Luo, C., Wei, Q., Du, B., Label-free electrochemical immunosensor based on graphene/methylene blue nanocomposite. *Anal. Biochem.*, 422, 1, 22–27, 2012.
112. Huang, Y., Dong, X., Shi, Y., Li, C.M., Li, L.-J., Chen, P., Nanoelectronic biosensors based on CVD grown graphene. *Nanoscale*, 2, 8, 1485–1488, 2010.
113. Kavitha, T., Gopalan, A.I., Lee, K.-P., Park, S.-Y., Glucose sensing, photocatalytic and antibacterial properties of graphene–ZnO nanoparticle hybrids. *Carbon*, 50, 8, 2994–3000, 2012.

114. Huang, Y., Dong, X., Liu, Y., Li, L.-J., Chen, P., Graphene-based biosensors for detection of bacteria and their metabolic activities. *J. Mater. Chem.*, 21, 33, 12358–12362, 2011.
115. Ang, P.K., Li, A., Jaiswal, M., Wang, Y., Hou, H.W., Thong, J.T.L., Lim, C.T., Loh, K.P., Flow sensing of single cell by graphene transistor in a microfluidic channel. *Nano Lett.*, 11, 12, 5240–5246, 2011.
116. Chen, F., Qing, Q., Xia, J., Li, J., Tao, N., Electrochemical gate-controlled charge transport in graphene in ionic liquid and aqueous solution. *J. Am. Chem. Soc.*, 131, 29, 9908–9909, 2009.
117. Schedin, F., Geim, A.K., Morozov, S.V., Hill, E.W., Blake, P., Katsnelson, M.I., Novoselov, K.S., Detection of individual gas molecules adsorbed on graphene. *Nat. Mater.*, 6, 9, 652, 2007.
118. Nomani, Md WK, Shishir, R., Qazi, M., Diwan, D., Shields, V.B., Spencer, M.G., Tompa, G.S., Sbrockey, N.M., Koley, G., Highly sensitive and selective detection of NO2 using epitaxial graphene on 6H-SiC. *Sens. Actuators B: Chem.*, 150, 1, 301–307, 2010.
119. Ang, W., Wei-Wei, L., Jing-Xia, W., Qing, L., Zhao, W., Li, X., Xiao-Chen, D., Wei, H., Single-Walled Carbon Nanotube Networked Field-Effect Transistors Functionalized with Thiolated Heme for NO2 Sensing. *Chin. Phys. Lett.*, 28, 12, 127304, 2011.
120. Kim, C.-H., Yoo, S.-W., Nam, D.-W., Seo, S., Lee, J.-H., Effect of Temperature and Humidity on $\hbox {NO} _{2} $ and $\hbox {NH} _{3} $ Gas Sensitivity of Bottom-Gate Graphene FETs Prepared by ICP-CVD. *IEEE Electron Device Lett.*, 33, 7, 1084–1086, 2012.
121. Wu, J., Li, H., Qi, X., He, Q., Xu, B., Zhang, H., Graphene Oxide Architectures Prepared by Molecular Combing on Hydrophilic-Hydrophobic Micropatterns. *Small*, 10, 11, 2239–2244, 2014.
122. Ren, Y., Zhu, C., Cai, W., Li, H., Ji, H., Kholmanov, I., Wu, Y., Piner, R.D., Ruoff, R.S., Detection of sulfur dioxide gas with graphene field effect transistor. *Appl. Phys. Lett.*, 100, 16, 163114, 2012.
123. Johnson, J.L., Behnam, A., Pearton, S.J., Ural, A., Hydrogen sensing using Pd-functionalized multi-layer graphene nanoribbon networks. *Adv. Mater.*, 22, 43, 4877–4880, 2010.
124. Chung, M.G., Kim, D.H., Lee, H.M., Kim, T., Choi, J.H., Seo, D. k., Yoo, J.-B., Hong, S.-H., Kang, T.J., Kim, Y.H., Highly sensitive NO2 gas sensor based on ozone treated graphene. *Sens. Actuators B: Chem.*, 166, 172–176, 2012.
125. Ang, P.K., Chen, W., Wee, A.T.S., Loh, K.P., Solution-gated epitaxial graphene as pH sensor. *J. Am. Chem. Soc.*, 130, 44, 14392–14393, 2008.
126. Cheng, Z., Li, Q., Li, Z., Zhou, Q., Fang, Y., Suspended graphene sensors with improved signal and reduced noise. *Nano Lett.*, 10, 5, 1864–1868, 2010.
127. K. Levon, A. Rahman, T. Sai, B. Zhao, Using floating gate field effect transistors for chemical and/or biological sensing, U.S. Patent Application 12/328,888, filed March 26, 2009.

128. Sohn, I.-Y., Kim, D.-J., Jung, J.-H., Yoon, O.J., Thanh, T.N., Quang, T.T., Lee, N.-E., pH sensing characteristics and biosensing application of solution-gated reduced graphene oxide field-effect transistors. *Biosens. Bioelectron.*, 45, 70–76, 2013.
129. Park, J.-U., Nam, S.W., Lee, M.-S., Lieber, C.M., Synthesis of monolithic graphene–graphite integrated electronics. *Nat. Mater.*, 11, 2, 120, 2012.
130. Li, J., Guo, S., Zhai, Y., Wang, E., High-sensitivity determination of lead and cadmium based on the Nafion-graphene composite film. *Anal. Chim. Acta*, 649, 2, 196–201, 2009.
131. Wen, Y., Li, F.Y., Dong, X., Zhang, J., Xiong, Q., Chen, P., The Electrical Detection of Lead Ions Using Gold-Nanoparticle-and DNAzyme-Functionalized Graphene Device. *Adv. Healthcare Mater.*, 2, 2, 271–274, 2013.
132. Gong, J., Zhou, T., Song, D., Zhang, L., Monodispersed Au nanoparticles decorated graphene as an enhanced sensing platform for ultrasensitive stripping voltammetric detection of mercury (II). *Sens. Actuators B: Chem.*, 150, 2, 491–497, 2010.
133. Zhang, T., Cheng, Z., Wang, Y., Li, Z., Wang, C., Li, Y., Fang, Y., Self-assembled 1-octadecanethiol monolayers on graphene for mercury detection. *Nano Lett.*, 10, 11, 4738–4741, 2010.
134. Mao, S., Lu, G., Yu, K., Bo, Z., Chen, J., Specific protein detection using thermally reduced graphene oxide sheet decorated with gold nanoparticle-antibody conjugates. *Adv. Mater.*, 22, 32, 3521–3526, 2010.
135. Maehashi, K., Sofue, Y., Okamoto, S., Ohno, Y., Inoue, K., Matsumoto, K., Selective ion sensors based on ionophore-modified graphene field-effect transistors. *Sens. Actuators B: Chem.*, 187, 45–49, 2013.
136. Fu, W., Nef, C., Tarasov, A., Wipf, M., Stoop, R., Knopfmacher, O., Weiss, M., Calame, M., Schönenberger, C., High mobility graphene ion-sensitive field-effect transistors by noncovalent functionalization. *Nanoscale*, 5, 24, 12104–12110, 2013.

7

Supercapacitor Electrodes Utilizing Graphene-Based Ternary Composite Materials

B. Saravanakumar[1]*, K. K. Purushothaman[2], S.Vadivel[3], A. Sakthivel[4], N. Karthikeyan[1] and P. A. Periasamy[5]

[1]*Department of Physics, Dr. Mahalinggam College of Engineering and Technology, Pollachi, India*
[2]*Department of Physics, Aringar Anna Government Arts and Science College, Karaikal, India*
[3]*Department of Chemistry, PSG College of Technology, Coimbatore, India*
[4]*Department of Electrical and Electronics Engineering, Dr. Mahalinggam College of Engineering and Technology, Pollachi, India*
[5]*Department of Chemistry, Dr. Mahalinggam College of Engineering and Technology, Pollachi, India*

Abstract
The rapid growth of electrical energy consumption over the globe urges to carry-out intense research on innovative electrode materials for supercapacitor electrode application. Till date, varieties of carbon-based nanocomposites have been tested to achieve high performance in supercapacitors. Among these, graphene is considered as a candidate utilized due to its unparallel physiochemical characteristics. The graphene has modified such as reduced graphene oxide to produce the better bonding with different materials including metal nanoparticles, metal oxides, conducting polymer, and some biopolymers. Recently, ternary composite materials (three different materials in a single system) evolved to produce enhancement in electrochemical features of supercapacitors. There are many methods that were adopted to blending three different materials for fabrication of ternary composite and tested for supercapacitor application. This chapter details the fabrication and characterization of graphene-based ternary composite materials towards supercapacitor electrode applications.

**Corresponding author*: saravanakumar123@gmail.com

Inamuddin, Rajender Boddula, Mohd Imran Ahamed and Abdullah M. Asiri (eds.)
Monoelements: Properties and Applications, (149–168) © 2020 Scrivener Publishing LLC

Keywords: Graphehe, supercapacitor, flexible electrode, ternary composite, conducting polymer

7.1 Introduction

The electrochemical energy storage systems (EES) are creating significant impact in the process of promoting the living style of human beings. The advancement in electronics industry and essentiality of electric vehicles (EVs) promoted large requirement for EES with superior charcteristics [1]. Among various varieties of EES, the supercapacitor (SC) could store and discharge huge electricity in a small duration of time [2]. SCs are attractive electrical energy storage device, which is utilized to develop high performance hybrid electric vehicles (HEV) owing to its advantageous features including high power, superior cycle stability, maintenance free, eco-friendliness. The SCs find use in electric gadgets and other instruments in our day routine. It provides supplementary power in EVs, wind turbine blade pitching system, trucks, industrial cranes, cell phones, camera, uninterruptable power supplies, etc. [3].

EES devices are categorized as electrochemical double layer capacitor (EDLC) and SCs using their charge storage principle. EDLC uses the splitting of electric charges at electrode-electrolyte interface. SCs utilizing Faradic reactions [4]. The oxidation-reduction reaction happening in the electrode and electrolyte leads to storage of electrical energy. The performance improvement in EES could be achieved through improvement of surface dependent process [5]. The electrode materials have to be engineered with more ion buffering reservoirs for creating higher charge accumulation. But getting the complete profit of the electroe material is a big challenging task [6].

There are many transition metal oxides (TMOs) including manganese oxide, ruthernium oxide, nickel oxide, vanadium oxide, iron oxide, etc., conducting polymers (PANI, PPy, and PEDOT) and their combinations are being tested for SC electrode applications. Even though TMOs and conducting polymers possess good redox properties, they suffer with poor electronic conductivity and agglomeration during synthesis. To overcome these critical concerns, a method of addition of nano dimensional carbon materials blended with metal oxides provides extremely good solution. The blending of nanostructured carbon with metal oxides facilitates many ion transferring channels for ion intercalation and deintercalation. Additionally, the higher electronic conductivity of carbon significantly improves the conductivity of the composite [7]. These attractive properties stress the requirement of adding nanostructured carbon-based material with metal oxide/conducting polymers. Here, it is valuable to note that the combination of more than two chemical functional groups (ternary) in a composite may also provide better solutions for existing issues. In this chapter, we discuss on charge storage in SCs, design of a SC electrode, functionalized forms of graphene, and the graphene-based ternary composite for supercapacitor electrode applications. The advancement in the performance of SCs using graphene-based ternary composite and their advantages are also discussed.

7.2 Charge Storage Mechanism of a Supercapacitor Device

In the year 1858, Hermann von Helmholtz was attempted firstly to elaborate the concept of electric double layer formation. More specifically, an electrochemical capacitor (Figure 7.1) possesses two different electrodes and these electrodes electrically isolated by a material named as separator. The electrodes with electrolyte solution captured the counter ions and oppose the co-ions under the external potential difference. The potential initiates the formation of the two different layers of polarized

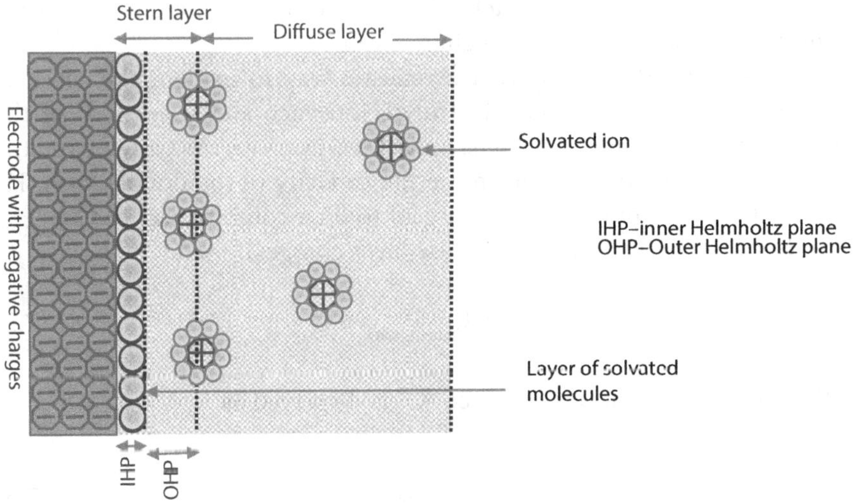

Figure 7.1 Charge storage mechanism in EDLC.

ions. The first layer is created over the surface of the electrode and another one with opposite polarity is created by solvated molecules present in the electrolyte.

A straight line present on solvated molecules is mentioned as inner Helmholtz plane (IHP). Further, the plane differentiates the polarized ions with opposite charge behaves as a molecular dielectric. Additionally, a line present in the center of the solvated ion layer indicating the boundary is considered as the outer Helmholtz plane (OHP) as displayed in Figure 7.1. The area that lies within this range is known as the electric double layer. According to Helmholtz considerations, the capacity C_h is given as

$$C_h = \frac{\varepsilon}{4\pi d} \tag{7.1}$$

In this equation, ε is dielectric constant and d is consider as thickness of the bilayer created. However, this model of Helmholtz fails to detail the relation between the capacitance, potential, and concentration of the electrolyte.

During the years between 1910 and 1913, David Leonard Chapman and Louis Georges Gouy explained the formation of the "diffuse layer" [8–10]. This model assumes that the solvated ions lie in the electrolyte as point charges are accumulated within a single diffuse layer. Further, this one details the relation between the capacity, concentration of the electrolyte, and the potential. But this model fails to describe the differences between the measured and estimated capacitance values.

In 1924, Otto Stern proposed a realistic way to explain the physical environment at the electrode/electrolyte interface and used both of the above-mentioned models [11]. His model used various electrolytes and the ionic radius of the ions to determine the capacity of the double layer. He assumed $Q_{solution}$ (solution side charges) to be a combination of $Q_{Helmholtz}$ (Helmholtz charge) and Q_{Gouy} (diffuse layer charges).

$$Q_{solution} = Q_{helmholtz} + Q_{gouy} \tag{7.2}$$

The total capacity at the interface "C" is displayed as

$$\frac{1}{C} = \frac{1}{C_h} + \frac{1}{C_g} \tag{7.3}$$

Here, C_h = Capacity due to Helmholtz layer
C_g = Capacity due to diffuse layer.

These two parallel layers act as a pair of capacitors and connected in series form. Furthermore, David Crockett Graham used an approach to describe about double layer [12]. He considered that some of the ionic moieties can penetrate the Stern layer and touches the IHP. He assumed these charge particles in attachment to the electrode as "specifically adsorbed ions" as presented in Figure 7.2.

Further, Bockris-Devanathan-Muller proposed the BDM model which we are utilizing commonly now [13]. In this model, they proposed that the solvent molecules possess a permanent alignment with the electrode. The inner Helmholtz plane was drawn through the center of the solvated ions. More accurately, the adsorbed ions could present in this area. The new layer created by the solvated ions appear next to the IHP is considered as OHP. Also, the diffuse layer is little away from the OHP. It's valuable to note that the electrical energy is accumulated by means of Faradic redox reactions on the electrode surface with electric double layer. Fundamentally, the pseudo capacitance arised due to exchange between the electrolyte and electrode from desolvated ions.

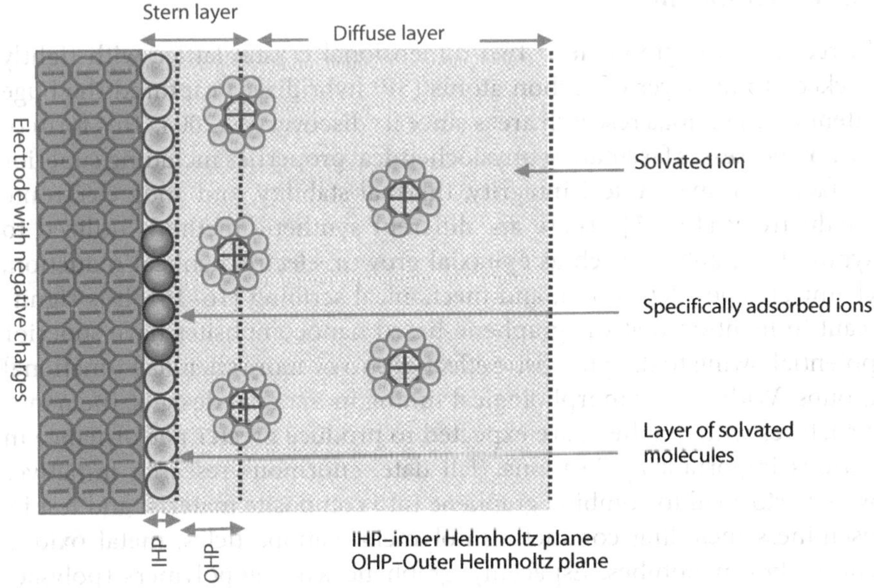

Figure 7.2 Charge storage mechanism via redox reactions-based pseudocapacitance.

Here, there are no chemical bonds arising between adsorbed ions and the atoms present. Only the charge transfer occurs in electrode and electrolyte interface.

7.2.1 Design of a Supercapacitor Electrode

Generally, a SC possesses four important components. It is composed of electrode active material, current collector, separator, and electrolyte. Among these, the innovative electrode active material holds the key to determine the performance of a SC system.

At present, the process of designing innovative electrode materials with higher electrochemical performance leads the better growth in SC technologies. In this SC technology, the electro-active material is the primary component to improve the performance of SC device. There are different varieties of electrode materials that are tested for SC electrode applications. It includes nanostrutured carbons, conducting polymers (CPs) transition metal oxides (TMOs), and binary/ternary composites.

7.3 Graphene and its Functionalized Forms

7.3.1 Graphene

In recent years, graphene, a two-dimensional crystal lattice with tightly packed atomic layer of carbon atoms (SP^2 hybridized) captured the huge attention in various research areas since its discovery in 2004. This form of carbon possesses fascinating physiochemical properties including superior surface area, mechanical integrity, thermal stability, and goog electronic conductivity [14, 15]. There are different synthetic methods utilized to synthesize graphene such as epitaxial growth, electrochemical reduction, chemical vapor deposition, and mechanical scribing [16–18]. It is significant to mention that the graphene-based nanocomposites have superior potential owing to its synergistic effect of two or more chemical functional groups. With proper morphological tuning in material designs, the nanostructures with graphene are expected to produce greater performance in various important applications. Till date, enormous research initiatives were performed to combine graphene into composite material and test its usefulness including conducting polymers, nanoparticles, metal oxides, and carbon nanotubes. Especially, graphene with biopolymers (polysaccharides, proteins, peptides, and nucleic acid) also has shown better results in achieving controllable multifunctionalities [19–22].

7.3.2 Graphene Oxide

The graphite powder was usually chemically exfoliated to synthesis of graphene oxide (GO). This chemical exfoliation can be done with mixed with strong oxidizers such as $KMnO_4$ in concentrated H_2SO_4. The process of chemical exfoliation separates the conjucation of the sheets of graphene in to SP^2 graphitic domains and also carbon vacancies [23]. The final product of this process provides the hydroxyl moities at the surface and carboxylic acid at the corners. Due to this, the sheets can able to form a suspension very easily with water [24]. This exfoliation process makes the insoluble graphite as processable material and which is much useful in the formation of many composite materials. Like graphene, the thickness of the GO is measured in molecular dimensions (ca. 1 nm) [24]. The lateral sizes colloidal particles are measured from nanometers upto micrometers [25]. Secondly, this can be analyzed as an unconventional material including anisotropic colloid, liquid crystal, membrane, and amphiphile. Owing to this, simple preparation of GO and its dispersion is very attractive in the applications including chemical sensors [26], biosensors [27], polymer nanocomposites [28], battery, and supercapacitors [29–30].

7.3.3 Reduced Graphene Oxide

The chemical treatment with reduction initators or high tempeature procedures on GO leads the formation of reduced graphene oxide (rGO). Primarily, hydrazine vapor is utilized for production of rGO. Due to explosive nature of the hydrazine, this process of reducing GO is needed much care [31]. Further, GO can be also reduced by utilizing sodium hydroxide, sodium carbonate, and sodium borohydride. The reducing process of GO leads the removal of oxygen functionalities on the exterior of GO. It significantly improves conductivity of the GO [32]. However, some of the defects and oxygen functionalities still remain in the GO [33]. The transmission electron microscopy (TEM) images of rGO are presented in the Figure 7.3.

7.4 Varieties of Graphene-Based Ternary Composite

The ternary composite materials utilizing graphene as a host material is expected to provide enhanced electrochemical properties based on the material composition. The graphene-based SC electrode active materials are categorized into following different types including graphene-conducting

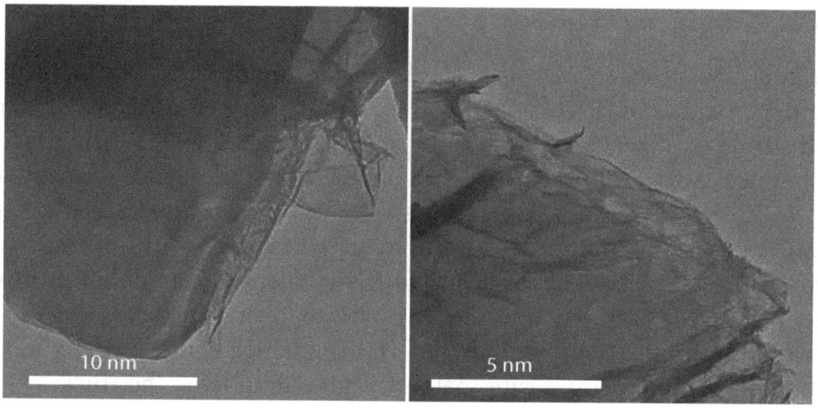

Figure 7.3 TEM images of reduced graphene oxide prepared by modified Hummers method.

polymer (CP)-TMO, graphene/other carbon/CP, graphene/other carbon/TMOs, and other graphene-based ternary materials.

7.4.1 Graphene-Conducting Polymer-Metal Oxide

The carbon family material possesses higher stability, electronic conductivity, and superior surface area. But it suffers from lower capacitance. The metal oxide has higher capacitance but possesses lower conductivity. The polymer may provide better flexibility and higher capacitance, but fails to provide higher stability. But, the combination of all three may address all the critical concerns for supercapacitor device fabrication. This synergetic combination is expected to provide better results. There are three important CPs including PEDOT (poly (3,4-ethylenedioxythiophene)), PPy (Polypyrole), and PANI (Poly aniline) are mostly utilized for design of flexible SC electro-active materials.

7.4.1.1 Graphene-PEDOT-Metal Oxide

PEDOT is a kind of conducting polymer based on 3,4-ethylenedioxylthiophene. This conducting polymer was explored for different energy storage applications owing to its high chemical better redox potential. Addition of this material with graphene and other metal oxides will facilitate better electrochemical properties. In this context, Azman *et al.* synthesized a ternary composite of poly (3,4-ethylenedioxythiophene)/GO/MnO_2 and investigated its usefulness towards supercapacitor applications [34].

This material exhibits capacitance of 239.4 Fg^{-1} at the voltage sweep rate of 25 mVs^{-1}. Further, this ternary composite shows 7.9 $Whkg^{-1}$ as energy density and power densitiy of 489 Wkg^{-1}. This material retains 95% of capacitance at 1,000 repeated cycles. Many authors synthesized a new variety of composite with a polymer, graphene, and mixed metal oxides. For instance, a new type of composite, namely, PEDOT/RGO/MnO_2/Fe_2O_3 was developed by Azman *et al.* [35] Here, MnO_2 and Fe_2O_3 particles were coated on sheets of PEDOT/RGO. This new material exhibits the capacitance of 287 Fg^{-1} with lower charge transfer resistance. This material diplays specific energy of 11 $Whkg^{-1}$ with specific power of 1,900 Wkg^{-1}.

A combination of MnO_2 nanorods, rGO, and poly (3,4-ethylenedioxythiophene)-poly (styrenesulfonate) (PEDOT:PSS) was fabricated by Hareesh *et al.* [36]. Authors used acetonitrile/lithium per chlorate organic electrolyte for electrochemical studies. This material exhibits capacitance of 633 Fg^{-1} at a current rate of 0.5 Ag^{-1}. Interestingly, this material holds 100% stability after 5,000 repeated continous cycles. Agnihotria *et al.* synthesized ternary composite of PEDOT-graphene-MnO_2 using one pot solvothermal procedure and analyzed the use of this materials for SC applications [37]. In this work, the binary composite was wrapped by PEDOT. This material exhibits highest capacitance of 633 Fg^{-1}.

A new kind of ternary composite, PEDOT:PSS/$NiFe_2O_4$/rGO has been fabricated by Hareesha *et al.* [38]. The electrochemical performance of this attractive composite material has compared with some other combinations including $NiFe_2O_4$/CNT, $NiFe_2O_4$/CNT/rGO, and $NiFe_2O_4$/rGO. This composite material displays capacitance 1,090 Fg^{-1}. A ternary polymer nanocomposite of graphene platelets (GnPs), MnO_2, and PEDOT onto a kitchen sponge [39]. Significantly, GnPs have been coated on sponge using esay dip/dry procedure. Further, it underwent polymerization of PEDOT. Further, MnO_2 was coated over the GnPs. The PEDOT layers by placing the sponge into potassium permanganate mixture. This combination has utilized for fabrication of flexible supercapacitor. This SC delivers capacitance of 802.99 Fg^{-1} with energy density of 55.76 $Whkg^{-1}$. Additionally, this device exhibits 99% retention of capacitnace after 1,000 charge-dischrge cycles.

7.4.1.2 Graphene-PANI-Metal Oxide

PANI is one among the promising conducting polymer utilized extensively for SC electrode applications due to its variable oxidation states, ease of doping and undoping leads the higher capacitance. In addition to this, this polymer material has better environmental stability with a higher theoritical capacitance of 2,000 Fg^{-1}. Making ternary composites with PANI is considered as a capable candidate to show better supercapacitive features. For instance, Shafi *et al.* synthesized $LaMnO_3$/RGO/PANI ternary composite and characterized its usefulness towards SC device fabrication [40]. An asymmetric supercapacitor device was assembled using $LaMnO_3$/rGO/PANI‖rGO that shows the specific capacitance of 111 Fg^{-1} at a current density of 2.5 Ag^{-1}. This device delivers energy density of 50 Wh kg^{-1} with a power density of 2.25 kW kg^{-1}. In addition, this SC retains 117% of capacitance after 100,000 cycles and showing excellent stability. A ternary composite of rGO, PANI, and nickel oxide prepared by Aranganathan *et al.* using *in situ* one step solution-based method [41]. While tested as SC electrode this composite showed maximum capacitance of 153.37 Fg^{-1} with a power density and energy density of 392.54 WKg^{-1} and 13.63 $Whkg^{-1}$, respectively. This material retains 100% capacitance after 10,000 charges-discharge cycles.

Palsaniya *et al.* fabricated a ternary composite of polyaniline (PANI), graphene and MoS_2 using polymerization of PANI [42]. This ternary composite electrode displays exhibits high specific capacitance (Cs, 142.30 Fg^{-1}). The authors reported that the cyclic stability of this ternary composite is significantly better than the pure PANI (~40%) and binary composites (~60%–96%). Peng *et al.* successfully designed a combination of graphene/nano-sulphur/PANI ternary material. This has utilized as an electro-active material in SC applications [43]. It exhibits outstanding supercapacitive features including specific capacitance of 619 Fg^{-1} based on cyclic voltammetry measurements (5 mVs^{-1}). This material retains 77.9% (457 F g^{-1}) of capacitance (586.8 Fg^{-1}) even at 500 repeated charge-discharge cycles.

A novel and low cost procedure to synthesis Fe/Co/Ni-graphene and PANI derived carbon/graphene was created by Zhao *et al.* Both the electrodes were used to assemble an asymmetric-type SC [44]. This device exhibits higher capacitance, superior energy density (16 $\mu Whcm^{-2}$), and good stability (87.5% after 8,000 repeated charge-discharge cycles).

Polyaniline-β-cyclodextrin-graphene nano sheet (PANI-β-CD-GNS) ternary composite was developed by Gandla *et al.* using *in situ* polymerization of aniline in presence of β-CD-GNS.PANI-β-CD-GNS [45]. This attractive material exhibits capacitance of 1,071 Fg^{-1}. In addition, this material shows cycling stability of 91.8% retention of capacitance after 5,000 cycles. Moreover, the authors reported that this material possesses higher radical cation stability that leads to better electrochemical features.

7.4.1.3 *Graphene-PPy-Metal Oxide*

The polypyrrole (PPy) is one among the good candidates for SC electrode applications. It possesses higher electronic conductivity, rapid charge-discharge, good thermal conductivity, and low cost. It is well demonstrated that the cross-linked PPy provides higher capacitnace, for instance, a ternary composite of titania-graphene-polypyrrole (TiO_2-G-PPy) films have synthesized by Jiang *et al.* using sol-gel procedure [46]. As a supercapacitor electrode, this materials delivers 201 Fg^{-1} with cyclic stability of 76.5% up to 100 charge-discharge cycles. Authors varied the content of titania and reported its electrochemical performance. Additionally, this report details the advantages and disadvantages of adding PPy in the composite.

Zhou *et al.* fabricated, RGO/MnO_2/PPy and tested its usefulness towards supercapacitor electrode applications [47]. This flexible electrode delivers capacitance of 682 Fg^{-1} and holds 86% of capacitance after 1,000 cycles. Further, authors reported the use of graphene sheets towards the reduction of agglomeration of PPy and MnO_2 during synthesis. Wang *et al.* synthesized a ternary supercapacitor electrode material with the combination of graphene, polypyrrole, SnO_2 using simple sol-gel procedure [48]. This material exhibits highest capacitance of 616 Fg^{-1}at the scan rate of 1 mVs^{-1}.

A ternary composite polypyrrole/Fe_2O_3/rGO were prepared by Moyseowicz *et al.* by hydrothermal and oxidative polymerization of pyrrole [49]. This kind of fabrication method involves preparation of Fe_2O_3/rGO composite using hydrothermal approach as first step. Further, PPy was pasted over the surface of this composite by using polymerization. This combination displays specific capacitance 140 Fg^{-1}. In addition, this electrode active material retains 93% of its capacitance after 5,000 charge-discharge cycles.

In a recent study, Han *et al.* prepared composite of rGO/MnO_2/PPy in a layer-by-layer structure [50]. The synergy between these materials displays higher capacitance as 404 Fg^{-1} for this material with better electrochemical features. Zheng *et al.* designed a ternary composite of rGO/SnO_2/PPy using reaction of Sn^{2+} ions and GO [51]. This process was followed by polymerization of pyrrole. The authors reported for better electrochemical features for this combination of ternary composite.

7.4.2 Graphene/Other Carbon/Conducting Polymer

The combination of graphene with some other carbon materials like mutil-wall carbon nanotube, single wall nanotube, and activated carbon with conducting polymers are also reported extensively to achive higher performance as supercapacitor electrode. For instance, Xu *et al.* synthesized a composite of graphene/AC/PPy [52]. In this report, the activated carbon is placed between graphene sheets. This attractive material shows the highest areal specific capacitance of 906 mF cm^{-2}. This material exhibits better cyclic stability (64.4% after 5,000 cycles). Zhang *et al.* synthesized a ternary graphene-based aerogel (RuO_2/g-C_3N_4/rGO) using combined hydrothermal and lyophilization process [53]. This ternary composite material exhibited specific capacitance of 704.3 Fg^{-1} at a current of 0.5Ag^{-1} and better electrochemical performances.

The rGO/MnO_2/CB ternary film synthesized by Chen *et al.* by vacuum filtration procedure. The authors explained the synergetic effect between MnO_2 nano particles, graphene, and carbon black [54]. Especially, CB acts as spacers between the graphene sheets and restricting the restacking. This attractive material exhibits the capacitance of 704.3 Fg^{-1}. The asymmetric capacitor based on this capacitor delivered energy density of 10 Wh kg^{-1} at the power density of 45 kW kg^{-1}. A ternary composite of molybdenum trioxide (α-MoO_3), thin layer of carbon, and graphene oxide is prepared by hydrothermal combined with ultrasonication method [55]. The authors reported the enhancement of electrical conductivity metal oxide by the addition of graphene sheets. This ternary composite (rGO/MoO_3@C) shows better electrochemical features including the higher capacitance and lower charge transfer resistance.

7.4.3 Graphene/Other Carbon Material/Metal Oxide

The other combination of forming ternary composite is blending graphene, other carbon-based material, and metal oxide. This type also showed some better results in terms of achieving enhanced electrochemical performance. For instance, Co_3O_4/N-doped graphene/CNTs was prepared using hydrothermal method by Bao *et al.* In this composite material, nitrogen doped graphene sheets were utilized as substrate with better electronic conductivity provides better electrochemical performance [56]. This Co_3O_4/NG, Co_3O_4/graphene ternary composite material displays a high specific capacitance of 456 Fg^{-1} at a current density of 1 Ag^{-1}. Interestingly, an increase of 23.2% of initial specific capacitance after 1,000 charge-discharge cycles for this material. A new kind of ternary composite of $FeNi_2S_4$-CNT-graphene nanocomposite was prepared by Tang *et al.* using simple solution-based method [57]. The authors reported that this ternary form performs better than the counterparts without CNT, the $FeNi_2S_4$-CNT-graphene in terms of electrochemical performance. This materials exhibits capacitance of 725 F g^{-1} at a current density of 10 Ag^{-1}, 3.63 Fcm^{-2} at a current density of 50 mA cm^{-2}. Further, $FeNi_2S_4$-CNT-graphene composite demonstrated a capacity retention of 88% after 2,500 cycles.

A novel ternary composite paper having rGO sheet, carbon nanotube (CNT), and MnO_2 was synthesized by Jin *et al.* using electrochemical deposition procedure [58]. The authors presented that the presence of graphene in this ternary composite paper remarkably enhancing the electrochemical performance. Further, an asymmetric supercapacitor was assembled using this flexible ternary composite paper operated in a 1.6V potential window exhibits the energy density at 24.8 $Whkg^{-1}$.

A ternary nanocomposite of nickel oxide (NiO)/multi-walled carbon nanotubes (CNTs)/reduced graphene oxide (RGO) was fabricated using solvothermal procedure by Rajendran *et al.* [59]. In this material, NiO nanostructures were uniformly deposited on the surface of RGO nanosheets and CNTs. This material displays 2.2 times higher capacitance than pristine NiO indicating the better synergy between RGO/CNT/NiO ternary nanocomposite. This material showed the cycling stability of 94% after 1,000 cycles. A highly porous $V_3O_7{\cdot}H_2O$ nanobelts/CNT/reduced graphene oxide ternary material using hydrothermal procedure by Hu *et al.* [60]. This attractive material possesses 3D micro-structure and exhibiting better supercapacitive properties when tested as a supercapacitor electrode. The authors reported that a notable increment of surface area (53.7 $m^2{\cdot}g^{-1}$) by the addition of graphene in this composite.The synergistic effect of these three materials exhibits capacitance of 685 Fg^{-1} at the current 0.5 Ag^{-1}. This material possesses good stability of 99.7% after 10,000 cycles.

7.4.4 Other Graphene-Based Ternary Materials

Ternary composite of $ZnCo_2O_4$/reduced graphene oxide/NiO nanowire arrays prepared by hydrothermal and thermal annealing method by Sahoo *et al.* [61]. This 3D ternary composite exhibits capacitance of 1,256 Fg^{-1} at a current density of 3 Ag^{-1} with a cyclic stability of 80% after 3,000 charge-discharge cycles. Additionally, authors reported the energy density of 62.8 $Whkg^{-1}$, maximum power density of 7492.5 Wkg^{-1} for this ternary composite. A different kind of porous ternary nanohybrid was fabricated using $NiMn_2O_4$, rGO, and polyaniline and investigates its usefulness for supercapacitor electrode by Sahoo *et al.* [62]. The hydrothermal-assisted thermal annealing method was adopted for synthesis of this ternary nanocomposite. The authors presented that proper wrapping of PANI in the composite significantly enhances the performance of this composite. This material shows specific capacitance of 757 Fg^{-1} with an energy density of 70 Wh kg^{-1}.

In addition, it retains the ~93% of its initial capacitance after 2,000 cycles. An rGO/Fe_2O_3/SnO_2 ternary composite material fabricated *via* precipitation of Fe_2O_3 on the surface of graphene oxide by Xia *et al.* [63]. Further, this material undergone reduction of GO with $SnCl_2$. The authors tested the effects of reduction temperature with time of reduction for GO with $SnCl_2$. In addition, Fe_2O_3 nanoparticles (~20 nm) were uniformly attached and surrounded by SnO_2 nanoparticles. Here, graphene is acting as a physical support matrix for relieving of stress during ion intercalation during cycling process. In addition, this material shows better electrochemical features.

7.5 Conclusion and Future Perspectives

An intensive research on supercapacitor electrodes materials has grown to next level, the researchers fabricating hybrid and ternary composite materials to take full advantage of individual material. Generally, binary materials cannot satisfy the need of advanced supercapacitor devices. Thus, the electrode active materials holding the key to determine the efficiency of a supercapacitor system. In this context, graphene and its different forms including graphene oxide, reduced graphene oxide-based ternary composites are considered better candidate owing to their signifigant chracteristics including enhanced specific surface area and good ionic conductivity. The use of polymeric materials in this composite provides higher capacitance and superior flexibility which is utilized to fabricate the flexible energy storage devices. However, we believe that there are many scope for further research to improve the supercapacitor features of the ternary composites using graphene.

References

1. Liu, W.W., Feng, Y.Q., Yan, X.B., Chen, J.T., Xue, Q.J., Superior Micro Supercapacitors Based on Graphene Quantum Dots. *Adv. Funct. Mater.*, 23, 4111–4122, 2013.
2. Choi, B.G., Hong, J., Hong, W.H., Hammond, P.T., Park, H., Facilitated Ion Transport in All-Solid-State Flexible Supercapacitors. *ACS Nano*, 5, 9, 7205–7213, 2011.
3. Hall, P.J., Mirzaeian, M., Fletcher, S.I., Sillars, F.B., Rennie, A.J.R., Shitta-Bey, G.O., Wilson, G., Cruden, A., Carter, R., Energy storage in electrochemical capacitors: Designing functional materials to improve performance. *Energy Environ. Sci.*, 3, 1238–1251, 2010.

4. Mishra, A.K. and Ramaprabhu, S.J., Functionalized Graphene-Based Nanocomposites for Supercapacitor Application. *J. Phys. Chem. C*, 115, 29, 14006–14013, 2011.
5. Khoo, E., Wang, J.M., Ma, J., Lee, P.S., Electrochemical energy storage in a β-$Na_{0.33}V_2O_5$nanobelt network and its application for supercapacitors. *J. Mater. Chem.*, 20, 8368–8374, 2010.
6. Mai, L., Xu, L., Han, C., Xu, X., Luo, Y., Zhao, S., Zhao, Y., Electrospun ultralong hierarchical vanadium oxide nanowires with high performance for lithium ion batteries. *Nano Lett.*, 10, 4750–4755, 2010.
7. Zhi, M., Xiang, C., Li, J., Li, M., Wu, N., Nanostructured carbonemetal oxide composite electrodes for supercapacitors: A review. *Nanoscale*, 5, 72–88, 2013.
8. Gouy, G., On the constitution of the electric charge on the surface of an electrolyte. *J. Phys. Radium*, 9, 457, 1910.
9. Gouy, G., Sur la constitution de la charge électrique a la surface d'un électrolyte, *Compt. Rend.*, 149, 654, 1910.
10. Chapman, D.L., A contribution to the theory of electrocapillarity. *Philos. Mag.*, 25, 475, 1913.
11. Stern, H.O., Zur theorie der elektrolytischen doppelschicht, *Electrochem.*, 30, 508, 1924.
12. Grahame, D.C., The electrical double layer and the theory of electrocapillarity. *Chem. Rev.*, 41, 441, 1947.
13. Bockris, J.O.M., Devanathan, M.A.V., Muller, K., On the structure of charged interfaces. *Proc. R. Soc., A*, 274, 55, 1966.
14. Geim, A.K. and Novoselov, K.S., The rise of graphene. *Nat. Mater.*, 6, 183–191, 2007.
15. Park, S. and Ruoff, R.S., Chemical methods for the production of graphenes. *Nat. Nanotechol.*, 4, 217–224, 2009.
16. Zhu, C., Guo, S., Fang, Y.Y., Dong, S., Reducing sugar: New functional molecules for the green synthesis of graphene nanosheets. *ACS Nano*, 4, 2429–2437, 2010.
17. Li, D., Muller, M.B., Gilje, S., Kaner, R.B., Wallace, G.G., Processable aqueous dispersions of graphene nanosheets. *Nat. Nanotechol.*, 3, 101–105, 2008.
18. Zhou, M., Wang, Y., Zhai, Y., Zhai, J., Ren, W., Wang, F., Dong, S., Controlled synthesis of large-area and patterned electrochemically reduced graphene oxide films. *Chem.-Eur. J.*, 15, 6116–6120, 2009.
19. Guo, Y., Guo, S., Ren, J., Zhai, Y., Dong, S., Wang, E., Cyclodextrin functionalized graphene nanosheets with high supramolecular recognition capability: Synthesis and host guest inclusion for enhanced electrochemical performance. *ACS Nano*, 4, 4001–4010, 2010.
20. Han, T.H., Lee, W., Lee, D.H., Kim, J.E., Choi, E.Y., Kim, S.O., Eptide/graphene hybrid assembly into core/shell nanowires. *Adv. Mater.*, 22, 2060–2064, 2010.
21. Liu, J., Fu, S., Yuan, B., Li, Y., Deng, Z., Toward a universal "adhesive nanosheet" for the assembly of multiple nanoparticles based on a protein-induced reduction/decoration of graphene oxide. *J. Am. Chem. Soc.*, 132, 7279–7281, 2010.

22. Lv, W., Guo, M., Liang, M.H., Jin, M.F., Cui, L., Yang, Q.H., Graphene-DNA hybrids: Self-assembly and electrochemical detection performance. *J. Mater. Chem.*, 20, 6668–6673, 2010.
23. Gao, W., Alemany, L.B., Ci, L., Ajayan, P.M., New insights into the structure and reduction of graphite oxide. *Nat. Chem.*, 1, 403–408, 2009.
24. Stankovich, S., Dikin, D.A., Piner, R.D., Kohlhaas, K.A., Kleinhammes, A., Jia, Y., Wu, Y., Nguyen, S.T., Ruoff, R.S., Synthesis of graphene-based nanosheets *via* chemical reduction of exfoliated graphite oxide. *Carbon*, 45, 1558–1565, 2007.
25. Su, C.Y., Xu, Y.P., Zhang, W.J., Zhao, J.W., Tang, X.H., Tsai, C.H., Li, L.J., Electrical and spectroscopic characterizations of ultra-large reduced graphene oxide monolayers. *Chem. Mater.*, 21, 5674–5680, 2009.
26. Malik, R., Tomer, V.K., Chaudhary, V., Functionalized Graphene Nanocomposites and their Derivatives, in: *Synthesis, Processing and Applications*, Micro and Nano Technologies, pp. 323–338, 2019.
27. Pandikumar, A. and Rameshkumar, P., *Graphene-Based Electrochemical Sensors for Biomolecules*, Elsevier Publishing, Amsterdam, Netherlands, 2019.
28. Zhang, D., Zhang, K., Wang, Y., Wang, Y., Yang, Y., Thermoelectric effect induced electricity in stretchable graphene-polymer nanocomposites for ultrasensitive self-powered strain sensor system. *Nano Energy*, 56, 25–32, 2019.
29. Kumar, N., Rodriguez, J.R., Pol, V.G., Sen, A., Facile synthesis of 2D graphene oxide sheet enveloping ultrafine 1D $LiMn_2O_4$ as interconnected framework to enhance cathodic property for Li-ion battery. *Appl. Surf. Sci.*, 463, 132–140, 2019.
30. Ultra-dispersed island-like Co_9S_8 nanoparticles composed of nanosheets *in-situ* grown on nitrogen-doped graphene for asymmetric supercapacitor. *Electrochim. Acta*, 293, 419–425, 2019.
31. Sundaram, R.S., *Graphene properties, preparation, characterization and devices*, pp. 50–80, Woodhead Publishing, Cambridge, United Kingdom, 2014.
32. Toh, S.Y., Loh, K.S., Kamarudin, S.K., Daud, W.R.W., Graphene production *via* electrochemical reduction of graphene oxide: Synthesis and characterisation. *Chem. Eng. J.*, 251, 422–434, 2014.
33. Kauppila, J., Kunnas, P., Damlin, P., Viinikanoja, A., Kvarnström, C., Electrochemical reduction of graphene oxide films in aqueous and organic solutions. *Electrochim. Acta*, 89, 84–89, 2013.
34. Azman, N.H.N., Lim, H.N., Mamat, M.S., Sulaiman, Y., Synergistic enhancement of ternary Poly(3,4-ethylenedioxythiophene)/graphene oxide/manganese oxide composite as a symmetrical electrode for supercapacitors. *Energies*, 11, 1510, 2018.
35. Azman, N.H.N., Sulaiman, Y., Mamat, M.S., Nazir, M., Lim, H., Novel poly (3,4-ethylenedioxythiophene)/reduced graphene oxide incorporated with manganese oxide/iron oxide for supercapacitor device. *J. Mater. Sci: Mater. Electron*, 30, 1458–1467, 2019.

36. Hareesh, K., Shateesh, B., Joshi, R.P., Williams, J.F., Phase, D.M., Haram, S.K., Dhole, S.D., Ultra high stable supercapacitance performance of conducting polymer coated MnO_2 nanorods/rGO nanocomposites. *RSC Adv.*, 7, 20027–20036, 2017.
37. Agnihotria, N., Senb, P., Dea, A., Mukherjee, M., Hierarchically designed PEDOT encapsulated graphene-MnO_2 nanocomposite as supercapacitors. *Mater. Res. Bull.*, 88, 218–225, 2017.
38. Hareesha, K., Shateeshb, B., Joshia, R.P., Dahiwalea, S.S., Bhoraskara, V.N., Haramb, S.K., Dholea, S.D., PEDOT:PSS wrapped $NiFe_2O_4$/rGO tertiary nanocomposite for the super-capacitor applications. *Electrochim. Acta*, 201, 106–116, 2016.
39. Moussa, M., Shi, G., Wua, H., Zhao, Z., Voelcker, N.H., Losic, D., Ma, J., Development of flexible supercapacitors using an inexpensive graphene/PEDOT/MnO_2 sponge composite. *Mater. Des.*, 125, 1–10, 2017.
40. Shafi, P.M., Ganesh, V., Bose, A.C., $LaMnO_3$/RGO/PANI ternary nanocomposite for supercapacitor electrode application and their outstandingperformance in all-solid-state asymmetrical device design. *ACS Appl. Energy Mater.*, 1, 6, 2802–2812, 2018.
41. Aranganathan V, V. and NityanandaShetty, A., Synthesis and characterization of reduced-graphene oxide/nickeloxide/polyaniline ternary nanocomposites for supercapacitors. *Mater. Today: Proc.*, 5, 8852–8861, 2018.
42. Palsaniya, S., Nemadea, H.B., Dasmahapatraa, A.K., Synthesis of Polyaniline/Graphene/MoS_2 nanocomposite for high performancesupercapacitor electrode. *Polymer*, 150, 150–158, 2018.
43. Peng, C., Liu, F., Chen, C., Synthesis of graphene/nano-sulphur/polyaniline ternary nanocomposite forhigh-performance supercapacitorelectrodes. *Polym. Polym. Compos.*, 27, 1, 3–10, 2019.
44. Zhao, J., Li, C., Zhang, Q., Zhang, J., Wang, X., Sun, J., Wang, J., Xie, J., Lin, Z., Li, Z., Lu, W., Lu, C., Yao, Y., Hierarchical ferric cobalt nickel ternary oxide nanowirearrays supported on graphene fibers as high-performanceelectrodes for flexible asymmetric supercapacitors. *Nano Res.*, 11, 1775–1786, 2018.
45. Gandla, D., Sarkar, S., Putta, C., Ghosh, S., Polyaniline-β-Cyclodextrin-Graphene nanocomposite for energystorage application: Efficiency enhancement through radicalcation stabilization. *J. Electrochem. Soc.*, 165, 11, A2549–A2556, 2018.
46. Jiang, L.L., Lu, X., Xie, C.M., Wan, G.J., Zhang, H.P., Youhong, T., Flexible, Free-Standing TiO_2–Graphene–Polypyrrole Composite Films as Electrodes for Supercapacitors. *J. Phys. Chem. C*, 119, 8, 3903–3910, 2015.
47. Zhou, H., Yan, Z., Yang, X., Lv, J., Kang, L., Liu, Z.H., RGO/MnO_2/polypyrrole ternary film electrode for supercapacitor. *Mater. Chem. Phys.*, 177, 40–47, 2016.
48. Wang, W., Hao, Q., Lei, W., Xia, X., Wang, X., Graphene/SnO_2/polypyrrole ternary nanocomposites as supercapacitor electrode materials. *RSC Adv.*, 2, 10268–10274, 2012.

49. Moyseowicz, Sliwak, A., Miniach, E., Gryglewicz, G., Polypyrrole/iron oxide/reduced graphene oxide ternary composite as a binder less electrode material with high cyclic stability for supercapacitors. *Compos. Part B-Eng.*, 109, 23–29, 2017.
50. Han, G., Liu, Y., Kan, E., Tang, J., Zhang, L., Wang, H., Tang, W., Sandwich-structured MnO_2/polypyrrole/reduced graphene oxide hybrid composites for high-performance supercapacitors. *RSC Adv.*, 4, 9898–9904, 2014.
51. Zheng, W., Li, S., Yu, X., Chen, C., Huang, H., Huang, Y., Li, L., Synthesis of hierarchical reduced graphene oxide/SnO_2/polypyrrole ternary composites with high electrochemical performance. *Mater. Res. Bull.*, 80, 303–308, 2016.
52. Xu, L., Jia, M., Li, Y., Zhang, S., Jin, X., Design and synthesis of graphene/activated carbon/polypyrrole flexible supercapacitor electrodes. *RSC Adv.*, 7, 31342, 2017.
53. Zhang, J., Ding, J., Li, C., Fabrication of novel ternary three dimensional RuO_2/graphitic-C_3N_4@reduced graphene oxide aerogel composites for supercapacitors. *ACS Sustainable Chem. Eng.*, 5, 6, 4982–4991, 2017.
54. Chen, J., Wang, Y., Cao, J., Flexible and solid-state asymmetric supercapacitor based on ternary graphene/MnO_2/carbon black hybrid film with high power performance. *Electrochim. Acta*, 182, 861–870, 2015.
55. Shaheen, W., Warsi, M.F., Shahid, M., Carbon coated MoO_3 nanowires/graphene oxide ternary nanocomposite for high-performance supercapacitors. *Electrochim. Acta*, 219, 330–338, 2016.
56. Bao, W., Yu, B., Li, W., Fan, H., Ren, Z., Co_3O_4/nitrogen-doped graphene/carbon nanotubes: An innovative ternary composite with enhanced electrochemical performance. *J. Alloys Compd.*, 647, 873–879, 2015.
57. Tang, X.F., Yang, Z.G., Liang, J.H., Synthesis of a ternary $FeNi_2S_4$/CNT/graphene nanocomposite with improved electrochemical properties. *RSC Adv.*, 6, 88168–88173, 2013.
58. Jin, Y., Chen, H., Chen, M., Liu, N., Wen, Q., Graphene-patched cnt/mno_2 nanocomposite papers for the electrode of high-performance flexible asymmetric supercapacitors. *ACS Appl. Mater. Interfaces*, 5, 8, 3408–3416, 2013.
59. Rajendran, R., Shrestha, L.K., Kumar, R.M., Composite Nanoarchitectonics for ternary systems of reduced graphene oxide/carbon nanotubes/nickel oxide with enhanced electrochemical capacitor performance. *J. Inorg. Organomet. Polym.*, 25, 267, 2015.
60. Hu, T., Liu, Y., Zhang, Y., Chen, M., Zheng, J., Tang, J., Meng, C., 3D hierarchical porous $V_3O_7{\cdot}H_2O$ nanobelts/CNT/reduced graphene oxide integrated composite with synergistic effect for supercapacitors with high capacitance and long cycling life. *J. Colloid Interface Sci.*, 1, 531, 382–393, 2018.
61. Sahoo, S. and Shim, J.J., Facile Synthesis of Three-Dimensional Ternary $ZnCo_2O_4$/Reduced Graphene Oxide/NiO composite film on nickel foam for next generation supercapacitor electrodes. *ACS Sustainable Chem. Eng.*, 5, 241–251, 2017.

62. Sahoo, S., Zhang, S., Shim, J.J., Porous ternary high performance supercapacitor electrode based on reduced graphene oxide, $NiMn_2O_4$, and Polyaniline. *Electrochim. Acta*, 216, 386–396, 2016.
63. Xia, G., Li, N., Li, D., Liu, R., Wang, C., Li, Q., Lü, X., Spendelow, J.S., Zhang, J., Wu, G., Graphene/Fe_2O_3/SnO_2 ternary nanocomposites as a high-performance anode for lithium ion batteries. *ACS Appl. Mater. Interfaces*, 5, 17, 8607–8614, 2013.

8

Graphene: An Insight Into Electrochemical Sensing Technology

Anantharaman Shivakumar[1]* and Honnur Krishna[2]

[1]PG Department of Chemistry, St. Philomena's College, Mysore, India
[2]Department of Chemistry, S.D.V.S. Sangh's S.S. Arts College and T.P. Science Institute, Belagum, India

Abstract

An interdisciplinary science, such as physics, material science, biochemistry, and biotechnology, microbiology environmental chemistry has attracted graphene in the synthesis and application due to features such as surface to volume ratio, strength to weight ratio, defects over these surfaces, fine tunable properties such as dimensions, lengths, and twisted form. The option is wide open due to sp^2 carbon for the creation for multidimensional characteristic properties such as electronic, presence of optical features.

The bipolar and nano nature in graphene synthesis methodology creates a conducible environment for biomolecules such as enzymes, nucleic acids, which in due course are responsible for the development of sensors for hydrogen peroxide, antioxidants, pharmaceutically important drugs and ammonia. Furthermore the commercialization has scaled up production in the industrial sector. Based on the extensive literature survey, reviewer intends to bring out some research related to graphene synthesis chemistry, physico and electrochemical properties, biotechnical functionalization and sensing technology.

Keywords: Graphene oxide, band gap, doping, marriage chemistry, Raman and IR study, chemistry of exfoliation, sensing techniques

Abbreviation Used

AgNps Silver Nanoparticles
AGNRs Armchair graphene nanoribbons

**Corresponding author*: shivakem77@gmail.com

Inamuddin, Rajender Boddula, Mohd Imran Ahamed and Abdullah M. Asiri (eds.)
Monoelements: Properties and Applications, (169–234) © 2020 Scrivener Publishing LLC

AuNps	Gold Nanoparticles
CF	carboxyfluorescein
CNT	Carbon Nanotubes
CV	Cyclic voltammetry
CVD	chemical vaporization deposition technique
ERGO	Electrochemically reduced graphene oxide
FAM	6-Carboxyfluorescein
GCE	glassy carbon electrode
GO	graphene oxide
GOD	glucose oxidase
HOPG	Highly oriented pyrolytic graphite
PSA	prostrate specific antigen
PVA	polyvinylalcohol
QDs	Quantum dots
QHE	quantum hall effect
RGO	reduced graphene oxide
ROX	Rhodamine-X
SCE	Saturated calomel electrode

8.1 Introduction

Carbon, even though not the most abundant element in the earth crust, is the fourth most common element after hydrogen, helium, and oxygen. Nevertheless, carbon constitutes the very fundamental building block of life. Carbon plays a unique role in nature. The carbon formation in stars as a result of merging of 3 α particles is a crucial process that leads to existence of all the relatively heavy elements in the universe [1]. Fundamental to organic chemistry and the basis for existence of life is the capability of carbon to form polymeric system due to catenation [2]. Carbon electrodes are frequently used in metallurgy for the production of silicon, yellow phosphorus, calcium carbide because of the tendency of amorphous carbon and graphite to remain as solids up to a temperature of 3,825 °C, just like how traditionally graphite and diamond have been known since ancient times. Nearly, three decades ago fullerenes and nanotubes were discovered [3–6].

Graphene as defined by the IUPAC is "a single layer of graphite structure, describing its nature by analogy to a polycyclic aromatic hydrocarbon of quasi-infinite size" [7]. It is difficult to set a sharp definition of graphene as the words are technically deviating due to modification of synthetic methodology. Yet again it has been retained to place graphene under single umbrella. Previously, descriptions such as graphite layers, carbon layers or carbon sheets have been used for the term graphene. Besides, graphite designates the modification of the element carbon in which planar sheets of carbon atoms, each atom bound to three neighbors in a honeycomb-like lattice network with interatomic C-C distance of 1.42 A°, it is not correct to use for a single layer a term which includes the term graphite, which would imply a three-dimensional structure. To be precise, the graphite and graphene interconversion physicochemical approach is nothing but a placement of layer by layer graphene one over the other which are held by weak van der Waal's forces separated by a distance of 3.35 A°. The term graphene should be used only when the reactions, structural relations, or other properties of individual layer is discussed. It is the property of graphite that makes mechanical or chemical exfoliation into smaller units, to ultimately obtain a single layer graphene layer. The exfoliation of graphite into smaller fragments, such as single and few layers graphene, increases the amount of basal plane per volume, but the amount of available adage plane remains the same [8]. Therefore, the exfoliation of graphite to graphene provides advantages for electrochemically active at the basal plane rather than the edge plane. The advantageous properties are shown in terms of physical, chemical, optical, electrical and mechanical change in the graphene conductance, a function of extent of surface adsorption [9], large specific surface area [10], and low Johnson noise [11]. Research demonstrated that monolayer graphene is a best contender to analyze a variety of molecules such as gases [12], biomolecules [13] involving the charge transfer between the adsorbed molecules and graphene. Ammonia [14], carbon monoxide [15], water [16], and nitrogen dioxide [17] have been assayed based on the donor/acceptor properties of graphene [18]. Good sensor properties in sensitivities at parts per billion for detection chemical warfare agents [19] and explosives [20] by the use of reduced graphene oxide has been achieved. Further the decrease in the overpotentials of hydrogen peroxide at Carbon nanotubes-based electrodes for a great deal of application in dual enzymatic system analyses of the biochemically important substrate [21].

Carbon might be the most promising and widely used material in sensing the electro analyzing and electrocatalyzing reactions. Carbon based nanomaterials especially CNT and graphene, are extremely attractive in the bioanalytical for electrode design as they can combine properties of high surface-volume ratio [22], high electrical conductivity [23], strong mechanical strength [24], acceptable biocompatibility [25], chemical and electrochemical stability [26, 27]. Carbon nanomaterials based sensors generally have higher sensitivities and lower detection limit than conventional ones [21, 28–30]. These fast electron transportation, high thermal conductivity, excellent mechanical flexibility and good biocompatibility make the graphene an ideal candidate for biosensing and its related application. The use of graphene can avoid the problems that are encountered in the transition metals like Fe, Ni, Co and such others in terms of low cost, solution processability, and unique structure to fabricate hydrophobic and lypophillic materials [31, 32]. In fact it possible to control the carbon–oxygen ratio of the graphene layers by carefully tuning the electrochemical properties. It has been demonstrated by Ambrosi that tuning C/O ratio of graphene oxide by application of fixed cathodic potential with a direct influence on the electrochemistry of graphene layers [33].

As a consequence of recent growth in the synthetic modification of the graphene related materials and its application to the sensing technology, the authors set out a logical frame work for the suitability of the application which has potential usefulness. A detailed survey of many graphene related sensors has been recently published [34] and there are earlier reviews [35] to complement and update the facts that the presented review article focuses on such as more widely used graphene sensors which are often ignored mainly on the synthesis, characterization and its application on biological and environmental sensing electrochemical probes.

8.2 Electronic Band Structure of Graphene

The unique electronic properties of the graphene based CNT are mainly because of the delocalization of π-bonds above and below the basal plane. These delocalization of π-electrons create high electrical conductivity and mobilities for graphene within the plane [36]. Traditionally, Fermi level in graphene falls within the conduction and valence band. Conduction in a semiconductor is normally carried through doping material while, charge carriers from the hole to electron takes place through a point which is called 'Dirac point'. Carbon has four valence electrons among which three are involved in the formation of strong three C–C sigma bond and remaining one electron is involved in formation of π-bond. The π-electrons are responsible for the electronic properties at low energies and sigma electrons form energy bands far away from the Fermi level. The electronic structure of graphene follows from a simple nearest-neighbor tight band approximation. The tight band approximation of the honey comb lattice energy band diagram of π-electrons was calculated by Wallace in 1947 [37].

The electronic structure can be explained by the reciprocal of the triangular lattice as shown in Figure 8.1. The atoms per unit cell results in two conical point per Brillouin zone, following a symmetry consideration [38].

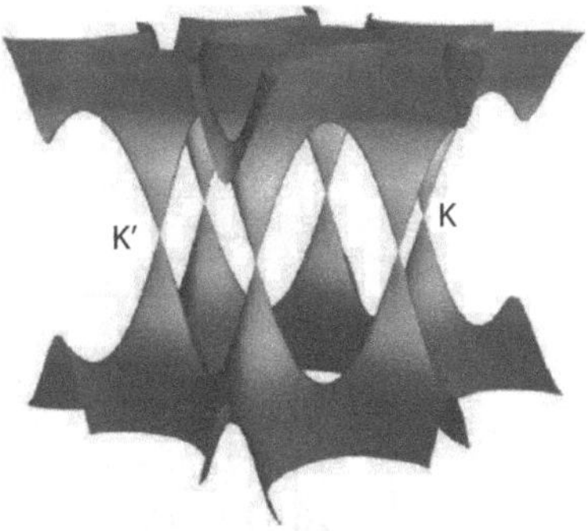

Figure 8.1 Band structure of graphene. The conductance band touches the valence band at the K and K′ points. Reproduced with permission from Graphene: carbon in two dimension [261].

The six corners in the zone form non-equivalent group of K-points, called K and K', called Dirac points. The essential feature of the Dirac spectrum converges from the basic principles of quantum mechanics and theory of relativity i.e. existence of antiparticles, more specifically in terms of positrons as anti particles for electrons, described as same spin or wave function. The wave function interaction of positron with electron leads to quantum mechanical hopping between the two leads to the formation of two energy bands of the Brillouin zone yielding a conical energy of spectrum. Because of this quasi particles in graphene behave differently from those in conventional metals and semiconductors where the energy spectrum can be approximated by a parabolic dispersion relation [39]. Hence, Graphene is called a zero gap semi-conductor because of meeting of the conduction and valence band at Dirac point. As a zero gap semiconductor, graphene plays an ambipolar electric field effect and charge carriers can be tuned between electrons and holes in concentration as high as 10^{13} $cm^2V^{-1}S^{-1}$ [40].

The other measure of which the electronic quality of the graphene can be confirmed is by observation of quantum Hall effect at room temperature [41]. When a magnetic field is applied in a direction perpendicular to the flow of current, a voltage is developed in a mutually perpendicular to the applied magnetic field and current flow due to the deflection of charge carriers to the edge of the samples by the magnetic field is called "QHE" [42]. For most of the semiconductors and metals, the Hall co-efficient varies linearly with temperature. Measured Hall conductivity of a single layer of graphene is quantized with odd multiples of the quantum $4e^2/h(N + 1/2)$ where N is the Landau level and 4 is a factor called double value and double spin degeneracy. Figure 8.2 depicts the QHE behavior of single

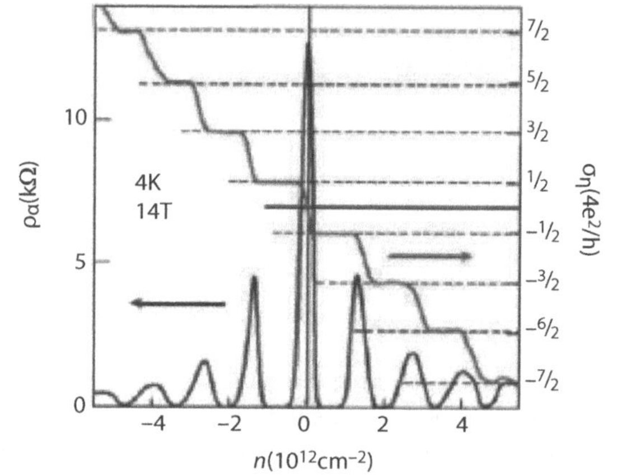

Figure 8.2 The hall mark of massless Dirac fermions is QHE plateau in σ_{xy} at half integers of $4e^2/h$. Reproduced with permission from Two-dimensional gas of massless Dirac fermions in graphene [262].

layer of graphene. A symmetrical graph showing the peak centered at n = 0 persisting through the neutrality point where charge carriers change from holes to electrons to holes. The sequence is shifted with respect to the standard QHE sequence by half, a unique observation when compared to normal semiconductors [43].

8.3 Electrochemical Influence of the Graphene Due to Doping Effect

The 'intrinsic corrugated' graphene in real 3D space can have other 'defects' including topological defects (such as pentagons, heptagons, or their combination), vacancies, adatoms, edges/cracks, adsorbed impurities and so on [40]. This internal property can be explored in the modification of graphene with doping materials. The doping character of the graphene is realized by adsorption of the atoms or molecules on its surface [44, 45]. In general the doping may be of the p-type or n-type for which the electrons can be extracted from or released to graphene layer. The challenge here is to make a conducible environment without altering the electronic structure by removal or addition of electron. Moreover, the enhanced electronic and ionic conductivity by graphene additives can also effectively benefit graphene performance by n and p doping [46]. The alkali metals release the valence electrons due to less electronegativity thus inducing an n-type doping. While atoms possessing high electronegativity forming a strong dimer bonds exhibits p-type doping. p-type doping characterization includes water [47], ammonia [48], nitrogen dioxide [49], tetrafluorotetracyanoquinodimethane. In pristine graphene, the apexes of the two conical bands meet at the fermi level, indicating a delicate balance between the electrons and holes. A schematic diagram for displaying of the pristine graphene with null doping and n/p-doping is depicted as shown in Figure 8.3.

Non-metallic doping of graphene such as nitrogen, boron, halogen, sulphur, and hydrogen can greatly influence the electrochemical properties such as catalytic potential or inhibitions of the redox reactions taking place at its surface. Electron donating dopants such as nitrogen exhibit a faster rate electron transfer towards ferricyanide/ferrocyanide reaction redox electrochemical probe compared with counter undoped graphene [50]. Nitrogen doped graphene can be prepared by carrying out exfoliation process in ammonia [50] or nitrogen atmosphere [51]. While, electron withdrawing dopants such as halogen prepared in halogen atmosphere can also influence the electrochemical analysis of ferricyanide/ferrocyanide system in counter way i.e. decreases the electron transfer rate. Also the size of the halogen counterpart plays a crucial role in the influence of electrochemical activity of rate transfer [52].

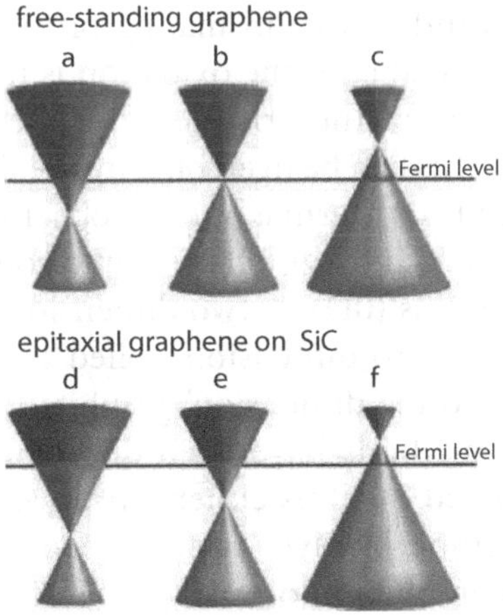

Figure 8.3 Doping graphene. Position of the Dirac point and the Fermi level of pristine and epitaxial graphene as a function of doping. The upper and lower panels show a free standing graphene layer and an epitaxial graphene layer on silicon carbide, respectively. The left and right panels visualize n-type and p-type doping, respectively, while the center panels show the pure graphene layers. For the epitaxial graphene, a natural substrate induced n-type doping is present. Reproduced with permission from Atomic hole doping of graphene [44].

8.4 Exfoliation of Graphite: Chemistry Behind Scientific Approach

Graphite oxide is an ancestor for the synthesis of Graphene oxide. GO is similar to that of graphite oxide in terms of its chemical structure, which is composed of number of oxygen functionalities in carbon basal plane. The graphite oxide structure contains stacked structure of graphene in layers similar to that of graphite. However, a general structure designed by Lerf *et al.* [53, 54] depicts that the hydroxyl and epoxy groups as functional groups residing at the basal plane while carbonyl and carboxyl group accommodating the edges of the GO sheets. Typically, the oxidation of graphite introduces oxygen functionalities at the basal plane which give rise to defect and wider spacing between the adjacent sheets. The oxygen functionalities in the graphite oxide introduces hydrophilic properties in the stacked structure, owing to the introduction of water molecules and some polar organic solvents such as ethyleneglycol, N-methylpyrrolidone, N,N-dimethylformamide to result in a colloidal suspension [55]. Thus increasing the spacing between the stacked sheets weakens the van der Waal's force of attraction between the sheets which in turn facilitates the exfoliation graphite oxide into GO upon ultrasonication or mechanical stirring in water or polar organic solvents. The disruption of sp^2 'π' network island by oxygen induces an insulating property for graphene. The graphene materials obtained after the oxidation is not expected to exhibit same conductivity as that of the pristine graphene due to oxygen functionalities and defects. This can be overcome by the deoxygenating process involving reducing agents or thermal or electrochemical process.

Exfoliation of graphite into graphene can produce dimension (3D, 2D, 1D) and dimension less (0D). A two dimensional graphite sheets can be wrapped to generate 'zero dimension', rolled to form 'one dimension' and staked to produce 'three dimension' graphite/graphene [56]. Several graphene preparation methods have been developed since its discovery [47]. The strategy adopted involves either top-down or bottom-up methods. The top down method involves cutting down large graphene sheets, CNT, carbon fibers or graphite into small pieces of graphene sheet. While in the bottom-up method, small molecules are used as starting materials to build graphene Quantum dots [57]. Top-down method generally includes the exfoliation of graphite as starting material through mechanical (scotch tape), chemical (solution based, graphite oxide) or electrochemical (oxidation/reduction exfoliation), nanolithography, acidic oxidation, hydrothermal, solvothermal, microwave assisted, sonication, photofenton reaction, selective plasma oxidation, chemical exfoliation, surfactant assisted liquid phase exfoliation, electrochemical expansion, formation of graphite intercalated compounds [58]. While for the bottom-up technique involves use of benzene derivatives, carbonization, fullerenes, unsubstituted hexaperihexabenzocaronene, catalytic (e.g., Chemical Vaporization Deposition technique, thermal, Silicon carbide decomposition).

Micromechanical exfoliation of highly oriented pyrolytic graphite is a top-down method which was proposed to isolate a single layer of graphene [47]. This method is commonly referred to as the "Scotch-tape method". The method involves repeated peeling of the layers of graphene from HOPG by means of adhesive tape to eventually give single layers that can be transferred successively onto an appropriate surface (e.g. silica wafers) through a wet or dry surface technique. The scotch-tape method enables the isolation of single graphene layer whose dimension is 1 mm [59]. Graphene prepared by scotch-tape method has extraordinary electronic [47, 60–62], mechanical [63] and thermal conductivity [64] properties that has triggered a burst of interest in various branches of science [42]. It has been documented that the electron transfer occur about 10^6 times faster at defects or edge-like portions of carbon materials as compared to basal or defect-free planes [65].

Bottom-up methods preparation of graphene powders by the use of gaseous phase is an important area of future research, which includes Chemical vaporization deposition technique that provides sufficient quantity for a host of likely applications, as well as for basic science [66]. Chemical vaporization decomposition is a high temperature disintegration, polymerization steps of a carbon feed stocks in the presence of transitional metals such as platinum [67], ruthenium [68], iridium [69], nickel and copper [70]. The carbon feed stock selected might be gaseous methane [71], ethane and propane [72] or solid source [73]. The carbon feedstock undergoes decomposition and rearranges into sp^2-hybridized graphene carbon structures. Sublimation of silicon, by heat treatment at high temperatures under ultra high vacuum conditions from silicon carbide to exposed layer of carbon atoms that rearranges to form epitaxial graphitic layers. Precise control of sublimation temperatures may lead to the formation of very thin graphitic layers. Surface segregation of carbon or decomposition of hydrocarbon on metallic surfaces like Cu and Ni has been also adopted for the preparation of graphene due to economical and easy availability. Non-uniform, multilayers of Ni and uniform mono layer of Cu in carbon has been realized. The very high solubility of Ni in carbon which is difficult suppress the C precipitation completely so that produced graphene layers vary from mono to ten layers whereas low solubility of Cu restricts the formation to monolayer

[74]. A diffusion limited kinetic growth and dendritic nature is accelerated due to the presence of surface oxygen on copper substrate. A fine tuning of temperature, pressure, cooling time and quality metal substrates have advantages of production of uni-layered graphene which is still an experimental research. The dimensions and heights of graphene sheets should be in the range 1.5 to 100 nm and 0.5 to 5.0 nm, respectively. On the whole the sizes of most graphene QDs range 3 to 20 nm and they consist of no more than 5 layers of graphene sheets [57, 75–78].

A general chemical exfoliation procedure adopted in the preparation of graphene from graphite involves different steps which are outlined as follows

- Oxidation of graphene under strong acidic conditions creating a large number of oxygen containing groups such as carboxyl, epoxy, and hydroxyl groups.
- These oxygen containing groups makes graphene hydrophilic in nature, which leads a tendency to be dissolved in water or organic solvents.
- The resulting GO can be reduced by the use of reducing agents such as hydrazine, hydroquinone, hydrogen sulfide, sodium borohydride, ascorbic acid or by heating in a reducing atmosphere in order to regain its actual structure and property. The intrinsic demerit of this procedure is that some quantity of oxygen carrying groups and significant carbon–oxygen groups are always present [79].

Chemical exfoliation of graphite based on oxidizable methods, require thermal or chemical reduction to partially restore the electronic properties of graphene [80]. Synthesis of GO either by Brodie [81], Staudenmaier [82], Hummers method [83] or variation of the above method. Staudenmaier and Brodie adopted potassium perchlorate with nitric acid, whereas Hummers viewed by oxidizing agent potassium permanganate and sulphuric acid. Graphitic salts with intercalation procedure involve sulphuric acid [84], nitric acid [85], or perchloric acid [86] as precursors and subsequent oxidation to GO. Figure below shows the GO proposed by He *et al.* [54], confirmed by solid state nuclear magnetic resonance and ^{13}C-labeled GO as shown in Figure 8.4.

RGO was prepared by several reducing agents that include hydrazine [87], hydroquinone [88], hydrogen sulphide [89], sodium borohydride [90], ascorbic acid [91], reduction via pyrolysis [92], and electrochemical reduction [93]. The detailing of chemical reduction or oxidation of graphite

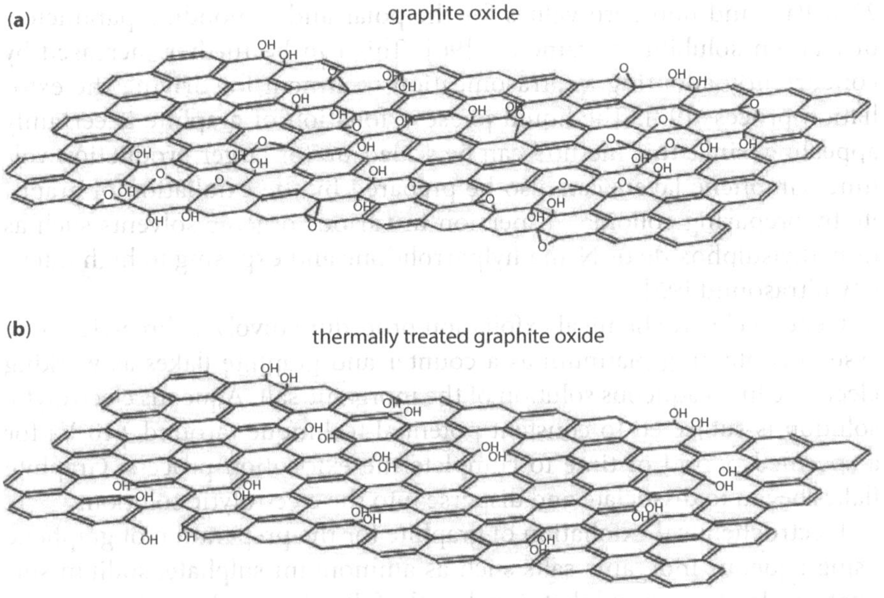

Figure 8.4 Structural model for (a) as-prepared GO; (b) GO after treatment at 100 °C for 24 h in vacuum. For clarity, the model indicates only the chemical connectivity and not the spatial orientation of the functional groups. Hydroxyl groups are perpendicular to the graphitic plane, and the interplane space contains also water molecules. Reproduced with permission from A new structural model for graphitic oxide [54].

in tuning the properties plays an active area of research in the future nanotechnology. The exfoliation of graphite involving oxidizing agents intends to separate the graphitic layers by the process of release of nascent oxygen, which enters the void volume to decrease the force of attraction, further the process can be efficiently improved by ultrasound waves treatment [54]. The oxidized graphite containing sp^3-hybridized carbon in conjunction with sp^2-hybridized carbon alters the intrinsic properties. GO exhibits low electrical conductivity due to disrupted sp^2-hybridized carbon network by oxygen functional groups and also exhibits photoluminescence due to band gap opening as a result of quantum confinement effect from small sp^2 domains [8]. Also, the abundance of oxygen functional groups provides GO with an inherent redox activity, as some of the oxygen functional groups can be reduced and regenerated electrochemically [94].

Solution based exfoliation of graphene sheets production that produces minimum alteration. Excellent solution based exfoliation can be obtained by minimizing the energy loss by using the organic solvents with surface tensions ~40 mJ m^{-2} [95] Hildebrand solubility parameter near to

23 $MPa^{1/2}$ and non-zero values for the polar and H-bonding parameters of Hansen solubility parameters [96]. This can be further increased by concurrently adopting a ultrasonication treatment for driving the exfoliation process [97]. The liquid phase exfoliation of graphite is certainly appealing, since this method can be scaled-up for larger production volume. Graphene layers can also be prepared by the exfoliation of graphite by preparing colloidal dispersion in various organic solvents such as dimethylsulphoxide or N-methylpyrrolidone and exposing to high intensity ultrasound [95].

General electrochemical exfoliation procedure involves three-electrode system containing platinum as a counter and graphite flakes as working electrode in an aqueous solution of the inorganic salt. Aqueous electrolytic solution is subjected to constant potential technique (around +10 V) for a specified period of time to complete the exfoliation process. Graphite flakes began to dissociate and disperse into the electrolytic solution.

Electrochemical exfoliation of graphite for the preparation of graphene using aqueous inorganic salts such as ammonium sulphate, sodium sulphate, and potassium sulphate involves the following mechanistic approach.

- Application of bias voltage results in the reduction of reducible species at cathode, creating a species that acts as a strong nucleophile.
- The nucleophilic attack of graphite by reduced ions initially occurs at the edge sites and grain boundaries.
- Oxidation at the edge sites and grain boundaries leads to depolarization and expansion of the graphite layers, thereby facilitating the intercalation of counter ions within the graphite layers.
- During this stage, solvent molecules may co-intercalate with counter ions with an inherent possibility of generation of hydroxyl and oxygen radicals by anodic oxidation if the process is carried out in aqueous medium.
- Reduction of counter anions and self oxidation of solvent produce gaseous species such as oxidized gaseous species and others molecules, as evidenced by vigorous gas evolution during electrochemical process [58, 98, 99].
- Radicals are powerful, highly reactive and non-selective oxidants. Thus, the oxidation of graphite edges planes by radicals opens up the edge sheets facilitating the intercalation of OH^- and O_2^2 and subsequently promoting depolarization and expansion graphite anode [100].

Production of anodic few layers graphene layers by electrochemical method should have significant advantages such as

a. An efficient, facile, fast, low concentration and cheap reagents
b. Simple reaction set up, ambient experimental condition and easy operational procedure
c. Single-step, environmental friendly, mild, soft, processing approach
d. Stable dispersion of high quality few layer graphene sheets.

The electrochemical exfoliation involves the application of cathodic or anodic potentials or current in aqueous or organic based HOPG working electrode in the presence counter (platinum) and reference electrode (SCE or Ag/AgCl) electrodes. When a positive potential is applied, the graphite is oxidized and the intercalation of negatively charged ions from the solution into the graphitic layers takes place. This is then followed by the application of negative potential that facilitates the exfoliation process. The use of high anodic potentials produces oxidized graphene, which is rich in oxygen functional groups and structural defects. Therefore, a cathodic reduction/intercalation method was attempted to generate nonoxidized graphene sheets with a more intact graphene structure. By first applying a negative potential, an intercalation process can be achieved with positive ions without any oxidation and formation of oxygen functionalities on the graphene sheets. Further microwave treatment can be performed to thermally expand the intercalated graphite to provide graphene [8]. Moreover, the use of anodic potentials to intercalate ions causes unwanted oxidation of graphite and introduction of oxygen functional groups disrupt the sp^2-hybridized carbon network. The demerit of this process is its irreversible nature, since neither electrochemical nor chemical processes are overruled for the regaining of sp^2 hybridized carbon. Also, the presence of both structural damage and oxygen functional groups has profound influence on the electrochemical properties of the resulting graphene. Similar to solution based exfoliation methods, the use of surfactants may cause an irreversible "functionalization" of the graphene flakes that may ultimately result in the alteration of the electrochemical behaviors [101 103].

In conclusion, both methods have their advantages and disadvantages, the bottom-up method is less defect free on graphene surfaces as compared to top-down approach, but the method is more difficult and non-economical. The biggest problem in top-down method is that imperfection of surface structure, significant crystallographic damage to processed pattern, internal stress and contamination. It is still an experimental research that are made to insisting high quality graphene by modification of presently known existing methods which in near future more methods will be published to challenge the older methods for future application. Table 8.1 outlines some of the merits and demerits of the available methods.

Table 8.1 Advantages and disadvantages of the graphene preparative methods.

Methods for graphene preparation	Advantages	Disadvantages
Mechanical	Easy and economical process	Low sample out put
Chemical process	Structural induction of functional groups are highly beneficial for binding of biomolecules such as enzymes, nucleic acids, avidin–biotin, bacteria, human immunoglobins, and cells	Removal of by products involves complicated chemical process, less purity
Microwave	Wide bandwidth, fast and homogeneous heating leading to uniform sized particles It does not require any passivating agent Size can fine tuned by variation in microwave heating time	Signal absorption by surrounding medium that may suffer attenuation
Sonication	Fast, Chemical less process	High intensity ultrasonic waves can explode violently that may cause destruction Costly instrumentation Decrease in lateral dimension and hole defects

(*Continued*)

Table 8.1 Advantages and disadvantages of the graphene preparative methods. (*Continued*)

Methods for graphene preparation	**Advantages**	**Disadvantages**
Nanolithography	High conductivity Well defined chirality A potential method for electronic application	Complicated procedural techniques in preparation and hence sophisticated instrumentation, low yield
Electrochemical	Facile, fast, scalable, economic environmental friendly Carbon to oxygen ratio can be controlled Coated directly on to substrate there by used for specific application	More than one reducing and oxidizing agents/ steps Agglomeration may take place
Solvothermal	Precise control over the size, shape distribution	Drastic experimental condition, side reaction
Photo Fenton reaction	Decreases degradation in comparison to Fenton reaction, negligible formation of iron sludge	Low pH confinement, handling iron sludge, slow regeneration of Fe(II)
Micromechanical	High quality properties	Low yield, not scalable
Epitaxial growth	Large continuous film area, high quality	Not transferable, high temperature, low vacuum
Reduction of GO	Solution processed, high yield	Poor electrical properties, small area flakes
CVD	High quality, large area, transferable	High temperature, low vacuum, different scalability
Solution exfoliation	High quality, good scalability, low temperature	Small area graphene flakes, colloidal stability, agglomeration upon reduction

8.5 Electrochemical Reduction of Oxidized Graphene

The electrochemical reduction may involve one or two steps. In one step electrochemical reduction process, the GO sheets are electrochemically reduced from an aqueous colloidal solution in the presence of buffer electrolyte such as phosphate buffer solution along with electrolytes like sodium chloride [104], sodium sulphate [105] to produce electrochemically reduced graphene oxide. This can be performed by cyclic voltammetry [106, 107], linear sweep voltammetry [108], differential pulse voltammetry [109] or constant potential mode [110] by a three-electrode system. The process of reduction takes place by diffusion controlled mechanism due to difference in concentration between electrode and electrolytic solution. The electrodeposition takes place due to difference in solubility between the GO and the resulting ERGO in the aqueous electrolyte [106].

For a process to be efficient, the conductivity and pH is a critical parameter. The conductivity of the electrolyte should be between 4 and 25 mS cm^{-1} and pH range between 1.5 and 12.5 [104]. pH greater than 10 may fail to undergo deposition of RGO on to electrodic surface because of ERGO solubility and instability in pH greater than 11.0 [106]. The solubility of ERGO is attributed for the presence of phenoxy group, carboxy group and alkoxy group on graphene basal plane [106]. During the process of reduction, a constant negative potential is applied—current decreases over time and approaches zero as reduction process is complete. The reduction of C=O and -OH, -C-O-C- can be carried out at a potential of −1.3 and −1.5 V respectively versus SCE [109]. Studies show that a potential greater than −1.5 V could lead to reduction of water, which competes in the reduction of ERGO making the process less than 100% efficient [104, 109].

In a two-step electrochemical approach, a thin film of GO is initially deposited on to a solid support, which is then dried out to form a GO-coated electrode. This GO-coated electrode is then subjected to electrochemical reduction using a three electrode system in the presence of buffer-electrolytic system as similar to one step reduction process. The solid support includes insulators such as flexible plastic, glass or conductors like indium tin oxide, glassy carbon, gold [111]. The mechanistic approach in the insulating and conducting supported is different. For insulated substrate systems, glassy carbon electrode is required for maintaining the conductivity, which acts as a working electrode in the electrochemical cell. In contrast to GO substrate coated on an insulator, GO-conducting material acts as working electrode in the electrochemical cell without the need of any additional conducting material. Various methods adopted in the assembly GO-substrate prior to the electrochemical reduction includes drop casting [93, 112], dip-coating [113], layered [114], spray coating [111, 115]. The well adhered films are produced because of the van der Waal's interaction between the solid support and the GO substrate [80]. Apart from this additional methods are also available that includes electrostatic interaction between the positively charged cystamine and negatively charged GO [114] with the inclusion of nafion as a binding agent [105, 115] to enhance the adherence of GO on to the surface of the electrode. Indeed the better method selection depends on the morphology, uniformity, thickness, coverage density of the GO substrate over the solid support and deposition techniques and parameters adopted [116]. Similar to one step reduction approach, the electrochemical deposition involving two step process includes phosphate buffer solution along with supporting electrolyte such as potassium chloride [117], potassium nitrate [114], sodium chloride [113]. But unlike one step electrochemical reduction process, pH of the electrolytic solution be preferably in acidic or neutral region, a low pH is generally not preferred due to the reduction of GO [118].

Both in one-step and two-step electrochemical reduction process selection of an appropriate cathodic reduction potential and time is made on the observed cyclic voltammetric peak potential or linear sweep voltammetry. The cathodic peak potential is function of pH. Hence it is required to select an appropriate pH value of the buffer medium. Zhou *et al.* has confirmed experimentally by linear sweep voltammetry in shifting of peak negatively with increase in pH of buffer medium [111]. Oxygen functionalities in GO varies as the method adopted which in turn influences the cathodic peak reduction potentials [119]. Researchers have claimed that possibility of non reducing character of oxygen functionalities by application of less negative cathodic potential despite the extended duration time. In contrary, if a more negative potential reduction rate relatively increases the defects in ERGO. Guo *et al.* suggested that these defects could be eliminated by performing electrochemical reduction at an elevated temperature or by annealing process [109].

The main process that is carried in ERGO is the recovery of sp^2 hybridized carbon. 100% complete efficiency of process is still a challenging process. However, oxygen functionalities can be controlled by selecting an appropriate peak potentials and duration of procedure. Graphically, (i – t) curve is significant in deciding complete or maximum efficiency. The electrochemical reduction is considered to be complete when current remains constant over time. This can also be evaluated visually by noticing change in color as GO is coated over the electrode. Table 8.2 summarizes various electrochemical reduction processes by either one-step or two-step process.

Table 8.2 Experimental parameters in the graphene sensing technology.

SlNo	Electrochemical technique employed	Reference electrode	Counter electrode	Working electrode	Assembled method	Supporting electrolyte	pH of the medium	Scan rate electrochemical cycle (100 mV/ S)	Potential range	Reference
1	CV	$AgClO_4$	Pt	Fluorine doped tin oxide/GO	Drop casting	Et_4NBF_4	7.0		-2.1 V to -1.1 V	[263]
2	CV	AgCl/Ag	Pt	SnO_2/GO	Spin casting	Et_4NBF_4	2	50	0 to -0.8 V	[264]
							12		0 to –1.5	
3	Constant potential	SCE	Pt	Au/GO	Drop casting	H2SO_4	-	-	-0.2 V	[265]
4	CV	SCE	Carbon rod	GCE/GO	Drop casting	$HAuCl_4$	-	25	-1.5 to 0.6 V	[266]
5	CV	SCE	Pt	GCE/GO	One step reduction	PBS	9.18	10	-1.5 to 0.5 V	[267]
7	CV	Ag/AgCl	Pt	GCE/1,6hexanediamine/ GO	Layer by layer	-	2.5	-	0 to -1.4 V	[268]
8	CV	SCE	Pt	Au/Cyst/GO	Layer by layer	KNO_3	2.5	10	0 to -1.0 V	[114]
9	CV	HgO/Hg	Pt	Graphite rod/GO	-	KOH	-	50	-1.45 to 0.0 V	[269]
10	CV	Ag/AgCl	Pt	GCE/GO	Drop casting	PBS	5.0	30	0 to -1.5 V	[270]
11	CV	SCE	Pt	Graphite/GO	-	PBS	6.9	50	0 to 3.0 V	[271]
12	CV	Ag/AgCl	Pt	Pencil graphite/GO	Dip coating	-	-	-	-1.4 to +0.3 V	[272]

Regardless of the electrochemical process adopted, obvious changes will be seen in the GO aqueous suspension. Guo *et al.* reported that the yellow-brown color of the native GO aqueous suspension turning to colorless with concomitant attachment of bulk amount of black precipitate onto the working electrode substrate during the electrochemical reduction process [109]. Table 8.2 chalks out some important experimental parameters adopted in the graphite sensing technique.

8.6 Spectroscopic Study of Graphene

Raman spectroscopy is widely used nondestructive method to characterize carbon-based materials to locate single layer graphene [120], to determine the diameter of single walled CNT [121], to study the electrochemical reduction of GO by observing the changes in the wavenumber of G band [114]. Raman spectra along with electrochemical studies can provide great details in extracting the fine structure of graphene. Raman spectrum of graphene or graphite is less complicated, consisting of only two bands and a few more very small bands. A HOPG and graphene have very similar Raman bands with variation in the intensity and position. The two main bands in the graphite spectrum are G-Band at ~1,582 cm^{-1} and 2D (G') band at ~2,685 cm^{-1}. A third, D band, indicating defect [122] at ~1350 cm^{-1} is also worth some discussion, although it is usually very weak in graphite [114]. The G-band represents thc in plane vibration mode of triagonal planar sp^2 bonded carbon that is highly sensitive to number of layers present in graphite. As thickness of layer increases position of band shifts towards lower energy demonstrating the softening of bonds. The empirical relation between band position and number of layers can be mathematically related [123] as

$$\omega_G = 1{,}581.6 + 11/(1 + n^{1.6})$$

It is worth noticing by spectral data that band position is affected by temperature, doping level and even strain accompanying the sample. Not only position of band but also intensity of 'G' band can give insight into number of layers. The intensity of band shows a ratio metric or linear trend as sample progress from single to multiple layer [123]. D-band is also called disorder or defect band representing ring breathing mode from sp^2 carbon rings. To be active, ring must be adjacent to graphene edge. The intensity of this band is proportional to degree of defect. Another important feature of this defect is presence of weak modes of vibration underlying this band as a consequence of its excitation laser. So, it is all important that same excitation laser should be used.

The G' is the second order band of D. It is result of two phonon lattice vibration process. The G' band will be present even in the absence of D band and it does not represent the defect unlike D band. For a single layer of graphene 2D is observed to be a symmetric peak. Addition of successive graphene layers causes 2D band to split into several overlapping modes due to lowering of symmetry with increase in layers of graphene [124]. Hence '2D' band is used to differentiate between single and multiple layer graphene. Similar to 'D' band, 2D band is resonant and exhibits strong dispersive behavior so that position and band shape can be significantly different with different excitation laser frequencies. Hence it is necessary to use the same excitation laser wavelength.

It is worth to point out that the ratio of I_{2D} to I_D can be used as a parameter for single or multiple graphene layer. This ratio, if the value is equal to 2 then it corresponds to defect free single layer of graphene. This ratio, lack of D band and sharp symmetric 2D is often characteristic defect free graphene existence of higher layered.

FT-IR is another nondestructive that can be used in complementarily to Raman spectroscopy to identify the oxygen functionalities and bonding configuration in the graphene related materials. The common oxygen functionalities in oxygenated graphene are -OH groups, C=O group, C=C skeletal vibration, C-O epoxy vibration in the region 3,200–3,500 cm^{-1} (broad and intense peak), 1,700 cm^{-1} (strong and intense peak), 1,600 cm^{-1} (weak peak) and 1,100 cm^{-1} (strong peak), respectively. The peak position and intensity depend on exclusively in different synthesis environments.

8.7 Biotechnical Functionalization of Graphene

Graphene have dimensions similar to those of biomolecules such as enzymes. Therefore, the combination of biomolecules with nanostructures produces functional nanostructured bio interfaces with synergistic functions and properties [125]. It is seen that the immobilization of the biosensing materials on the surface is the vital step in biosensors production, particularly when dealing with enzymes and antibodies [126]. The marriage chemistry of graphene with electrochemical techniques has created many indigenous biosensing strategies for its application in the areas of clinical diagnosis and food safety [127]. The unique structures, components and properties of graphene are providing fascinating opportunities for the biotechnological and biochemical development [128]. The high surface area and ease of functionalizing its surface can accommodate highly active probes and targets of interest which makes a favorable condition for development of novel bio-sensing interface [129]. The absorption mechanism of molecules on graphene and how these interactions affect transport properties of graphene, these studies will provide us further understanding of graphene and its interaction with molecules, which may lead to great advancement of graphene science and its various application such as catalysts and sensors [130]. Moreover, the integration principle adopted recently involves graphene with different nanomaterials such as metals, metal oxides, and QDs provides a abundant opportunities for developing novel biosensors with enhanced performance [131]. This speculation was further enhanced by different groups working on nanoparticles by decoration of nanoparticles on the layer of graphene which showed enhanced electrical properties resulting in a improved performance of electrochemical biosensors [8, 132–135].

GO due to its stable dispersion in aqueous medium can be dissolved in water due to ester, epoxide, hydroxyl and carboxyl functionalities which have strong affinity towards water molecules [136]. The ideal behavior for the substrate adsorption by 2D graphene is due to its compatible aromatic nature [137]. Today, in the modern world, the intersection of the biotechnology and nanotechnology by conceptualization of modern nanomaterials has made impressive achievements for the generation of mechanical, optical, electronic and biosensors devices [36, 138]. Fabrication of graphene based nanomaterials with functionalized biosystems integrated with nucleic acids, peptides and even cells are the center point of discussion in the adsorption of single stranded DNA on to the graphene sheets, its ability in quenching of the electron donors, protection of biomolecules from the enzymatic cleavage and transportation *in vivo* systems has interpreted the potential application in the field of biochemical and biological studies [137].

Recently, the research in the area of graphene nanocomposites, coupled with electroanalytical technique has given growth different in frontier areas. Among them it includes organic, inorganic, biochemical analytes such as glucose, cysteine, proteins, DNA, metals, non-metals, gas, pharmaceuticals, anti-oxidants, proteins and peptides [139], nucleic acids [140], avidin–biotin [141], aptamers [142], bacteria [143], tumor cells [144], human immunoglobin [145], and small molecules such as ammonia, water [146] through physical adsorption or chemical conjugation.

The molecular sensors so far developed have essential features such as probe target interaction to trigger the transducer for electronic readout. The minimum requirements for molecular biosensors include a molecular recognition layer and signal transducer that works in a mutually constructive way, coupled to an appropriate read out device. The outstanding physicochemical properties indicate graphene as a potential application in sensor based research fields. However, association of graphene to form graphite further decreases the surface area and thus reducing the adsorption capacity.

8.8 Graphene Technology in Sensors

8.8.1 Glucose Sensors

A direct electrochemistry of the electrode–enzyme chemistry refers mediator less involving the direct electronic communication between the electrode and the active center of the enzyme without the participation of the mediator either in solution or solid phase.

Disadvantages of preceding generation biosensors have been an issue in the development of subsequent biosensor. Each of the disadvantages is categorized as follows

- An issue concerning the first generation biosensors for the development of glucose assay is the electroactive interference for the generated hydrogen peroxide due to the requirement of high potential during its assay. This high potential constraint leads to reduction of endogenous species such as ascorbic acid, uric acid and some drugs like acetaminophen [147]. The efforts were being made to minimize this by
  - ✓ Coating electrode with a layer of nafion and cellulose acetate.
  - ✓ Changing the operative potentials to nullify the electroactivity of interferents.
- Another issue in the first generation is the dependency of oxygen during the process. Efforts are addressed by introduction of mass transport limiting films such as polyurethane or polycarbonate [148, 149] and internal source of oxygen can also be restored by use of the carbon based electrode which have a high oxygen solubility which can act as an internal source of oxygen due to the presence π-electron system [150].

The potential cause of the development of second generation biosensors is oxygen deficit. This has been resolved by employing a nonphysiological electron acceptor in shuttling the electron between FAD and $FADH_2$ of glucose. In overtaking oxygen deficit, different redox polymers like poly(vinylpyridine), poly(vinylimidazole) that are covalently linked with osmium complex that are able to reduce the distance between the polymer redox center and FAD of glucose oxidase leading high current and fast sensing response [151]. Artificial mediators such as ferrocene, ferricyanide, transitional metal complexes, single walled CNT on gold surfaces can be bonded by self assembly using spontaneous chemical reactions via Au–S bonds due to similar size gold nanoparticles and CNT [152].

The present third generation sensor enriches the elimination of leachable artificial mediator and biomolecules. One possible way is to attach or adsorb the mediator or biomolecules covalently/non-covalently with electrodes of interest.

A non-physiological mediator should satisfy following critical conditions: (i) Process of interaction between the reduced form of GOD should be instantaneous compared to oxygen, (ii) Electrochemical potential of the mediator should be low enough, having low solubility in aqueous medium, and (iii) Non-toxicity and chemically stable in both reduced and oxidized forms.

Sensitive determination of hydrogen peroxide is of great significant importance in biological, clinical and environmental and many other fields because of its unique feature. Many analytical methods have been designed which includes electrochemical, luminescence, conventional titrimetry. Among these, it is observed that voltammetric and amperometric methods are best recognized due to its sensitivity, selectivity [153]. Table 8.3 discusses sensitivity and detection limit of the some of the available methods in the determination of glucose.

Table 8.3 Graphene based assay for glucose analyte.

Fabricated graphene probe	Linear range	Detection limit	References
GOD/graphene/chitosan	0.08–12.0 mM	0.02 mM	[22]
Graphene/polyethyleneimine/ ionic liquid/GOD	2.0 14.0 mM	–	[273]
GOD/RGO/GCE	0.1–27.0 mM	–	[274]
GOD/Pt/graphene/chitosan	0.15–4.2 mM	0.6 μM	[275]

(Continued)

Table 8.3 Graphene based assay for glucose analyte. (*Continued*)

Fabricated graphene probe	Linear range	Detection limit	References
GCE/3-aminopropyltriethoxysilane/RGO/GOD	–	–	[113]
GO/GOD	28.0 mM		[276]
Polypyrrole/graphene/GOD	–	–	[277]
Graphene nanosheets/nafion	0.2–1.4 mM	–	[278]
Graphene oxide	1.0–20.0 μM	1.0 μM	[51]
Graphene/cobalt oxide	80.0 μM	100.0 nM	[279]
Graphene/AuNps/chitosan	2.0–10.0 mM at −0.2 V	180.0 μM	[280]
	2 .0–14.0 mM at 0.5 V		
Graphene/AgNp's	1×10^{-4}–0.01 M	7 μM	[281]

8.8.2 DNA and Aptamer Sensors

Mutations of DNA and specific sequential order are related to many diseases. Due to its unique base pairing phenomenal approach, DNA plays a pivotal role in the life sciences which offers a great potential application in the medical field after a great accomplishment in human genome project [154]. DNA gets adsorbed on the surface of the graphene oxide through π–π stacking interaction [155]. The presence of the hole density of the order $5.61 \times 10^{12}/cm^2$ in the fabrication of ssDNA on graphene opens up the designing of Field Effect Transistor for reversible detection of ssDNA [156]. The adsorption of ssDNA on both sides of the graphene takes place through π–π stacking [157]. The 2D biointerface of thiolated DNA-graphene oxide sheet is decorated with gold nanoparticles, with maintaining optical properties after assembly. Today, in the modern scientific and technical world, DNA biosensors have a promising application in the field of clinical, pharmacogenomics, drug screening, medical diagnosis, bioterrorism, pollution and environmental, food safety, forensic investigation. In recent times, acceleration of biological and biomedical fields that can perform DNA sequencing have been developed [158]. In a typical DNA sensor, ssDNA molecules are immobilized on to the surface of electrode, called recognition layer involving either chemical or physical method. The base-pairing interaction recruit the complementary base pair, called the target DNA to the surface leading to a process called hybridization. This hybridization signal is transduced into a recognizable electronic signal for data interpretation. Indeed this system can be captured for the point care of testing in clinical diagnostics and more crucially environmental monitoring of hazardous chemicals [159]. Nucleic acids and particularly DNA have been analyzed with liquid mercury, solid carbon, boron-doped diamond, gold, copper, silver, platinum electrodes. The operation of the solid electrode potential in the region beyond +1 V makes the system advantageous over the mercury electrodes which operate in the negative potential region against the standard calomel electrode. Thus the oxidations of DNA bases can be studied by the solid electrodes while the reduction with the help of mercury electrodes [160]. Positively charged carbon electrodes with electrostatically bound probe DNA can be used in DNA hybridization sensors. On the contrary, strong hydrophobic interaction of bases of the probe DNA with the mercury surface may prevent an efficient DNA hybridization at the mercury electrode. Various electrochemical methods that have been applied in the analysis include differential pulse polarography, differential pulse voltammetry, square wave voltammetry, cyclic voltammetry, constant current chronopotentiometry. It is also shown that purine and pyrimidine bases of nucleic acids produce sparingly soluble salts of mercury [161]. Groups that are responsible for the formation of sparingly soluble compounds of mercury may involve exocyclic and ring nitrogens dependent on the nature of the base and on the experimental conditions [161]. Due to the hidden nature of double stranded involving hydrogen bonds between the base pairs, this strongly influences the peak heights. High sensitivities of the electrochemical signals to changes in the DNA structure were utilized for local conformational changes and DNA helix–coil transitions [162]. Relatively small changes in DNA structure may significantly affect the electrochemical signal of DNA at electrodes. Damage to DNA may lead to altered availability of its electroactive sites, resulting in changes in the measured electrochemical signal [163].

Palecek *et al.* [181] was first to propose a method to distinguish between double stranded and single stranded DNA through reduction. Later, the same group placed an electrochemical technique to show a distinguishing electrochemical signal between ssDNA and dsDNA by adsorptive stripping voltammetry [164, 165]. The electroactive nature of the purine and pyrimidine bases at the electrodic system makes a system to carry out electrocatalytic analysis to study the oxidation or reduction [166].

A simple method for the preparation of DNA-modified electrode involves the immersion of bare electrode such as pyrolytic graphite electrode, carbon paste electrode, screen printed electrode except gold electrode into a small drop of a DNA solution for a short period of time and the electrochemical measurements was performed in a usual electrochemical cell with a DNA modified electrode against the corresponding blank without adsorbed DNA. This method involving the adsorption of DNA was adopted for both ssDNA and dsDNA. As the DNA is negatively charged due to the phosphate backbone structure, positively charged carbon electrode can be used for the hybridization. On the contrary, a strong hydrophobic interaction of bases of the DNA probe with the mercury surface may prevent an efficient DNA hybridization at the mercury electrode. The structure of the DNA is subject to change as the small potential is need to be concentrated on the electrode [167].

Mediator chemistry applied to DNA involves the investigation of the intervention of transitional metal complexes, hydroxyl radicals, singlet oxygen and alkylating agents for the oxidation of guanine in DNA at ITO electrodes. Addition of DNA to a solution of $[Ru(bipy)_3]^{2+}$ resulted in a catalytic improvement of the $[Ru(bipy)_3]^{2+}$ oxidation peak [168]. The two step mechanism suggested is as depicted below

$$[Ru(bipy)_3]^{2+} \longrightarrow [Ru(bipy)_3]^{3+} + e^-$$

$$[Ru(bipy)_3]^{2+} + DNA \longrightarrow DNA_{ox} + [Ru(bipy)_3]^{2+}$$

Adsorption of DNA solution on to a modified polymer coated glassy carbon electrode is another approach. The electrochemical measurement was performed for both the complementary and non-complementary with a manifestation of the faradays signal [169].

The fluorescence, quenching and enhancing property of binding of DNA to the GO shows the disturbance of external materials with that of the dye-labeled DNA with GO is the basis for various detection using frequency resonance energy transfer, in which another molecule transfers energy to GO [170, 171]. The chemiluminiscence assay can also be explored in similar dimension for the DNA adsorbed GO involving either enhancement or depriving of the chemiluminiscence intensity.

Metals ions (like Mg^{2+}, Ca^{2+}, K^+, and Na^+) decrease the force of repulsion between the ds DNA and GO thus facilitating the π–π interaction between the aromatic region of GO and nucleobases which protects from enzymatic digestion. Whereas the nonionic surfactant like triton X-100, strongly interacts with GO which restores the enzymatic digestion of dsDNA [172].

GO-DNA sensors have been facilitated for the determination of various proteins involving restoring/quenching property of fluorescence dye. This mechanism is adopted in the determination of protein such as enzymes. Xu *et al.* performed the quenching property of dye based fluorescence attached to dsDNA adsorbed to the surface of GO by DNA polymerase [173]. Furthermore, the He *et al.* reported a tethering of oligonucleotides with small protein molecules for the quantification of protein [174]. Addition of DNAase for the adsorbed dsDNA over the GO for the restoring of fluorescence intensity was adopted by He *et al.* [175]. The discussion regarding the assay of DNA and protein based on GO is detailed in Table 8.4.

Table 8.4 Fluoroscence sensing technology in nucleic acid/aptamer assay.

Sl. No.	Fabricated ss DNA/ GO probe	Target	Ex wavelength (nm)	Em wavelength (nm)	Reference
1	Fluorescein based dye	HIV-1	480	525	[282]
2	Mercaptoacetic acid capped Cd-Te QD-cyclin Molecular beacon	–	–	590	[283]
3	4-(1-pyrenylvinyl)-N-butylpyridinium bromide	Salmontested-DNA Baker's yeast-RNA BSA Hemoglobin Lens proteins Alpha-cymotrypsinogen-A D(+)-glucose	595	–	[284]
4	FAM	Bleomycin-Fe(II)	–	–	[285]
5	Aptamer/FAM	Thrombin	470	517	[286]
6	Aptamer/FAM	ATP	–	525	[142]
7	Multiplexed aptamer/ FAM or carboxy-X-rhodamine (ROX)	Exonuclease-III	Ex 520-FAM Ex 610-ROX	–	[287]

(*Continued*)

Table 8.4 Fluoroscence sensing technology in nucleic acid/aptamer assay. (*Continued*)

Sl. No.	Fabricated ss DNA/ GO probe	Target	Ex wavelength (nm)	Em wavelength (nm)	Reference
8	CF	Exonuclease	–	–	[288, 289]
9	Fluorescein isothiocyanate Indocarbocyanine indodicarbocyanine	Thrombin	480 520 630	519 570 670	[290]
12	CF	–	494	–	[291]
	Cyanine-5	–	643	–	
	6-carboxy-X-Rhodamine	–	587	–	

A SELEX (Systematic evolution of ligands by exponential enrichment) is the procedure adopted in the selection of aptamers from the random library of 10^{13}–10^{16} single stranded DNA or RNA sequences. The SELEX procedure involves cycles of absorption, recovery and amplification. Among the three steps adopted the recovery of the DNA/RNA is a critical step for ensuring the purity and selectivity. Aptamer isolation associates different procedures such as affinity chromatography [176], flow cyclometry [177], capillary gel electrophoresis [178], surface plasmon resonance [179], atomic force microscopy [180].

Aptamers are the artificial ssDNA or RNA or peptides, selected artificially by combinatorial procedure, consists of sequences that fold into secondary and tertiary structures making them to bind to certain targets with extremely high specificity [181, 182]. Aptamer is thought to resemble chemical antibodies [183], with its dissociation constant ranging from nanomolar to picomolar. Advantages of aptamers lies in the invitro selection from a synthetic library, making it possible target any (from inorganic nature to intact cells), overcoming the limit of having only to use cell lines or animals, as it is necessary for antibodies. High purity aptamers synthesis by polymerase chain reaction makes it possible to produce large quantity. Thirdly, the simple chemical structure makes it easily amenable to further modification of functional groups as per the need. Finally, aptamers are much more stable than antibodies, making them suitable in applications requiring harsh conditions such as high temperature or extreme pH [184]. Therefore, aptamers have sensitivity and selectivity comparable to antibodies, and additionally they have extended shelf life [185]. Wide range dynamic application in the field of biosensors, diagnostics, therapeutics makes aptamers to be explored technically. Due to their stability, easy chemical modification and prone to nanostructures device construction, aptamer based sensor have been incorporated in a variety of application facilitating the development of electrochemical devices. Also, the graphene based cytosensor can be considered as an economical process due to low production cost. The common surface fuctionalization include covalent, gold–thiol, or biotin–streptavidin attachments. Approaches that has been adopted in designing the sensing platforms include covalent, non-covalent, QDs, magnetic nanoparticles. Basic strategies that are adopted in the fabrication of aptasensors include target-induced structural switching, or displacement, sandwich and competitive modes [186].

Aptamers can be immobilized on graphene nano composites via covalent bonds developing reusable sensors or via physical adsorption of aptamers on graphene to obtain one time use sensors. 3,4,9,10-perylenetetracarboxylic acid, a archetypal perylene, due to its chemical and photostability has been explored in the adsorption on graphene through π–π stacking thus preventing the aggregation of graphene for covalent attachment of aptamers, with a well-defined cathodic-peak for the detection of thrombin [187].

A complex containing a gold nanoparticle-graphene was immobilized on glassy carbon electrode and finally aptamer, consisting of ferrocene modified catalytic DNA were immobilized on the gold surface through gold–thiol bonds. The approach of ferrocene to the surface increases the electrical signal with a detection limit of 0.1 pM [188].

8.8.3 Pollutant Sensors

Environmental pollution especially toxic gases, heavy metallic ions and organic pollutants in air and water, caused by industry and agricultural activities severely threaten ecological imbalance and human health and have received extensive worldwide attention. For example heavy metal ions in bodies accumulated from the food chains will cause various chronic diseases. Heavy metal ions such as lead, cadmium, chromium, mercury, arsenic, copper, organic pollutants in waste water, such as oils and solvents, phenolic compounds, pesticides have severe risk to the ecological and biological and need to be removed from the soil and water for better living of the future generation of the human community. Hence the detection and determination of all these is necessary in many fields especially in the environmental monitoring due to its risk and toxicity [189].

Many fluorescence based recognition tools are designed due to the molecular flexibility of the structure. Determination of the metal ion by Fluorescence resonance energy transfer (FRET) involves two steps first the intercalation of GO between the base pairs of DNA with peripheral carboxylate or alcohol or epoxide linkage of GO attached by metal ion. Second the cleavage of DNA by GO involves oxidative and hydrolytic. FRET based sensors are explored by number of research groups due to the ability of graphene to quench fluorescence graphene signal. A rapid, sensitive, method for the detection of mercury based on the induced quenching of A-15 labeled with CF reported by Zhang *et al.* [190]. The method is based on the non-hybridization of complementary bases A-15 and T-15. The absence of mercury made the CF fluoresced. Li *et al.* adopted energy donor properties of GO in the quantification of mercury by FRET sensor formatted with the thymine–mercury–thymine on the surface of GO sheets with ssDNA aptamers. The presence of mercury quenched the emission under 600 nm by the application of 488 nm as excitation wavelength. Mercury was quantified with detection limit of 0.92 nM [191]. A G-quadraduplex structure associated with acridine orange captured on the surface of GO is used in the quantification of silver ion. The quantitative destroying property of silver ion caused AO to be released with the enhancement of the fluorescence intensity. The detection limit that was achieved was 58 nM [192]. A CF labeled DNA sequence in the absence of silver ion, containing cytosine rich nucleic acids-tin a single stranded preferentially quenched by the GO. The presence of silver augmented the fluoro property with a detection limit 5 nM for silver detection. [193]. A selective and sensitive detection for potassium based on the disturbance of the fluoroscence property of positively charged ruthenium complex with negatively charged GO. A potassium aptamer, G-quadraduplex formation was promoted by potassium ions. The fluoroscence property was restored in the absence of the potassium [194]. Fu *et al.* developed a 'turn on' fluorescence sensor based on the leaching of gold nanoparticles from the graphene surface which results in the retaining of the fluorescence property. Graphene adsorbed by gold nanoparticles dissolved thiosulphate, which on treating with Pb^{2+} and 2-mercaptoethanol, leached gold nanoparticles from the graphene which leads to increase in the reappearing and increasing in the relative fluorescence intensity [195]. Kundu *et al.* reported a tool for selective sensing of gold in aqueous media using enhanced fluorescence property of GO–PVA. The GO–PVA system has augmented fluorescence system due to the presence of hydrogen bonding between the alcohol group of the polyvinyl system with that of the alcohol group of the GO. The gold caused the GO–PVA fluorescence system to be quenched by approximately around 83%, whereas all the metal ions (such as Cu^{2+}, Ag^{+}, Mg^{2+}, Ca^{2+}, Zn^{2+}, K^{+}, Pb^{2+}, Co^{2+}, Ni^{2+}, Pd^{2+}, Fe^{2+}, Fe^{3+}, Cr^{3+}) have a significantly smaller effect on the fluorescence hybrid. The detection limit of gold is 1,400 nM and its high selectivity is due to the high value of the reduction potential than that of the other metal ions [196]. A lead sensor was designed based on the curbing interaction between GO with DNAzyme-carboxyfluorescein solution. The strong bonding through π–π stacking interaction between the ssDNA loop of DNAzyme-CF and GO results in the quenching of the fluoro property, which in turn retained the fluorescence ability of CF in the presence of the Pb^{2+} ion [197]. The sensor showed a linear relationship between the range 1 and 100 nM with a detection limit 300 pM. A graphene decorated with polypyrrole system for the selective determination of Hg^{2+} was proposed by Zhao *et al.* [198]. In this technique, RGO was fabricated with polypyrrole system for the adsorption of mercury and after this preconcentration step the RGO/Ppy/Hg^{2+} was anodically stripped for quantitative analysis with a limit of detection 15 nM. Latter, similar principle was extended by Zhou *et al.* for the same metal ion using cysteamine with a detection limit 3 nM [199]. The fascinating feature of the developed sensor is that it demonstrated a excellent selectivity to Hg^{2+} in the presence of Cu^{2+}, Co^{2+}, Fe^{2+}, Zn^{2+}, and Mn^{2+}.

GO based sensors offer a better advantage over other related materials such as carbon nano tubes due to its unique properties as

1. The GO system has an excellent solubility in aqueous, non-aqueous system by ionic and covalent nature due to the presence of sp^3 hybridized oxygen and sp^2 hybridized carbon. Furthermore, the fine tuning of the ratio of sp^2 to sp^3 fractions by reduction chemistry offers a powerful way to engineer the band gap of GO through reduction which in turn influences the tailoring of electrical and optical properties.
2. The effectiveness in quenching clearly influences the advantages of the GO materials over the carbon materials such as diamond.

Hence the tunable fluorescence property from GO could be used to develop GO based devices for practical analytical application. RGO has high advantages over the conventional macro electrodes such as high signal to noise ratio, low power consumption, ideal in accommodating for metal ion electron transfer, can determine individual ion as well as multiple ions with low detection limit, and its commercialization lies in on-site measurement of the metal ions concentration [200].

8.8.4 Gas Sensors

Graphene in the sensors technology has a new paradigm shift in the rapid growth both in research and commercial applications. Our daily lives can be greatly influenced by the development and advancement of portable gas sensors capable of detecting analytes in real-time. Rapid advancement in the nano engineering and production of integrated electronic components in supplementation can allow for achieving these goals [201–205]. Graphene holds a pivotal place in potential for application in variety of sensors because of dangling π-bonds that react with chemical elements [206]. It is hard to generate fully saturated edges without any dangling bonds. The chemical reactivity of the dangling bonds in CNT plays a very important role [207, 208].

Sensing gas molecules is critical to environmental monitoring, control of chemical processes, space missions, and agricultural and medical applications. New generations of gas sensors have demonstrated using CNT and semiconductor nanowires. Graphene sensors that are developed for gases are based on the modulation of the electronic structure in diverse ways. The adsorption of gases on the surface converts the parent graphene into either n-type or p-type semiconductor. It is practically seen that adsorption of gases like carbon di oxide and oxygen converts the system into p-type, while the adsorption of ammonia leads to n-type performer. These rich adsorption effects are caused by the intrinsic property of gas molecules and the bonding characteristics between the gas molecules and graphene which is charge transfer in nature [209]. The p-and n-type semiconducting behavior can be detected by application of gate voltage and is also useful from the application view point. It has been practically demonstrated that all the gas molecules such as NO_2, CO, NO, O_2, CO_2, NH_3 can influence the electronic structure of arm chair graphene nano ribbons but only ammonia molecule can modify the conductance of AGNR while the other molecule has a little effect on conductance [209]. The defects in the graphene sheets are very important in sensing the gas molecules. It is also seen that nitrogen-doped graphene has a strong interaction with NO_2 [17] and boron-doped graphene exhibits an enhanced interaction with CO, CO_2, NO, NH_3 [210].

The adsorption energies of CO, NO, NO_2, and O_2 are all larger than 1 eV, corresponding to strong chemisorptions. The adsorption energies for NH_3 and CO_2 on Armchair graphene nanoribbons are −0.18 and −0.31 eV, respectively indicating that the adsorptions are between weak chemisorptions and strong physisorption [209].

Table 8.5 Bond order and bond length of gas molecules.

Molecule	Bond type	Bond length (A⁰)		Change
		Isolated	Adsorbed	
O_2	O-O	1.23	1.38	0.15
CO_2	C-O	1.16	1.26	0.10
NO	N-O	1.17	1.24	0.07
NO_2	N-O	1.18	1.25	0.07
CO	C-O	1.13	1.18	0.05
NH_3	N-H	1.01	1.03	0.02

The decrease in the bond order with reference to the adsorbed species such as CO, NO, NO_2, O_2, CO_2 can be accounted in terms of charge transfer concept of the molecular orbital theory. The bond length of all the species is longer than that of isolated molecules, indicating the weakening of adsorption process of original molecules. Based on the charge transfer it can be seen that CO, NO, NO_2, O_2, and CO_2 have electron withdrawing and NH_3 has a balancing of electron withdrawing and donating capacity. A balancing act in the case of ammonia has been suggested due to the fact that the value of the adsorbed system are almost close to unadsorbed species, might be due to the lack of detection with high sensitivity. The electron withdrawing tendency is of the order $O_2 > CO_2 > NO_2 = NO > CO > NH_3$ (Table 8.5). The much greater increase in bond length of oxygen indicates that electron withdrawing capacity of oxygen is highest towards antibonding level of oxygen and least in ammonia points out its least accepting capacity as detailed in Figure 8.5.

8.8.5 Pharmaceutical Sensors and Antioxidant Sensors

The real time pharmaceutical and antioxidant sensors are very important in the pharma industry and research laboratories. The non-degradable sensor methods are significant in projecting the purity of the sample. Graphene based electrodes have gone through significant improvements in the past decades. Thus graphene based sensors have been successfully applied for *in situ* detection of a plethora of pharmaceutical analytes in a wide range of sample mixture due to their advantages such as disposability,

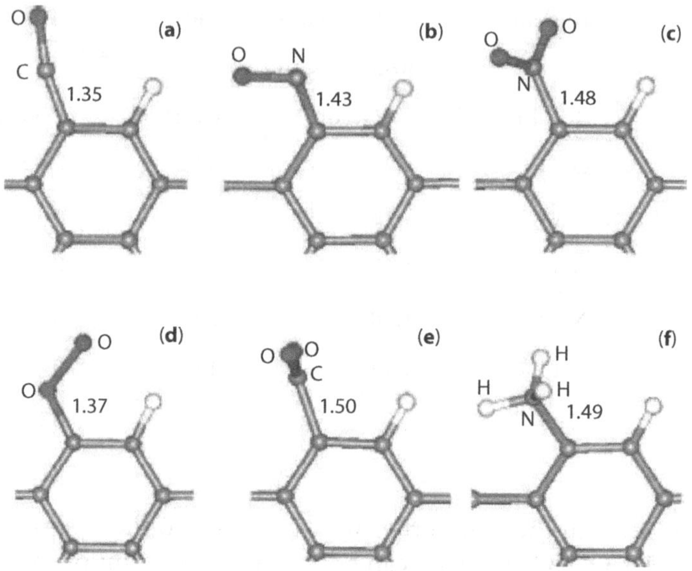

Figure 8.5 Optimized structure of 10-AGNRs with gas molecule adsorption: (a) CO, (b) NO, (c) NO_2, (d) O_2, (e) CO_2, and (f) NH_3. We only show the structure around the adsorbed molecule. The bond lengths (in units of A°) between the molecule and the ribbon are also given. Reproduced with permission from Adsorption of gas molecules on graphene nanoribbons and its implication for nanoscale molecule sensor [209].

simplicity and rapid responses. In particular development of electrochemical sensors based on graphene for pharmaceutical analysis has received massive consideration since they enable rapid screening of pharmaceuticals in matrix requiring small volume samples and no pretreatment steps. Non-enzymatic based biosensors has been widely used in diagnostics for the detection of wide range of pharmaceutically important analytes and antioxidants such as paracetamol, salbutamol, ractopamine, clenbuterol, β-agonists, isoniazid, uric acid, isoquercitin, baicillin, taxifollin, prostrate specific antigen, glucose, honokiol, hydroquinone, catechol, rutin, caffeic acid, ascorbic acid. The immense potential application is based on direct/indirect electron transfer between electrode surface and analyte of interest.

Graphene sheets/methylene blue/glutaraldehyde/anti PSA/bovine serum albumin was proposed by Mao *et al.* [211] for the PSA. The chitosan provides acts as a dispersant for graphene-methylene blue nanocomposite. The high surfaces areas graphene sheets and vast amino groups and hydroxyl group of chitosan provides a platform for crosslinking of antibody. A glassy carbon electrode coated with electrochemically reduced graphene doped with gold nanoparticles and PSA antibody solution was considered in PSA assay by Jang research group [212]. Glassy carbon electrode along with chemically RGO and polymeric α-D-glucopyranose, called β-cyclodextrins in which cyclodextrin enables to capture pharmaceutically important drugs such as imidocloprid into their cavity by host–guest interaction complexes was suggested by Chen group. A casting procedure was demonstrated by Kang research group [214] for paracetamol. The π–π conjugation between the oxidized product of paracetamol and sp^2 carbon in graphene sheets facilitates higher stability compared to p–π conjugation of the reactant. Suspension of TiO_2-graphene–nafion over the surface was chosen for paracetamol analysis by Fan and his co-workers [215]. A chemiluminiscence probe for sulphamethoxazole was proposed by Huamin and his co-workers [216] by fabricating a molecular imprinted polymer composed of acrylamide, ethyleneglycoldimethylacrylate and 2,2-azobisisobutyronitrile as monomer, cross linker and radical initiator, respectively. An inkjet layer containing poly(3,4-ethylenedioxythiophene): poly (styrene-sulphonate) was printed over an area 3 mm× 5 mm for the electrochemical determination of salbutamol was proposed by Karuwan and his research group [217]. An electrochemical polymerization of carbazole over chemically RGO has been adopted in the imidacloprid assay by Lei and his research workers [218]. Graphene sheets/nafion/gold magnetic particles coated with bovine serum albumin was chosen by Yang *et al.* for the clenbuterol, a pharmaceutically important drug [219]. A cyclic voltammetric and differential pulse voltammetry was optimized for isoniazid and uric acid quantification by the use of electrode fabricated involving poly(sulphosalicyclic acid)/carboxylated graphene over glassy carbon electrodic system was studied by Yan *et al.* [220]. A Rhodium nano sensors was developed by Cheemalapati group using 1 mM Rh (III) chloride in sulphuric acid for isoniazid drug [221]. RGO with 1-ethyl-3(3-dimethylaminopropyl) carbodiimide and N-hydroxysuccinimide along with artificial antigen for Ractopamine/salbutamol/clenbuterol was proposed by wang and co-workers [222]. A nano-Au/prussian blue/nano-Au/glassy carbon electrode was accomplished by $FeCl_3$, potassium ferricyanide in acidic medium by application of constant potential of 400 mV to Nano-Au/GCE for 40 s. Further, 1% $HAuCl_4$ at −0.2 V for 300 s allows the deposition of gold nano particle. Electropolymerization of aniline allows the coating of polyaniline over the gold nano. Then it is polished with salbutamol, salbutamol antibody along with free salbutamol and 0.5% horseradish peroxidase. Finally once again layered by gold nanoparticles for the assay of salbutamol by Huang group workers [223]. Terbiumhexacyanoferrate, chemically RGO has been adopted in the quantifying method for rutin involving electrochemical process at the working electrode, which was

Table 8.6 Sensitivity of the drug and antioxidant sensors.

Sensing material	Analyte	Methods	Linearity	Detection limit	References
Graphene-modified glassy carbon electrode	Paracetamol	Cyclic voltammetry and square wave voltammetry	–	32 nM	[214]
Nafion-TiO_2-graphene	Paracetamol	Cyclic voltammetry	1–100 μM	210 nM	[215]
Prussian Blue, polyaniline/poly (acrylic acid) and Au-hybrid graphene nanocomposite	Salbutamol	Electrochemical immunosensor		0.04 ng mL^{-1}	[223]
Graphene-poly(3,4-ethyl enedioxythiophene):poly(styrene-sulfonate)	Salbutamol	Electrochemical sensing	1–500 μM	1.25 μM	[217]
RGO and silver–palladium alloy nanoparticles	Ractopamine, Salbutamol and Clenbuterol	Screen-printed carbon electrode	0.01–100 ng/ml	1.52 pg mL^{-1}	[222]
				1.44 pg mL^{-1}	
				1.38 pg mL^{-1}	

(*Continued*)

Table 8.6 Sensitivity of the drug and antioxidant sensors. (*Continued*)

Sensing material	Analyte	Methods	Linearity	Detection limit	References
4,5-dihydroxy-3-[(2-hydroxy-5-sulfophenyl) azo]-2,7-naphthalenedisulfonic acid trisodium salt (acid chrome blue K) at a graphene oxide -nafion modified glassy carbon electrode	Clenbuterol, salbutamol, terbutaline, ractopamine, dopamine, dobutamine, adrenaline, and isoprenaline,	Linear sweep voltammetry	1.0–36 ng/mL	0.58–1.46 ng mL^{-1}	[292]
Poly(sulfosalicylic acid) and carboxylated graphene modified glassy carbon electrode	Isoniazid and uric acid	Cyclic voltametry and differential pulse voltammetry	0.05–15 μM isoniazid and 0.02–15 μM for Uric acid	12 nM	[220]
Rhodium modified glassy carbon electrode	Isoniazid	Linear sweep voltammetry	70–1300 μM	13 μM	[221]
2,6-dimethyl-β-cyclodextrin functionalized graphene nanosheets	Isoquercitrin and baicalin	–	10 nM–3.0 μM and 0.04 μM–3.0 μM	4 nM and 10 nM	[293]

(*Continued*)

Table 8.6 Sensitivity of the drug and antioxidant sensors. (*Continued*)

Sensing material	Analyte	Methods	Linearity	Detection limit	References
Chitosan/graphene oxide-molecularly imprinted polymers	Sulfamethoxazole	Chemiluminiscence	10^{-7} M–2.3 × 10^{-3} M	29 nM	[216]
RGO modified with water-soluble β-cyclodextrin polymer	Imidacloprid	Cyclic voltammetry and differential pulse voltammetry	1 × 10^{-6} to 1.5 × 10^{-4} M	100 nM	[213]
Graphene nanosheets and carbon nanospheres/ chitosan	Mebendazole	Cyclic and differential pulse voltammetry	0.02–1 μM	10.5 nM	[294]
Guanine on graphene nanoribbon modified glassy carbon electrode	–	–	0.1 to 4 mgl^{-1}	0.05 mgL^{-1}	[229]
RGO film with preferred vertical orientation was fabricated on a glassy carbon electrode	Taxifolin	–	1.0 × 10^{-8}–1.0 × 10^{-6}M	2.0 nM	[295]

(*Continued*)

Table 8.6 Sensitivity of the drug and antioxidant sensors. (*Continued*)

Sensing material	Analyte	Methods	Linearity	Detection limit	References
Graphene sheets–methylene blue–chitosan	Protsate specific antigen	Cyclic voltammetry	0.05–5.00 $ngml^{-1}$	13 pg mL^{-1}	[211]
(GOD/Nafion/(copper phthalocyanine-3,4′,4″,4‴-tetrasulfonic acid tetrasodium salt or alcian blue pyridine variant)/sulfanilic acid S/Glassy carbon electrode	Glucose	Cyclic voltammetry	8 mM	0.05 mM	[296]
Graphene oxide–ferroferric oxide–hemoglobin	Carbofuran	Electrochemical	–		[297]
Polycarbazole/chemically reduced graphene oxide/GCE	Imidacloprid insecticides	Cyclic voltammetry and differential pulse voltammetry	–	0.22 and 0.44 μM	[218]
Graphene based nanocomposite material	Carbamate pesticides	–	0.1–50 ng mL^{-1}	0.02–0.04 $ngmL^{-1}$	[298]

further applied in pharmaceutical tablets by Yang and his co-workers [224]. GCE/CNT system for total phenolic content by developing standard optimized methods for gallic acid, caffeic acid, ferlic acid, p-coumaric acid was elaborated by Arribas research group [225]. The higher peak and capacitative current that those observed for bare glassy carbon electrode due to improved electrochemical characteristics of nanotubes and increased active surface area. A dropsen carbon/Pt-nanoparticles/RGO/laccase coated with nafion involving caffeic acid as a substrate for laccase enzyme was chosen by Eremia *et al.* [226], which was further applied in the estimation of antioxidant capacity in tea infusion. A yttriumhexacyanoferrate system coated over chemically RGO involves the redox cycle of rutin was applied to determination in pharmaceutical samples containing rutin by Yang and his co-workers [227]. GO deposited over GCE by pulse potentiostatic method has been adopted for taxifollin solution by CV, chronocoulometry, square wave voltammetry [219]. Oxidation of rutin was studied by fabricated GCE containing poly(3,4-ethylenedioxythiophene) and GO system by Zhang and his co-workers [228]. The decrease in guanine oxidation peak was considered as a measure of antioxidant capacity by GCE/graphene nano ribbon/guanine by a Fenton solution (Fe(II) + EDTA + hydrogen peroxide) involving co workers of Yang research group [229], which was extended in the antioxidant parameter in fruit juices. A multiwalled CNT coated over poly(diallyldimethyl ammonium chloride) and reduced graphene system was chosen for the simultaneous detection of catechol and hydroquinone by song and his research colleagues [230]. A screen printed electrode coated with covalently attached 1-pyrenebutanoic acid and succinimidyl ester along with amines of tyrosinase protected gold nanoparticles was chosen by song and his co-workers as a representative method for catechol [231]. Below Table 8.6 lists some of the pharmaceutical and antioxidant sensors that are adopted along with graphene technology with detection limit and its corresponding linearity range.

8.9 Conclusion

Graphene in sensing technology has a unique place due to its high surface area, electrical conductivity and its ease of augmentation of biomolecules with the help of marriage chemistry between the graphene and its substrate. This review article closely highlights the recent past developments in the preparation of graphene by its mother graphite initially involving an oxidation process for diminishing van der Waal's interaction between the adjoining graphene layers in graphite. Despite, its potential application, the main difficulty is still the production of high quality graphene. Since its unique nanostructure and remarkable synergistic effect allows a research analyst in selecting it as a potential candidate for sensing process towards determination of hydrogen peroxide, metals, pharmaceutics, antioxidants, DNA, aptamers, biomarkers. Although the research towards the sensing technology from date to date is increasing in a promising way for graphene based biosensors, the application in the real samples is still at the infant stage. Recently, sister graphene materials like C_3N_4, BN [232], MoS_2 [233] have been investigated in two-dimensional nanomaterials based biosensors [234–239]. Recently, MoS_2 sensors has been explored in various gas sensors such as ammonia, nitric oxide, DNA, water and chemical vapors [240–246]. Also, high electrical conductivity and optical transparency makes it as promising candidate for developing liquid crystal display [247], touchscreens [248], organic light emitting diode photovoltaic cells [26], spintronics [249], supercapacitors [250], radioactive waste removal [251, 252]. Solar cells in the energy production, automobile, aero planes. In automobiles and aero planes, it can replace copper wire by much lighter carbon material that could save substantially on weight and use of fuel. A class of graphene in planar morphology has emerged and made significant attraction in sensing technology due to diverse structural features and molding capability in its composition by introduction extra sensing process. These types of nanomaterials have promoted great efforts to improve its application, reliability, durability, and economical nature of electrochemical biosensors. Looking into these parameters the reviewer feels that synthesis and functionalization for planar graphene morphology are highly desirable. The bilayer graphene have also been shown to have extreme potential candidate for optoelectronics [253] and nanoelectronics [254] application due to tunable conduction and valence band gap, QHE. A cost effective oxidation of ethanol [255] has been developed for the beneficiary of organic chemists. The graphene filters are in the move for desalination process [256]. The graphene forms such as nanoribbons and QDs have been shown to have tremendous application for ultrasensitive detection of analytes of biochemical importance. The toxicity and biocompatibility is a concern in the field of medical in the evaluation of *in vivo* results [257]. Graphene technology could also contribute to medication monitoring devices for patients with diabetes [258, 259] or cancer [260]. Therefore, compliance of international regulatory guidelines for assessing the toxicity of graphene based materials is critical for the development of commercial viable sensors. Keeping all the views from different research fraternity it can be essenced that analytical performance of graphene based biosensors and devices have significant contribution in environmental assessment, health monitoring, bio and non-bio based industries. Despite all these facts about carbon based material it is necessary that for essential commercialization multi industrial sector should keep an eye on the cost of the graphene for effective research.

Acknowledgements

Dr. Anantharaman Shivakumar (A.S.K) thanks St. Philomena's College facilities for the research investigation.

References

1. Fowler, W.A., Experimental and theoretical nuclear astrophysics: The quest for the origin of the elements. *Rev. Mod. Phys.*, 56, 2, 149–179, 1984.
2. Pauling, L., *The Nature of the Chemical Bond*, Cornell University Press, Ithaca, NY, 1960.
3. Curl, R.F., Dawn of the fullerenes: Experiment and conjecture. *Rev. Mod. Phys.*, 69, 3, 691–702, 1997.
4. Kroto, H., Symmetry, space, stars and C_{60}. *Rev. Mod. Phys.*, 69, 3, 703–722, 1997.
5. Smalley, R.E., Discovering the fullerenes. *Rev. Mod. Phys.*, 69, 3, 723–730, 1997.
6. Iijima, S., Helical microtubules of graphitic carbon. *Nature*, 354, 6348, 56–58, 1991.
7. Fitzer, E., Kochling, K.-H. H.P.B., Marsh, H., Recommended terminology for the description of carbon as a solid. *Pure Appl. Chem.*, 67, 3, 473–506, 1995.
8. Ambrosi, A., Chua, C.K., Bonanni, A., Pumera, M., Electrochemistry of Graphene and Related Materials. *Chem. Rev.*, 114, 14, 7150–7188, 2014.
9. Li, Y., Yan, H., Farmer, D.B., Meng, X., Zhu, W., Osgood, R.M., Heinz, T.F., Avouris, P., Graphene Plasmon Enhanced Vibrational Sensing of Surface-Adsorbed Layers. *Nano Lett.*, 14, 3, 1573–1577, 2014.
10. Qian, Y., Ismail, I.M., Stein, A., Ultralight, high-surface-area, multifunctional graphene-based aerogels from self-assembly of graphene oxide and resol. *Carbon*, 68, 221–231, 2014, http://dx.doi.org/10.1016/j.carbon.2013.10.082.
11. Shayeganfar, F. and Rochefort, A., Electronic Properties of Self-Assembled Trimesic Acid Monolayer on Graphene. *Langmuir*, 30, 32, 9707–9716, 2014.

12. Fathalian, A., Jalilian, J., Shahidi, S., Effects of gas adsorption on the electronic properties of graphene nanoribbons. *Physica B: Condens. Matter*, 417, 75–78, 2013, http://dx.doi.org/10.1016/j.physb.2013.02.018.
13. Jang, H., Lee, J., Min, D.-H., Graphene oxide for fluorescence-mediated enzymatic activity assays. *J. Mater. Chem. B*, 2, 17, 2452–2460, 2014.
14. Tang, S. and Cao, Z., Adsorption and Dissociation of Ammonia on Graphene Oxides: A First-Principles Study. *J. Phys. Chem. C*, 116, 15, 8778–8791, 2012.
15. Zheng, J., Wang, L., Katsiev, K., Losovyj, Y., Vescovo, E., Goodman, D.W., Dowben, P., Lu, J., Mei, W.-N., Adsorption configurations of carbon monoxide on gold monolayer supported by graphene or monolayer hexagonal boron nitride: A first-principles study. *Eur. Phys. J. B*, 86, 10, 1–7, 2013.
16. Rangel, E., Ruiz-Chavarria, G., Magana, L.F., Water molecule adsorption on a titanium–graphene system with high metal coverage. *Carbon*, 47, 2, 531–533, 2009, http://dx.doi.org/10.1016/j.carbon.2008.11.037.
17. Ko, G., Kim, H.Y., Ahn, J., Park, Y.M., Lee, K.Y., Kim, J., Graphene-based nitrogen dioxide gas sensors. *Curr. Appl. Phys.*, 10, 4, 1002–1004, 2010, http://dx.doi.org/10.1016/j.cap.2009.12.024.
18. Imani, K., Jafari, G., Abolhasani, M.R., Electronic Structure Calculation of Adsorbate Gas Molecules on an Armchair Graphene Nanoribbon. *ISRN Condens. Matter Phys.*, 2012, 5, 2012.
19. Singh, V.V., Nigam, A.K., Yadav, S.S., Tripathi, B.K., Srivastava, A., Boopathi, M., Singh, B., Graphene oxide as carboelectrocatalyst for *in situ* electrochemical oxidation and sensing of chemical warfare agent simulant. *Sens. Actuators B: Chem.*, 188, 1218–1224, 2013, http://dx.doi.org/10.1016/j.snb.2013.08.013.
20. Chen, T.-W., Sheng, Z.-H., Wang, K., Wang, F.-B., Xia, X.-H., Determination of Explosives Using Electrochemically Reduced Graphene. *Chem. – Asian J.*, 6, 5, 1210–1216, 2011.
21. Zhu, Z., Garcia-Gancedo, L., Flewitt, A.J., Xie, H., Moussy, F., Milne, W.I., A Critical Review of Glucose Biosensors Based on Carbon Nanomaterials: Carbon Nanotubes and Graphene. *Sensors*, 12, 5, 5996, 2012.
22. Kang, X., Wang, J., Wu, H., Aksay, I.A., Liu, J., Lin, Y., Glucose Oxidase–graphene–chitosan modified electrode for direct electrochemistry and glucose sensing. *Biosens. Bioelectron.*, 25, 4, 901–905, 2009, http://dx.doi.org/10.1016/j.bios.2009.09.004.
23. Worsley, M.A., Pauzauskie, P.J., Olson, T.Y., Biener, J., Satcher, J.H., Baumann, T.F., Synthesis of Graphene Aerogel with High Electrical Conductivity. *J. Am. Chem. Soc.*, 132, 40, 14067–14069, 2010.
24. Chen, H., Müller, M.B., Gilmore, K.J., Wallace, G.G., Li, D., Mechanically Strong, Electrically Conductive, and Biocompatible Graphene Paper. *Adv. Mater.*, 20, 18, 3557–3561, 2008.

25. Pinto, A.M., Gonçalves, I.C., Magalhães, F.D., Graphene-based materials biocompatibility: A review. *Colloids Surf., B: Biointerfaces*, 111, 188–202, 2013, http://dx.doi.org/10.1016/j.colsurfb.2013.05.022.
26. Wang, X., Zhi, L., Müllen, K., Transparent, Conductive Graphene Electrodes for Dye-Sensitized Solar Cells. *Nano Lett.*, 8, 1, 323–327, 2008.
27. Chatterjee, S., Carter, R., Oakes, L., Erwin, W.R., Bardhan, R., Pint, C.L., Electrochemical and Corrosion Stability of Nanostructured Silicon by Graphene Coatings: Toward High Power Porous Silicon Supercapacitors. *J. Phys. Chem. C*, 118, 20, 10893–10902, 2014.
28. Liu, F., Piao, Y., Choi, K.S., Seo, T.S., Fabrication of free-standing graphene composite films as electrochemical biosensors. *Carbon*, 50, 1, 123–133, 2012, http://dx.doi.org/10.1016/j.carbon.2011.07.061.
29. Sheng, Q., Wang, M., Zheng, J., A novel hydrogen peroxide biosensor based on enzymatically induced deposition of polyaniline on the functionalized graphene–carbon nanotube hybrid materials. *Sens. Actuators B: Chem.*, 160, 1, 1070–1077, 2011, http://dx.doi.org/10.1016/j.snb.2011.09.028.
30. YeoHeung, Y., Zhongyun, D., Vesselin, N.S., Mark, J.S., Electrochemical impedance measurement of prostate cancer cells using carbon nanotube array electrodes in a microfluidic channel. *Nanotechnology*, 18, 46, 465505, 2007.
31. Kemp, K.C., Seema, H., Saleh, M., Le, N.H., Mahesh, K., Chandra, V., Kim, K.S., Environmental applications using graphene composites: Water remediation and gas adsorption. *Nanoscale*, 5, 8, 3149–3171, 2013.
32. Lingamdinne, L.P., Koduru, J.R., Karri, R.R., A comprehensive review of applications of magnetic graphene oxide based nanocomposites for sustainable water purification. *J. Environ. Manage.*, 231, 622–634, 2019, https://doi.org/10.1016/j.jenvman.2018.10.063.
33. Ambrosi, A. and Pumera, M., Precise Tuning of Surface Composition and Electron-Transfer Properties of Graphene Oxide Films through Electroreduction. *Chem.—Eur. J.*, 19, 15, 4748–4753, 2013.
34. Nag, A., Mitra, A., Mukhopadhyay, S.C., Graphene and its sensor-based applications: A review. *Sens. Actuators A: Phys.*, 270, 177–194, 2018, https://doi.org/10.1016/j.sna.2017.12.028.
35. Mohan, V.B., Lau, K.-t., Hui, D., Bhattacharyya, D., Graphene-based materials and their composites: A review on production, applications and product limitations. *Compos. Part B: Eng.*, 142, 200–220, 2018, https://doi.org/10.1016/j.compositesb.2018.01.013.
36. Yang, W., Ratinac, K.R., Ringer, S.P., Thordarson, P., Gooding, J.J., Braet, F., Carbon Nanomaterials in Biosensors: Should You Use Nanotubes or Graphene? *Angew. Chem. Int. Ed.*, 49, 12, 2114–2138, 2010.
37. Wallace, P.R., The Band Theory of Graphite. *Phys. Rev.*, 71, 9, 622–634, 1947.

38. Slonczewski, J.C. and Weiss, P.R., Band Structure of Graphite. *Phys. Rev.*, 109, 2, 272–279, 1958.
39. Reese, C. and Bao, Z., Organic single-crystal field-effect transistors. *Mater. Today*, 10, 3, 20–27, 2007, http://dx.doi.org/10.1016/S1369-7021(07)70016-0.
40. Zhu, Y., Murali, S., Cai, W., Li, X., Suk, J.W., Potts, J.R., Ruoff, R.S., Graphene and Graphene Oxide: Synthesis, Properties, and Applications. *Adv. Mater.*, 22, 35, 3906–3924, 2010.
41. Novoselov, K.S., Jiang, Z., Zhang, Y., Morozov, S.V., Stormer, H.L., Zeitler, U., Maan, J.C., Boebinger, G.S., Kim, P., Geim, A.K., Room-Temperature Quantum Hall Effect in Graphene. *Science*, 315, 5817, 1379, 2007.
42. Geim, A.K. and Novoselov, K.S., The rise of graphene. *Nat. Mater.*, 6, 3, 183–191, 2007.
43. Miller, D.L., Kubista, K.D., Rutter, G.M., Ruan, M., de Heer, W.A., First, P.N., Stroscio, J.A., Observing the Quantization of Zero Mass Carriers in Graphene. *Science*, 324, 5929, 924–927, 2009.
44. Gierz, I., Riedl, C., Starke, U., Ast, C.R., Kern, K., Atomic Hole Doping of Graphene. *Nano Lett.*, 8, 12, 4603-4607, 2008.
45. Wehling, T.O., Novoselov, K.S., Morozov, S.V., Vdovin, E.E., Katsnelson, M.I., Geim, A.K., Lichtenstein, A.I., Molecular Doping of Graphene. *Nano Lett.*, 8, 1, 173–177, 2008.
46. Wang, L., Wei, Z., Mao, M., Wang, H., Li, Y., Ma, J., Metal oxide/graphene composite anode materials for sodium-ion batteries. *Energy Storage Mater.*, 16, 434–454, 2019, https://doi.org/10.1016/j.ensm.2018.06.027.
47. Novoselov, K.S., Geim, A.K., Morozov, S.V., Jiang, D., Zhang, Y., Dubonos, S.V., Grigorieva, I.V., Firsov, A.A., Electric Field Effect in Atomically Thin Carbon Films. *Science*, 306, 5696, 666–669, 2004.
48. Jijun, Z., Alper, B., Jie, H., Jian Ping, L., Gas molecule adsorption in carbon nanotubes and nanotube bundles. *Nanotechnology*, 13, 2, 195, 2002.
49. Chang, H., Lee, J.D., Lee, S.M., Lee, Y.H., Adsorption of NH3 and NO2 molecules on carbon nanotubes. *Appl. Phys. Lett.*, 79, 23, 3863–3865, 2001, http://dx.doi.org/10.1063/1.1424069.
50. Poh, H.L., Simek, P., Sofer, Z., Tomandl, I., Pumera, M., Boron and nitrogen doping of graphene *via* thermal exfoliation of graphite oxide in a BF3 or NH3 atmosphere: Contrasting properties. *J. Mater. Chem. A*, 1, 42, 13146–13153, 2013.
51. Wang, Y., Shao, Y., Matson, D.W., Li, J., Lin, Y., Nitrogen-Doped Graphene and Its Application in Electrochemical Biosensing. *ACS Nano*, 4, 4, 1790–1798, 2010.
52. Poh, H.L., Šimek, P., Sofer, Z., Pumera, M., Halogenation of Graphene with Chlorine, Bromine, or Iodine by Exfoliation in a Halogen Atmosphere. *Chem.—Eur. J.*, 19, 8, 2655–2662, 2013.
53. Lerf, A., He, H., Forster, M., Klinowski, J., Structure of Graphite Oxide Revisited. *J. Phys. Chem. B*, 102, 23, 4477–4482, 1998.

54. He, H., Klinowski, J., Forster, M., Lerf, A., A new structural model for graphite oxide. *Chem. Phys. Lett.*, 287, 1–2, 53–56, 1998, http://dx.doi.org/10.1016/S0009-2614(98)00144-4.
55. Paredes, J.I., Villar-Rodil, S., Martínez-Alonso, A., Tascón, J.M.D., Graphene Oxide Dispersions in Organic Solvents. *Langmuir*, 24, 19, 10560–10564, 2008.
56. Rao, C.N.R., Sood, A.K., Subrahmanyam, K.S., Govindaraj, A., Graphene: The New Two-Dimensional Nanomaterial. *Angew. Chem. Int. Ed.*, 48, 42, 7752–7777, 2009.
57. Sun, H., Wu, L., Wei, W., Qu, X., Recent advances in graphene quantum dots for sensing. *Mater. Today*, 16, 11, 433–442, 2013, http://dx.doi.org/10.1016/j.mattod.2013.10.020.
58. Parvez, K., Wu, Z.-S., Li, R., Liu, X., Graf, R., Feng, X., Müllen, K., Exfoliation of Graphite into Graphene in Aqueous Solutions of Inorganic Salts. *J. Am. Chem. Soc.*, 136, 16, 6083–6091, 2014.
59. Geim, A.K., Graphene: Status and Prospects. *Science*, 324, 5934, 1530–1534, 2009.
60. Novoselov, K.S., Jiang, D., Schedin, F., Booth, T.J., Khotkevich, V.V., Morozov, S.V., Geim, A.K., Two-dimensional atomic crystals. *Proc. Natl. Acad. Sci. U.S.A.*, 102, 30, 10451–10453, 2005.
61. Wu, J., Pisula, W., Müllen, K., Graphenes as Potential Material for Electronics. *Chem. Rev.*, 107, 3, 718–747, 2007.
62. Guinea, F., Katsnelson, M.I., Geim, A.K., Energy gaps and a zero-field quantum Hall effect in graphene by strain engineering. *Nat. Phys.*, 6, 1, 30–33, 2010, http://www.nature.com/nphys/journal/v6/n1/suppinfo/nphys1420_S1.html.
63. Lee, C., Wei, X., Kysar, J.W., Hone, J., Measurement of the Elastic Properties and Intrinsic Strength of Monolayer Graphene. *Science*, 321, 5887, 385–388, 2008.
64. Balandin, A.A., Ghosh, S., Bao, W., Calizo, I., Teweldebrhan, D., Miao, F., Lau, C.N., Superior Thermal Conductivity of Single-Layer Graphene. *Nano Lett.*, 8, 3, 902–907, 2008.
65. Kampouris, D.K. and Banks, C.E., Exploring the physicoelectrochemical properties of graphene. *Chem. Commun.*, 46, 47, 8986–8988, 2010.
66. Dato, A., Radmilovic, V., Lee, Z., Phillips, J., Frenklach, M., Substrate-Free Gas-Phase Synthesis of Graphene Sheets. *Nano Lett.*, 8, 7, 2008.
67. Sun, J., Nam, Y., Lindvall, N., Cole, M.T., Teo, K.B.K., Woo Park, Y., Yurgens, A., Growth mechanism of graphene on platinum: Surface catalysis and carbon segregation. *Appl. Phys. Lett.*, 104, 15, 152107, 2014, http://dx.doi.org/10.1063/1.4871978.
68. Sutter, P.W., Flege, J.-I., Sutter, E.A., Epitaxial graphene on ruthenium. *Nat. Mater.*, 7, 5, 406–411, 2008.

69. Šrut, I., Mikšić Trontl, V., Pervan, P., Kralj, M., Temperature dependence of graphene growth on a stepped iridium surface. *Carbon*, 56, 193–200, 2013, http://dx.doi.org/10.1016/j.carbon.2013.01.026.
70. Losurdo, M., Giangregorio, M.M., Capezzuto, P., Bruno, G., Graphene CVD growth on copper and nickel: Role of hydrogen in kinetics and structure. *Phys. Chem. Chem. Phys.*, 13, 46, 20836–20843, 2011.
71. Li, W.-N., Ding, Y.-S., Suib, S.L., DiCarlo, J.F., Galasso, F.S., Controlling the growth of CVD carbon from methane on transition metal substrates. *Surf. Coat. Technol.*, 190, 2–3, 366–371, 2005, http://dx.doi.org/10.1016/j.surfcoat.2004.06.026.
72. Wassei, J.K., Mecklenburg, M., Torres, J.A., Fowler, J.D., Regan, B.C., Kaner, R.B., Weiller, B.H., Chemical Vapor Deposition of Graphene on Copper from Methane, Ethane and Propane: Evidence for Bilayer Selectivity. *Small*, 8, 9, 1415–1422, 2012.
73. Sun, Z., Yan, Z., Yao, J., Beitler, E., Zhu, Y., Tour, J.M., Growth of graphene from solid carbon sources. *Nature*, 468, 7323, 549–552, 2010, http://www.nature.com/nature/journal/v468/n7323/abs/nature09579.html#supplementary-information.
74. Wang, B., Guo, W., Ling, L., Zhang, R., C2 oxygenates formation from syngas over the promoter M(M = Rh, Co) monolayer-modified Cu(111) surface: Probing into the role of monolayer promoter on the selectivity. *Appl. Surf. Sci.*, 488, 434–444, 2019, https://doi.org/10.1016/j.apsusc.2019.05.033.
75. Shen, J., Zhu, Y., Yang, X., Li, C., Graphene quantum dots: Emergent nanolights for bioimaging, sensors, catalysis and photovoltaic devices. *Chem. Commun.*, 48, 31, 3686–3699, 2012.
76. Zhu, S., Tang, S., Zhang, J., Yang, B., Control the size and surface chemistry of graphene for the rising fluorescent materials. *Chem. Commun.*, 48, 38, 4527–4539, 2012.
77. Zhang, Z., Zhang, J., Chen, N., Qu, L., Graphene quantum dots: An emerging material for energy-related applications and beyond. *Energy Environ. Sci.*, 5, 10, 8869–8890, 2012.
78. Li, L., Wu, G., Yang, G., Peng, J., Zhao, J., Zhu, J.-J., Focusing on luminescent graphene quantum dots: Current status and future perspectives. *Nanoscale*, 5, 10, 4015–4039, 2013.
79. Tung, V.C., Allen, M.J., Yang, Y., Kaner, R.B., High-throughput solution processing of large-scale graphene. *Nat. Nanotechnol.*, 4, 1, 25–29, 2009, http://www.nature.com/nnano/journal/v4/n1/suppinfo/nnano.2008.329_S1.html.
80. Eda, G., Fanchini, G., Chhowalla, M., Large-area ultrathin films of reduced graphene oxide as a transparent and flexible electronic material. *Nat. Nanotechnol.*, 3, 5, 270–274, 2008, http://www.nature.com/nnano/journal/v3/n5/suppinfo/nnano.2008.83_S1.html.

81. Brodie, B.C., On the Atomic Weight of Graphite. *Philos. Trans. R. Soc. London*, 149, 249–259, 1859.
82. Staudenmaier, L., Verfahren zur Darstellung der Graphitsäure. *Ber. Dtsch. Chem. Ges.*, 31, 2, 1481–1487, 1898.
83. Hummers, W.S. and Offeman, R.E., Preparation of Graphitic Oxide. *J. Am. Chem. Soc.*, 80, 6, 1339–1339, 1958.
84. Hong, Y., Wang, Z., Jin, X., Sulfuric Acid Intercalated Graphite Oxide for Graphene Preparation. *Sci. Rep.*, 3, 3439, 2013, 10.1038/srep03439, http://www.nature.com/articles/srep03439#supplementary-information.
85. Forsman, W.C., Vogel, F.L., Carl, D.E., Hoffman, J., Chemistry of graphite intercalation by nitric acid. *Carbon*, 16, 4, 269–271, 1978, http://dx.doi.org/10.1016/0008-6223(78)90040-4.
86. Besenhard, J.O., Minderer, P., Bindl, M., Hydrolysis of perchloric acid and sulfuric acid graphite intercalation compounds. *Synth. Met.*, 34, 1–3, 133–138, 1989, http://dx.doi.org/10.1016/0379-6779(89)90376-7.
87. Park, S., An, J., Potts, J.R., Velamakanni, A., Murali, S., Ruoff, R.S., Hydrazine-reduction of graphite- and graphene oxide. *Carbon*, 49, 9, 3019–3023, 2011, http://dx.doi.org/10.1016/j.carbon.2011.02.071.
88. Sheshmani, S. and Fashapoyeh, M.A., Suitable chemical methods for preparation of graphene oxide, graphene and surface fuctionalised graphene nanosheets. *Acta Chim. Slov.*, 60, 4, 813–825, 2013.
89. Nethravathi, C., Nisha, T., Ravishankar, N., Shivakumara, C., Rajamathi, M., Graphene–nanocrystalline metal sulphide composites produced by a one-pot reaction starting from graphite oxide. *Carbon*, 47, 8, 2054–2059, 2009, http://dx.doi.org/10.1016/j.carbon.2009.03.055.
90. Shin, H.-J., Kim, K.K., Benayad, A., Yoon, S.-M., Park, H.K., Jung, I.-S., Jin, M.H., Jeong, H.-K., Kim, J.M., Choi, J.-Y., Lee, Y.H., Efficient Reduction of Graphite Oxide by Sodium Borohydride and Its Effect on Electrical Conductance. *Adv. Funct. Mater.*, 19, 12, 1987–1992, 2009.
91. Zhu, X., Liu, Q., Zhu, X., Li, C., Xu, M., Liang, Y., Reduction of Graphene Oxide Via Ascorbic Acid and Its Application for Simultaneous Detection of Dopamine And Ascorbic Acid. *Int. J. Electrochem. Sci.*, 7, 5172–5184, 2012.
92. Endo, M., Takeuchi, K., Igarashi, S., Kobori, K., Shiraishi, M., Kroto, H.W., The production and structure of pyrolytic carbon nanotubes (PCNTs). *J. Phys. Chem. Solids*, 54, 12, 1841–1848, 1993, http://dx.doi.org/10.1016/0022-3697(93)90297-5.
93. Zhang, X., Zhang, D., Chen, Y., Sun, X., Ma, Y., Electrochemical reduction of graphene oxide films: Preparation, characterization and their electrochemical properties. *Chin. Sci. Bull.*, 57, 23, 3045–3050, 2012.
94. Chng, E.L.K. and Pumera, M., Solid-State Electrochemistry of Graphene Oxides: Absolute Quantification of Reducible Groups using Voltammetry. *Chem.—Asian J.*, 6, 11, 2899–2901, 2011.

95. Hernandez, Y., Nicolosi, V., Lotya, M., Blighe, F.M., Sun, Z., De, S., McGovern, I.T., Holland, B., Byrne, M., Gun'Ko, Y.K., Boland, J.J., Niraj, P., Duesberg, G., Krishnamurthy, S., Goodhue, R., Hutchison, J., Scardaci, V., Ferrari, A.C., Coleman, J.N., High-yield production of graphene by liquid-phase exfoliation of graphite. *Nat. Nanotechnol.*, 3, 9, 563–568, 2008, http://www.nature.com/nnano/journal/v3/n9/suppinfo/nnano.2008.215_S1.html.
96. Hernandez, Y., Lotya, M., Rickard, D., Bergin, S.D., Coleman, J.N., Measurement of Multicomponent Solubility Parameters for Graphene Facilitates Solvent Discovery. *Langmuir*, 26, 5, 3208–3213, 2010.
97. Godoy, A.P., Ecorchard, P., Beneš, H., Tolasz, J., Smržová, D., Seixas, L., Pedrotti, J.J., de Souza, E.A.T., El Seoud, O.A., and Donato, R.K., Ultrasound exfoliation of graphite in biphasic liquid systems containing ionic liquids: A study on the conditions for obtaining large few-layers graphene. *Ultrason. Sonochem.*, 55, 279–288, 2019, https://doi.org/10.1016/j.ultsonch.2019.01.016.
98. Beck, F., Jiang, J., Krohn, H., Potential oscillations during galvanostatic overoxidation of graphite in aqueous sulphuric acids. *J. Electroanal. Chem.*, 389, 1-2, 161-165, 1995, http://dx.doi.org/10.1016/0022-0728(95)03870-M.
99. Beck, F., Junge, H., Krohn, H., Graphite intercalation compounds as positive electrodes in galvanic cells. *Electrochim. Acta*, 26, 7, 799–809, 1981, http://dx.doi.org/10.1016/0013-4686(81)85038-4.
100. Rao, K.S., Senthilnathan, J., Liu, Y.-F., Yoshimura, M., Role of Peroxide Ions in Formation of Graphene Nanosheets by Electrochemical Exfoliation of Graphite. *Sci. Rep.*, 4, 4237, 2014, 10.1038/srep04237, http://www.nature.com/articles/srep04237#supplementary-information.
101. Brownson, D.A.C. and Banks, C.E., Graphene electrochemistry: Surfactants inherent to graphene inhibit metal analysis. *Electrochem. Commun.*, 13, 2, 111–113, 2011, http://dx.doi.org/10.1016/j.elecom.2010.11.024.
102. Brownson, D.A.C., Metters, J.P., Kampouris, D.K., Banks, C.E., Graphene Electrochemistry: Surfactants Inherent to Graphene Can Dramatically Effect Electrochemical Processes. *Electroanalysis*, 23, 4, 894–899, 2011.
103. Bonanni, A. and Pumera, M., Surfactants used for dispersion of graphenes exhibit strong influence on electrochemical impedance spectroscopic response. *Electrochem. Commun.*, 16, 1, 19–21, 2012, http://dx.doi.org/10.1016/j.elecom.2011.12.012.
104. Hilder, M., Winther-Jensen, B., Li, D., Forsyth, M., MacFarlane, D.R., Direct electro-deposition of graphene from aqueous suspensions. *Phys. Chem. Chem. Phys.*, 13, 20, 9187–9193, 2011.
105. Shao, Y., Wang, J., Engelhard, M., Wang, C., Lin, Y., Facile and controllable electrochemical reduction of graphene oxide and its applications. *J. Mater. Chem.*, 20, 4, 743–748, 2010.

106. Liu, C., Wang, K., Luo, S., Tang, Y., Chen, L., Direct Electrodeposition of Graphene Enabling the One-Step Synthesis of Graphene–Metal Nanocomposite Films. *Small*, 7, 9, 1203–1206, 2011.
107. Jiang, Y., Lu, Y., Li, F., Wu, T., Niu, L., Chen, W., Facile electrochemical codeposition of "clean" graphene–Pd nanocomposite as an anode catalyst for formic acid electrooxidation. *Electrochem. Commun.*, 19, 21–24, 2012, http://dx.doi.org/10.1016/j.elecom.2012.02.035.
108. Ping, J., Wang, Y., Fan, K., Wu, J., Ying, Y., Direct electrochemical reduction of graphene oxide on ionic liquid doped screen-printed electrode and its electrochemical biosensing application. *Biosens. Bioelectron.*, 28, 1, 204–209, 2011, http://dx.doi.org/10.1016/j.bios.2011.07.018.
109. Guo, H.-L., Wang, X.-F., Qian, Q.-Y., Wang, F.-B., Xia, X.-H., A Green Approach to the Synthesis of Graphene Nanosheets. *ACS Nano*, 3, 9, 2653–2659, 2009.
110. Chang, H.-H., Chang, C.-K., Tsai, Y.-C., Liao, C.-S., Electrochemically synthesized graphene/polypyrrole composites and their use in supercapacitor. *Carbon*, 50, 6, 2331–2336, 2012, http://dx.doi.org/10.1016/j.carbon.2012.01.056.
111. Zhou, M., Wang, Y., Zhai, Y., Zhai, J., Ren, W., Wang, F., Dong, S., Controlled Synthesis of Large-Area and Patterned Electrochemically Reduced Graphene Oxide Films. *Chem.—Eur. J.*, 15, 25, 6116–6120, 2009.
112. Li, W., Liu, J., Yan, C., Reduced graphene oxide with tunable C/O ratio and its activity towards vanadium redox pairs for an all vanadium redox flow battery. *Carbon*, 55, 313–320, 2013, http://dx.doi.org/10.1016/j.carbon.2012.12.069.
113. Wang, Z., Zhou, X., Zhang, J., Boey, F., Zhang, H., Direct Electrochemical Reduction of Single-Layer Graphene Oxide and Subsequent Functionalization with Glucose Oxidase. *J. Phys. Chem. C*, 113, 32, 14071–14075, 2009.
114. Ramesha, G.K. and Sampath, S., Electrochemical Reduction of Oriented Graphene Oxide Films: An *in Situ* Raman Spectroelectrochemical Study. *J. Phys. Chem. C*, 113, 19, 7985–7989, 2009.
115. Zhou, Y.-G., Chen, J.-J., Wang, F.-b., Sheng, Z.-H., Xia, X.-H., A facile approach to the synthesis of highly electroactive Pt nanoparticles on graphene as an anode catalyst for direct methanol fuel cells. *Chem. Commun.*, 46, 32, 5951–5953, 2010.
116. Eda, G. and Chhowalla, M., Chemically Derived Graphene Oxide: Towards Large-Area Thin-Film Electronics and Optoelectronics. *Adv. Mater.*, 22, 22, 2392–2415, 2010.
117. Liu, S., Ou, J., Wang, J., Liu, X., Yang, S., A simple two-step electrochemical synthesis of graphene sheets film on the ITO electrode as supercapacitors. *J. Appl. Electrochem.*, 741, 7, 881–884, 2011.

118. Compton, O.C. and Nguyen, S.T., Graphene Oxide, Highly Reduced Graphene Oxide, and Graphene: Versatile Building Blocks for Carbon-Based Materials. *Small*, 6, 6, 711–723, 2010.
119. Grimshaw, J., *Electrochemical Reactions and Mechanisms in Organic Chemistry*, Elsevier, Amsterdam, The Netherlands, 2000.
120. Ferrari, A.C., Meyer, J.C., Scardaci, V., Casiraghi, C., Lazzeri, M., Mauri, F., Piscanec, S., Jiang, D., Novoselov, K.S., Roth, S., Geim, A.K., Raman Spectrum of Graphene and Graphene Layers. *Phys. Rev. Lett.*, 97, 18, 187401, 2006.
121. Rao, A.M., Richter, E., Bandow, S., Chase, B., Eklund, P.C., Williams, K.A., Fang, S., Subbaswamy, K.R., Menon, M., Thess, A., Smalley, R.E., Dresselhaus, G., Dresselhaus, M.S., Diameter-Selective Raman Scattering from Vibrational Modes in Carbon Nanotubes. *Science*, 275, 5297, 187–191, 1997.
122. Kalbac, M., Reina-Cecco, A., Farhat, H., Kong, J., Kavan, L., Dresselhaus, M.S., The Influence of Strong Electron and Hole Doping on the Raman Intensity of Chemical Vapor-Deposition Graphene. *ACS Nano*, 4, 10, 6055–6063.
123. Wang, H., Wang, Y., Cao, X., Feng, M., Lan, G., Vibrational properties of graphene and graphene layers. *J. Raman Spectrosc.*, 40, 12, 1791–1796, 2009.
124. Poncharal, P., Ayari, A., Michel, T., Sauvajol, J.L., Raman spectra of misoriented bilayer graphene. *Phys. Rev. B*, 78, 11, 113407, 2008.
125. Chen, D., Wang, G., Li, J., Interfacial Bioelectrochemistry: Fabrication, Properties and Applications of Functional Nanostructured Biointerfaces. *J. Phys. Chem. C*, 111, 6, 2351–2367, 2007.
126. Rad, A.S., Mirabi, A., Binanian, E., Tayebi, H., A review on glucose and hydrogen peroxide biosensor based on modified electrode include silver nanoparticles. *Int. J. Electrochem. Sci.*, 6, 3671–3683, 2011.
127. Song, Y., Luo, Y., Zhu, C., Li, H., Du, D., Lin, Y., Recent advances in electrochemical biosensors based on graphene two-dimensional nanomaterials. *Biosens. Bioelectron.*, 76, 195–212, 2016, https://doi.org/10.1016/j.bios.2015.07.002.
128. Martin, C.R. and Kohli, P., The emerging field of nanotube biotechnology. *Nat. Rev. Drug Discovery*, 2, 1, 29–37, 2003.
129. Fang, Y. and Wang, E., Electrochemical biosensors on platforms of graphene. *Chem. Commun.*, 49, 83, 9526–9539, 2013.
130. Shao, Y., Wang, J., Wu, H., Liu, J., Aksay, I.A., Lin, Y., Graphene Based Electrochemical Sensors and Biosensors: A Review. *Electroanalysis*, 22, 10, 1027–1036, 2010.
131. Mao, H.Y., Laurent, S., Chen, W., Akhavan, O., Imani, M., Ashkarran, A.A., Mahmoudi, M., Graphene: Promises, Facts, Opportunities, and Challenges in Nanomedicine. *Chem. Rev.*, 113, 5, 3407–3424, 2013.
132. Du, D., Yang, Y., Lin, Y., Graphene-based materials for biosensing and bioimaging. *MRS Bull.*, 37, 12, 1290–1296, 2012.

133. Kamat, P.V., Graphene-Based Nanoarchitectures. Anchoring Semiconductor and Metal Nanoparticles on a Two-Dimensional Carbon Support. *J. Phys. Chem. Lett.*, 1, 2, 520–527, 2010.
134. Zhou, Y., Fang, Y., Ramasamy, R.P., Non-Covalent Functionalization of Carbon Nanotubes for Electrochemical Biosensor Development. *Sensors*, 19, 2, 392, 2019.
135. Atchudan, R., Muthuchamy, N., Edison, T.N.J.I., Perumal, S., Vinodh, R., Park, K.H., Lee, Y.R., An ultrasensitive photoelectrochemical biosensor for glucose based on bio-derived nitrogen-doped carbon sheets wrapped titanium dioxide nanoparticles. *Biosens. Bioelectron.*, 126, 160–169, 2019, https://doi.org/10.1016/j.bios.2018.10.049.
136. Du, D., Wang, L., Shao, Y., Wang, J., Engelhard, M.H., Lin, Y., Functionalized Graphene Oxide as a Nanocarrier in a Multienzyme Labeling Amplification Strategy for Ultrasensitive Electrochemical Immunoassay of Phosphorylated p53 (S392). *Anal. Chem.*, 83, 3, 746–752, 2011.
137. Wang, Y., Li, Z., Wang, J., Li, J., Lin, Y., Graphene and graphene oxide: Biofunctionalization and applications in biotechnology. *Trends Biotechnol.*, 29, 5, 205–212, 2011.
138. Ulijn, R.V. and Smith, A.M., Designing peptide based nanomaterials. *Chem. Soc. Rev.*, 37, 4, 664–675, 2008.
139. Yan, Z., Congyu, W., Shouwu, W., Jingyan, Z., Interactions of graphene and graphene oxide with proteins and peptides. *Nanotechnol. Rev.*, 2, 1, 27–45, 2013.
140. Tang, L., Wang, Y., Li, J., The graphene/nucleic acid nanobiointerface. *Chem. Soc. Rev.*, pp. 6954–6980, 2015.
141. Liu, Z., Jiang, L., Galli, F., Nederlof, I., Olsthoorn, R.C.L., Lamers, G.E.M., Oosterkamp, T.H., Abrahams, J.P., A Graphene Oxide·Streptavidin Complex for Biorecognition—Towards Affinity Purification. *Adv. Funct. Mater.*, 20, 17, 2857–2865, 2010.
142. Wang, Y., Li, Z., Hu, D., Lin, C.-T., Li, J., Lin, Y., Aptamer/Graphene Oxide Nanocomplex for *in Situ* Molecular Probing in Living Cells. *J. Am. Chem. Soc.*, 132, 27, 9274–9276, 2010.
143. Hernández, R., Vallés, C., Benito, A.M., Maser, W.K., Xavier Rius, F., Riu, J., Graphene-based potentiometric biosensor for the immediate detection of living bacteria. *Biosens. Bioelectron.*, 54, 553–557, 2014, http://dx.doi.org/10.1016/j.bios.2013.11.053.
144. Yoon, H.J., Kim, T.H., Zhang, Z., Azizi, E., Pham, T.M., Paoletti, C., Lin, J., Ramnath, N., Wicha, M.S., Hayes, D.F., Simeone, D.M., Nagrath, S., Sensitive capture of circulating tumour cells by functionalized graphene oxide nanosheets. *Nat. Nanotechnol.*, 8, 10, 735–741, 2013, 10.1038/nnano.2013.194, http://www.nature.com/nnano/journal/v8/n10/abs/nnano.2013.194.html#supplementary-information.

145. Zhao, H., Chang, Y., Liu, M., Gao, S., Yu, H., Quan, X., A universal immunosensing strategy based on regulation of the interaction between graphene and graphene quantum dots. *Chem. Commun.*, 49, 3, 234–236, 2013.
146. Böttcher, S., Weser, M., Dedkov, Y.S., Horn, K., Voloshina, E.N., Paulus, B., Graphene on ferromagnetic surfaces and its functionalization with water and ammonia. *Nanoscale Res. Lett.*, 6, 214–214, 2011.
147. Wang, J., Electrochemical Glucose Biosensors. *Chem. Rev.*, 108, 2, 814–825, 2008.
148. Gough, D.A., Lucisano, J.Y., Tse, P.H.S., Two-dimensional enzyme electrode sensor for glucose. *Anal. Chem.*, 57, 12, 2351–2357, 1985.
149. Armour, J.C., Lucisano, J.Y., McKean, B.D., Gough, D.A., Application of Chronic Intravascular Blood Glucose Sensor in Dogs. *Diabetes*, 39, 12, 1519–1526, 1990.
150. Wang, J. and Lu, F., Oxygen-Rich Oxidase Enzyme Electrodes for Operation in Oxygen-Free Solutions. *J. Am. Chem. Soc.*, 120, 5, 1048–1050, 1998.
151. Pishko, M.V., Katakis, I., Lindquist, S.-E., Ye, L., Gregg, B.A., Heller, A., Direct Electrical Communication between Graphite Electrodes and Surface Adsorbed Glucose Oxidase/Redox Polymer Complexes. *Angew. Chem. Int. Ed. Engl.*, 29, 1, 82–84, 1990.
152. Liu, Z., Shen, Z., Zhu, T., Hou, S., Ying, L., Shi, Z., Gu, Z., Organizing Single-Walled Carbon Nanotubes on Gold Using a Wet Chemical Self-Assembling Technique. *Langmuir*, 16, 8, 3569–3573, 2000.
153. Gu, T. and Hasebe, Y., DNA–Cu(II) poly(amine) complex membrane as novel catalytic layer for highly sensitive amperometric determination of hydrogen peroxide. *Biosens. Bioelectron.*, 21, 11, 2121–2128, 2006, http://dx.doi.org/10.1016/j.bios.2005.10.007.
154. Alphey, L., *DNA Sequencing: From Experimental Methods to Bioinformatics*, BIOS Scientific Publishers, Oxford, UK, 1997.
155. Tang, Z., Wu, H., Cort, J.R., Buchko, G.W., Zhang, Y., Shao, Y., Aksay, I.A., Liu, J., Lin, Y., Constraint of DNA on Functionalized Graphene Improves its Biostability and Specificity. *Small*, 6, 11, 1205–1209, 2010.
156. Ohno, Y., Maehashi, K., Yamashiro, Y., Matsumoto, K., Electrolyte-Gated Graphene Field-Effect Transistors for Detecting pH and Protein Adsorption. *Nano Lett.*, 9, 9, 3318–3322, 2009.
157. Patil, A.J., Vickery, J.L., Scott, T.B., Mann, S., Aqueous Stabilization and Self-Assembly of Graphene Sheets into Layered Bio-Nanocomposites using DNA. *Adv. Mater.*, 21, 31, 3159–3164, 2009.
158. Kahvejian, A., Quackenbush, J., Thompson, J.F., What would you do if you could sequence everything? *Nat. Biotechnol.*, 26, 10, 1125–1133, 2008.
159. Lee, X.J., Hiew, B.Y.Z., Lai, K.C., Lee, L.Y., Gan, S., Thangalazhy-Gopakumar, S., Rigby, S., Review on graphene and its derivatives: Synthesis methods and

potential industrial implementation. *J. Taiwan Inst. Chem. Eng.*, 98, 163–180, 2019, https://doi.org/10.1016/j.jtice.2018.10.028.

160. Fojta, M., Vetterl, V., Tomschik, M., Jelen, F., Nielsen, P., Wang, J., Palecek, E., Adsorption of peptide nucleic acid and DNA decamers at electrically charged surfaces. *Biophys. J.*, 72, 5, 2285–2293, 1997, http://dx.doi.org/10.1016/S0006-3495(97)78873-8.
161. Paleček, E., Determination of pseudouridine at submicromolar concentrations by cathodic stripping voltammetry at a mercury electrode. *Anal. Chim. Acta*, 174, 103–113, 1985, http://dx.doi.org/10.1016/S0003-2670(00)84369-X.
162. Palecek, E., Fojta, M., Jelen, F., Vetterl, V., *In the Encyclopedia of Electrochemistry*, Wiley-VCH, Weinheim, 2002.
163. Muti, M. and Muti, M., Electrochemical monitoring of the interaction between anticancer drug and DNA in the presence of antioxidant. *Talanta*, 178, 1033–1039, 2018, https://doi.org/10.1016/j.talanta.2017.08.089.
164. Paleček, E., Adsorptive transfer stripping voltammetry: Determination of nanogram quantities of DNA immobilized at the electrode surface. *Anal. Biochem.*, 170, 2, 421–431, 1988, http://dx.doi.org/10.1016/0003-2697(88)90654-9.
165. Palecek, E., Oscillographic Polarography of Highly Polymerized Deoxyribo nucleic Acid. *Nature*, 188, 4751, 656–657, 1960.
166. Palecek, E., *Topics in bioelectrochemistry and bioenergetics*, John Wiley, Chichester, NY, 1983.
167. Palecek, E. and Fojta, M., *Electrochemical DNA Sensors*, Weinheim, Wiley-VCH Verlag GmbH & Co. KGaA, Germany, 2005.
168. Popovich, N.D. and Thorp, H.H., New strategies for electrochemical nuclei acid detection. *Electrochem. Soc. Interface*, 11, 30–34, 2002, https://www.electrochem.org/dl/interface/wtr/wtr02/IF12-02-Pages30-34.pdf.
169. Tarlov, M.J. and Steel, A.B., *Biomolecular films. Design, functions and applications*, Marcell Dekker, New York, 2003.
170. Lu, C.-H., Yang, H.-H., Zhu, C.-L., Chen, X., Chen, G.-N., Graphene Platform for Sensing Biomolecules, *Angew. Chem. Int. Ed.*, 48, 26, 4785–4787, 2003.
171. Lu, C.-H., Li, J., Liu, J.-J., Yang, H.-H., Chen, X., Chen, G.-N., Increasing the Sensitivity and Single-Base Mismatch Selectivity of the Molecular Beacon Using Graphene Oxide as the "Nanoquencher". *Chem.—Eur. J.*, 16, 16, 4889–4894, 2010.
172. Lei, H., Mi, L., Zhou, X., Chen, J., Hu, J., Guo, S., Zhang, Y., Adsorption of double-stranded DNA to graphene oxide preventing enzymatic digestion. *Nanoscale*, 3, 9, 3888–3892, 2011.
173. Xu, F., Shi, H., He, X., Wang, K., Ye, X., Yan, L.A., Wei, S., A facile graphene oxide-based DNA polymerase assay. *Analyst*, 137, 17, 3989–3994, 2012.
174. He, Y., Xing, X., Tang, H., Pang, D., Graphene Oxide-Based Fluorescent Biosensor for Protein Detection *via* Terminal Protection of Small-Molecule-Linked DNA. *Small*, 9, 12, 2097–2101, 2013.

175. Zhou, J., Lu, Q., Tong, Y., Wei, W., Liu, S., Detection of DNA damage by using hairpin molecular beacon probes and graphene oxide. *Talanta*, 99, 625–630, 2012, http://dx.doi.org/10.1016/j.talanta.2012.06.049.
176. Zhao, Q., Wu, M., Chris Le, X., Li, X.-F., Applications of aptamer affinity chromatography. *TrAC Trends Anal. Chem.*, 41, 46–57, 2012, http://dx.doi.org/10.1016/j.trac.2012.08.005.
177. Meyer, M., Scheper, T., Walter, J.-G., Aptamers: Versatile probes for flow cytometry. *Appl. Microbiol. Biotechnol.*, 97, 16, 7097–7109, 2013.
178. Mosing, R. and Bowser, M., Isolating Aptamers Using Capillary Electrophoresis–SELEX (CE–SELEX), in: *Nucleic Acid and Peptide Aptamers*, G. Mayer (Ed.), pp. 33–43, Humana Press, Totowa, NJ, 2009.
179. Ashley, J. and Li, S.F.Y., An aptamer based surface plasmon resonance biosensor for the detection of bovine catalase in milk. *Biosens. Bioelectron.*, 48, 126–131, 2013, http://dx.doi.org/10.1016/j.bios.2013.04.009.
180. Jiao, F., Fan, H., Yang, G., Zhang, F., He, P., Directly investigating the interaction between aptamers and thrombin by atomic force microscopy. *J. Mol. Recognit.*, 26, 12, 672–678, 2013.
181. Ellington, A.D. and Szostak, J.W., *In vitro* selection of RNA molecules that bind specific ligands. *Nature*, 346, 6287, 818–822, 1990.
182. Robertson, D.L. and Joyce, G.F., Selection *in vitro* of an RNA enzyme that specifically cleaves single-stranded DNA. *Nature*, 344, 6265, 467–468, 1990.
183. Hernandez, F.J. and Ozalp, V.C., Graphene and Other Nanomaterial-Based Electrochemical Aptasensors. *Biosensors*, 2, 1, 1, 2012.
184. Iliuk, A.B., Hu, L., Tao, W.A., Aptamer in Bioanalytical Applications. *Anal. Chem.*, 83, 12, 4440–4452, 2011.
185. Yao, C., Zhu, T., Qi, Y., Zhao, Y., Xia, H., Fu, W., Development of a Quartz Crystal Microbalance Biosensor with Aptamers as Bio-recognition Element. *Sensors*, 10, 6, 5859, 2010.
186. Han, K., Liang, Z., Zhou, N., Design Strategies for Aptamer-Based Biosensors. *Sensors*, 10, 5, 4541, 2010.
187. Yuan, Y., Gou, X., Yuan, R., Chai, Y., Zhuo, Y., Ye, X., Gan, X., Graphene-promoted 3,4,9,10-perylenetetracarboxylic acid nanocomposite as redox probe in label-free electrochemical aptasensor. *Biosens. Bioelectron.*, 30, 1, 123–127, 2011, http://dx.doi.org/10.1016/j.bios.2011.08.041.
188. Liang, J., Chen, Z., Guo, L., Li, L., Electrochemical sensing of l-histidine based on structure-switching DNAzymes and gold nanoparticle-graphene nanosheet composites. *Chem. Commun.*, 47, 19, 5476–5478, 2011.
189. Lü, M., Li, J., Yang, X., Zhang, C., Yang, J., Hu, H., Wang, X., Applications of graphene-based materials in environmental protection and detection. *Chin. Sci. Bull.*, 58, 22, 2698–2710, 2013.
190. Zhang, Y. and Jiang, W., Decorating graphene sheets with gold nanoparticles for the detection of sequence-specific DNA. *Electrochim. Acta*, 71, 239–245, 2012, http://dx.doi.org/10.1016/j.electacta.2012.03.136.

191. Li, M., Zhou, X., Ding, W., Guo, S., Wu, N., Fluorescent aptamer-functionalized graphene oxide biosensor for label-free detection of mercury(II). *Biosens. Bioelectron.*, 41, 889–893, 2013, http://dx.doi.org/10.1016/j.bios.2012.09.060.
192. Liu, J.-J., Song, X.-R., Wang, Y.-W., Chen, G.-N., Yang, H.-H., A graphene oxide (GO)-based molecular beacon for DNA-binding transcription factor detection. *Nanoscale*, 4, 12, 3655–3659, 2012.
193. Wen, Y., Xing, F., He, S., Song, S., Wang, L., Long, Y., Li, D., Fan, C., A graphene-based fluorescent nanoprobe for silver(i) ions detection by using graphene oxide and a silver-specific oligonucleotide. *Chem. Commun.*, 46, 15, 2596–2598, 2010.
194. Sun, W., Shi, S., Yao, T., Graphene oxide-Ru complex for label-free assay of DNA sequence and potassium ions viafluorescence resonance energy transfer. *Anal. Methods*, 3, 11, 2472–2474, 2011.
195. Fu, X., Lou, T., Chen, Z., Lin, M., Feng, W., Chen, L., "Turn-on" Fluorescence Detection of Lead Ions Based on Accelerated Leaching of Gold Nanoparticles on the Surface of Graphene. *ACS Appl. Mater. Interfaces*, 4, 2, 1080–1086, 2012.
196. Kundu, A., Layek, R.K., Kuila, A., Nandi, A.K., Highly Fluorescent Graphene Oxide-Poly(vinyl alcohol) Hybrid: An Effective Material for Specific Au3+ Ion Sensors. *ACS Appl. Mater. Interfaces*, 4, 10, 5576–5582, 2012.
197. Zhao, X.-H., Kong, R.-M., Zhang, X.-B., Meng, H.-M., Liu, W.-N., Tan, W., Shen, G.-L., Yu, R.-Q., Graphene–DNAzyme Based Biosensor for Amplified Fluorescence "Turn-On" Detection of Pb2+ with a High Selectivity. *Anal. Chem.*, 83, 13, 5062–5066, 2011.
198. Zhao, Z.-Q., Chen, X., Yang, Q., Liu, J.-H., Huang, X.-J., Selective adsorption toward toxic metal ions results in selective response: Electrochemical studies on a polypyrrole/reduced graphene oxide nanocomposite. *Chem. Commun.*, 48, 16, 2180–2182, 2012.
199. Zhou, H., Wang, X., Yu, P., Chen, X., Mao, L., Sensitive and selective voltammetric measurement of Hg2+ by rational covalent functionalization of graphene oxide with cysteamine. *Analyst*, 137, 2, 305–308, 2012.
200. Chang, J., Zhou, G., Christensen, E., Heideman, R., Chen, J., Graphene-based sensors for detection of heavy metals in water: A review. *Anal. Bioanal. Chem.*, 406, 16, 3957–3975, 2014.
201. Xu, K., Tian, X., Wu, C., Liu, J., Li, M., Sun, Y., Wei, F., Fabrication of single-walled carbon nanotube-based highly sensitive gas sensors. *Sci. China Technol. Sci.*, 56, 1, 32–35, 2013.
202. Mohammed Javad Kiani, M.T., Akbari, E., Karimi, H., Haran, F.K.C., Graphene nanoribbon based gas sensor. *Key Eng. Mater.*, 553, 7–11, 2013.
203. Amiri, I.S., Ahsan, R., Shahidinejad, A., Ali, J., Yupapin, P.P., Characterisation of bifurcation and chaos in silicon microring resonator. *Commun., IET*, 6, 16, 2671–2675, 2012.

204. Amiri, I.S. and Ali, J., Nano Particle Trapping by Ultra-Short Tweezer and Wells Using Microring Resonator Interferometer System for Spectroscopy Application. *Nanosci. Nanotechnol. Lett.*, 5, 8, 850–856, 2013.
205. Rahmani, M., Ahmadi, M.T., Karimi, H.F.A., Kiani, M.J., Akbari, E., Ismail, R., Analytical Modeling of Monolayer Graphene-based NO2 Sensor. *Sens. Lett.*, 11, 2, 270–275, 2013.
206. Akbari, E., Arora, V.K., Enzevaee, A., Ahmadi, M.T., Saeidmanesh, M., Khaledian, M., Ka, rimi, H., and Yusof, R., An analytical approach to evaluate the performance of graphene and carbon nanotubes for NH3 gas sensor applications. *Beilstein J. Nanotechnol.*, 5, 726–734, 2014.
207. Robinson, J.A., Snow, E.S., Bădescu, Ş.C., Reinecke, T.L., Perkins, F.K., Role of Defects in Single-Walled Carbon Nanotube Chemical Sensors. *Nano Lett.*, 6, 8, 1747–1751, 2006.
208. Basiuk, V.A., Reactivity of Carboxylic Groups on Armchair and Zigzag Carbon Nanotube Tips: A Theoretical Study of Esterification with Methanol. *Nano Lett.*, 2, 8, 835–839, 2002.
209. Huang, B., Li, Z., Liu, Z., Zhou, G., Hao, S., Wu, J., Gu, B.-L., Duan, W., Adsorption of Gas Molecules on Graphene Nanoribbons and Its Implication for Nanoscale Molecule Sensor. *J. Phys. Chem. C*, 112, 35, 13442–13446, 2008.
210. El-Barbary, A.A., Eid, K.M., Kamel, M.A., Taha, H.O., Ismail, G.H., Adsorption of CO, CO2, NO and NO2 on Boron Nitride Nanotubes: DFT Study. *J. Surf. Eng. Mater. Adv. Technol.*, 5, 3, 154–161, 2015.
211. Mao, K., Wu, D., Li, Y., Ma, H., Ni, Z., Yu, H., Luo, C., Wei, Q., Du, B., Label-free electrochemical immunosensor based on graphene/methylene blue nanocomposite. *Anal. Biochem.*, 422, 1, 22–27, 2012, http://dx.doi.org/10.1016/j.ab.2011.12.047.
212. Jang, H.D., Kim, S.K., Chang, H., Choi, J.-W., 3D label-free prostate specific antigen (PSA) immunosensor based on graphene–gold composites. *Biosens. Bioelectron.*, 63, 546–551, 2015, http://dx.doi.org/10.1016/j.bios.2014.08.008.
213. Chen, M., Meng, Y., Zhang, W., Zhou, J., Xie, J., Diao, G., β-Cyclodextrin polymer functionalized reduced-graphene oxide: Application for electrochemical determination imidacloprid. *Electrochim. Acta*, 108, 1–9, 2013, http://dx.doi.org/10.1016/j.electacta.2013.06.050.
214. Kang, X., Wang, J., Wu, H., Liu, J., Aksay, I.A., Lin, Y., A graphene-based electrochemical sensor for sensitive detection of paracetamol. *Talanta*, 81, 3, 754–759, 2010, http://dx.doi.org/10.1016/j.talanta.2010.01.009.
215. Fan, Y., Liu, J.-H., Lu, H.-T., Zhang, Q., Electrochemical behavior and voltammetric determination of paracetamol on Nafion/TiO2–graphene modified glassy carbon electrode. *Colloids Surf., B: Biointerfaces*, 85, 2, 289–292, 2011, http://dx.doi.org/10.1016/j.colsurfb.2011.02.041.

216. Huamin, Q., Lulu, F., Li, X., Li, L., Min, S., Chuannan, L., Determination sulfamethoxazole based chemiluminescence and chitosan/graphene oxide-molecularly imprinted polymers. *Carbohydr. Polym.*, 92, 1, 394–399, 2013, http://dx.doi.org/10.1016/j.carbpol.2012.09.092.
217. Karuwan, C., Sriprachuabwong, C., Wisitsoraat, A., Phokharatkul, D., Sritongkham, P., Tuantranont, A., Inkjet-printed graphene-poly(3,4-ethylenedioxythiophene):poly(styrene-sulfonate) modified on screen printed carbon electrode for electrochemical sensing of salbutamol. *Sens. Actuators B: Chem.*, 161, 1, 549–555, 2012, http://dx.doi.org/10.1016/j.snb.2011.10.074.
218. Lei, W., Wu, Q., Si, W., Gu, Z., Zhang, Y., Deng, J., Hao, Q., Electrochemical determination of imidacloprid using poly(carbazole)/chemically reduced graphene oxide modified glassy carbon electrode. *Sens. Actuators B: Chem.*, 183, 102–109, 2013, http://dx.doi.org/10.1016/j.snb.2013.03.118.
219. Yang, X., Wu, F., Chen, D.-Z., Lin, H.-W., An electrochemical immunosensor for rapid determination of clenbuterol by using magnetic nanocomposites to modify screen printed carbon electrode based on competitive immunoassay mode. *Sens. Actuators B: Chem.*, 192, 529–535, 2014, http://dx.doi.org/10.1016/j.snb.2013.11.011.
220. Yan, H., Xiao, H., Xie, Q., Liu, J., Sun, L., Zhou, Y., Zhang, Y., Chao, L., Chen, C., Yao, S., Simultaneous electroanalysis of isoniazid and uric acid at poly(-sulfosalicylic acid)/electroreduced carboxylated graphene modified glassy carbon electrode. *Sens. Actuators B: Chem.*, 207, Part A, 167–176, 2015, http://dx.doi.org/10.1016/j.snb.2014.10.002.
221. Cheemalapati, S., Chen, S.-M., Ali, M.A., Al-Hemaid, F.M.A., Enhanced electrocatalytic oxidation of isoniazid at electrochemically modified rhodium electrode for biological and pharmaceutical analysis. *Colloids Surf., B: Biointerfaces*, 121, 444–450, 2014, http://dx.doi.org/10.1016/j.colsurfb.2014.06.035.
222. Wang, H., Zhang, Y., Li, H., Du, B., Ma, H., Wu, D., Wei, Q., A silver–palladium alloy nanoparticle-based electrochemical biosensor for simultaneous detection of ractopamine, clenbuterol and salbutamol. *Biosens. Bioelectron.*, 49, 14–19, 2013, http://dx.doi.org/10.1016/j.bios.2013.04.041.
223. Huang, J., Lin, Q., Zhang, X., He, X., Xing, X., Lian, W., Zuo, M., Zhang, Q., Electrochemical immunosensor based on polyaniline/poly (acrylic acid) and Au-hybrid graphene nanocomposite for sensitivity enhanced detection of salbutamol. *Food Res. Int.*, 44, 1, 92–97, 2011, http://dx.doi.org/10.1016/j.foodres.2010.11.006.
224. Yang, S., Wang, G., Li, G., Du, J., Qu, L., Decoration of graphene modified carbon paste electrode with flower-globular terbium hexacyanoferrate for nanomolar detection of rutin. *Electrochim. Acta*, 144, 268–274, 2014, http://dx.doi.org/10.1016/j.electacta.2014.08.050.
225. Sánchez Arribas, A., Martínez-Fernández, M., Moreno, M., Bermejo, E., Zapardiel, A., Chicharro, M., Analysis of total polyphenols in wines by FIA with highly stable amperometric detection using carbon nanotube-modified

electrodes. *Food Chem.*, 136, 3–4, 1183–1192, 2013, http://dx.doi.org/10.1016/j.foodchem.2012.09.027.

226. Eremia, S.A.V., Vasilescu, I., Radoi, A., Litescu, S.-C., Radu, G.-L., Disposable biosensor based on platinum nanoparticles-reduced graphene oxide-laccase biocomposite for the determination of total polyphenolic content. *Talanta*, 110, 164–170, 2013, http://dx.doi.org/10.1016/j.talanta.2013.02.029.
227. Yang, S., Li, G., Wang, G., Zhao, J., Qiao, Z., Qu, L., Decoration of chemically reduced graphene oxide modified carbon paste electrode with yttrium hexacyanoferrate nanoparticles for nanomolar detection of rutin. *Sens. Actuators B: Chem.*, 206, 126–132, 2015, http://dx.doi.org/10.1016/j.snb.2014.09.027.
228. Zhang, K., Xu, J., Zhu, X., Lu, L., Duan, X., Hu, D., Dong, L., Sun, H., Gao, Y., Wu, Y., Poly(3,4-ethylenedioxythiophene) nanorods grown on graphene oxide sheets as electrochemical sensing platform for rutin. *J. Electroanal. Chem.*, 739, 66–72, 2015, http://dx.doi.org/10.1016/j.jelechem.2014.12.013.
229. Yang, Y., Zhou, J., Zhang, H., Gai, P., Zhang, X., Chen, J., Electrochemical evaluation of total antioxidant capacities in fruit juice based on the guanine/graphene nanoribbon/glassy carbon electrode. *Talanta*, 106, 206–211, 2013, http://dx.doi.org/10.1016/j.talanta.2012.12.030.
230. Song, D., Xia, J., Zhang, F., Bi, S., Xiang, W., Wang, Z., Xia, L., Xia, Y., Li, Y., Xia, L., Multiwall carbon nanotubes-poly(diallyldimethylammonium chloride)-graphene hybrid composite film for simultaneous determination of catechol and hydroquinone. *Sens. Actuators B: Chem.*, 206, 111–118, 2015, http://dx.doi.org/10.1016/j.snb.2014.08.084.
231. Song, W., Li, D.-W., Li, Y.-T., Li, Y., Long, Y.-T., Disposable biosensor based on graphene oxide conjugated with tyrosinase assembled gold nanoparticles. *Biosens. Bioelectron.*, 26, 7, 3181–3186, 2011, http://dx.doi.org/10.1016/j.bios.2010.12.022.
232. Bernard, S. and Miele, P., Polymer-Derived Boron Nitride: A Review on the Chemistry, Shaping and Ceramic Conversion of Borazine Derivatives. *Materials*, 7, 11, 7436, 2014.
233. Wang, X., Nan, F., Zhao, J., Yang, T., Ge, T., Jiao, K., A label-free ultrasensitive electrochemical DNA sensor based on thin-layer MoS2 nanosheets with high electrochemical activity. *Biosens. Bioelectron.*, 64, 386–391, 2015, http://dx.doi.org/10.1016/j.bios.2014.09.030.
234. Pumera, M. and Loo, A.H., Layered transition-metal dichalcogenides (MoS2 and WS2) for sensing and biosensing. *TrAC Trends Anal. Chem.*, 61, 49–53, 2014, http://dx.doi.org/10.1016/j.trac.2014.05.009.
235. Farimani, A.B., Min, K., Aluru, N.R., DNA Base Detection Using a Single-Layer MoS2. *ACS Nano*, 8, 8, 7914–7922, 2014.
236. Huang, X., Zeng, Z., Bao, S., Wang, M., Qi, X., Fan, Z., Zhang, H., Solution-phase epitaxial growth of noble metal nanostructures on dispersible single-layer molybdenum disulfide nanosheets. *Nat. Commun.*, 4, 1444, 2013, http://www.nature.com/ncomms/journal/v4/n2/suppinfo/ncomms2472_S1.html.

237. Huang, X., Zeng, Z., Zhang, H., Metal dichalcogenide nanosheets: Preparation, properties and applications. *Chem. Soc. Rev.*, 42, 5, 1934–1946, 2013.
238. Hou, Y., Wen, Z., Cui, S., Guo, X., Chen, J., Constructing 2D Porous Graphitic C3N4 Nanosheets/Nitrogen-Doped Graphene/Layered MoS2 Ternary Nanojunction with Enhanced Photoelectrochemical Activity. *Adv. Mater.*, 25, 43, 6291–6297, 2013.
239. Linares, F.M.M., Nanotechnology: An urgent need. *Soft Nanosci. Lett.*, 3, 45–45, 2013.
240. Yao, Y., Tolentino, L., Yang, Z., Song, X., Zhang, W., Chen, Y., Wong, C.-P., High-Concentration Aqueous Dispersions of MoS2. *Adv. Funct. Mater.*, 23, 28, 3577–3583, 2013.
241. Li, H., Yin, Z., He, Q., Li, H., Huang, X., Lu, G., Fam, D.W.H., Tok, A.I.Y., Zhang, Q., Zhang, H., Fabrication of Single- and Multilayer MoS2 Film-Based Field-Effect Transistors for Sensing NO at Room Temperature. *Small*, 8, 1, 63–67, 2012.
242. Liu, K., Feng, J., Kis, A., Radenovic, A., Atomically Thin Molybdenum Disulfide Nanopores with High Sensitivity for DNA Translocation. *ACS Nano*, 8, 3, 2504–2511, 2014.
243. Zhang, S.-L., Choi, H.-H., Yue, H.-Y., Yang, W.-C., Controlled exfoliation of molybdenum disulfide for developing thin film humidity sensor. *Curr. Appl. Phys.*, 14, 3, 264–268, 2014, http://dx.doi.org/10.1016/j.cap.2013.11.031.
244. Perkins, F.K., Friedman, A.L., Cobas, E., Campbell, P.M., Jernigan, G.G., Jonker, B.T., Chemical Vapor Sensing with Monolayer MoS2. *Nano Lett.*, 13, 2, 668–673, 2013.
245. Late, D.J., Huang, Y.-K., Liu, B., Acharya, J., Shirodkar, S.N., Luo, J., Yan, A., Charles, D., Waghmare, U.V., Dravid, V.P., Rao, C.N.R., Sensing Behavior of Atomically Thin-Layered MoS2 Transistors. *ACS Nano*, 7, 6, 4879–4891, 2013.
246. Lee, K., Gatensby, R., McEvoy, N., Hallam, T., Duesberg, G.S., High-Performance Sensors Based on Molybdenum Disulfide Thin Films. *Adv. Mater.*, 25, 46, 6699–6702, 2013.
247. Blake, P., Brimicombe, P.D., Nair, R.R., Booth, T.J., Jiang, D., Schedin, F., Ponomarenko, L.A., Morozov, S.V., Gleeson, H.F., Hill, E.W., Geim, A.K., Novoselov, K.S., Graphene-Based Liquid Crystal Device. *Nano Lett.*, 8, 6, 1704–1708, 2008.
248. Wang, J., Liang, M., Fang, Y., Qiu, T., Zhang, J., Zhi, L., Rod-Coating: Towards Large-Area Fabrication of Uniform Reduced Graphene Oxide Films for Flexible Touch Screens. *Adv. Mater.*, 24, 21, 2874–2878, 2012.
249. Han, W., Kawakami, R.K., Gmitra, M., Fabian, J., Graphene spintronics. *Nat. Nano*, 9, 10, 794–807, 2014.
250. Guo, C.X. and Li, C.M., A self-assembled hierarchical nanostructure comprising carbon spheres and graphene nanosheets for enhanced supercapacitor performance. *Energy Environ. Sci.*, 4, 11, 4504–4507, 2011.

251. Sun, Y., Shao, D., Chen, C., Yang, S., Wang, X., Highly Efficient Enrichment of Radionuclides on Graphene Oxide-Supported Polyaniline. *Environ. Sci. Technol.*, 47, 17, 9904–9910, 2013.
252. Yang, H., Sun, L., Zhai, J., Li, H., Zhao, Y., Yu, H., *In situ* controllable synthesis of magnetic Prussian blue/graphene oxide nanocomposites for removal of radioactive cesium in water. *J. Mater. Chem. A*, 2, 2, 326–332, 2014.
253. Wang, Q.H., Kalantar-Zadeh, K., Kis, A., Coleman, J.N., Strano, M.S., Electronics and optoelectronics of two-dimensional transition metal dichalcogenides. *Nat. Nanotechnol.*, 7, 11, 699–712, 2012.
254. Qin, Z., Taylor, M., Hwang, M., Bertoldi, K., Buehler, M.J., Effect of Wrinkles on the Surface Area of Graphene: Toward the Design of Nanoelectronics. *Nano Lett.*, 14, 11, 6520–6525, 2014.
255. Singh, R.N. and Awasthi, R., Graphene support for enhanced electrocatalytic activity of Pd for alcohol oxidation. *Catal. Sci. Technol.*, 1, 5, 778–783, 2011.
256. Cohen-Tanugi, D. and Grossman, J.C., Water Desalination across Nanoporous Graphene. *Nano Lett.*, 12, 7, 3602–3608, 2012.
257. Yang, K., Li, Y., Tan, X., Peng, R., Liu, Z., Behavior and Toxicity of Graphene and Its Functionalized Derivatives in Biological Systems. *Small*, 9, 9–10, 1492–1503, 2013.
258. Weichung, C.W., Siang, C.S., Mei, L.P.S., SP, C., Pheng, S., Effects of a pharmaceutical care model on medication adherence and glycemic control of people with type 2 diabetes. *Patient Prefer. Adherence*, 8, 1185–1194, 2014.
259. DiSanto, R.M., Subramanian, V., Gu, Z., Recent advances in nanotechnology for diabetes treatment. *Wiley Interdiscip. Rev.: Nanomed. Nanobiotechnol.*, 7, 4, 548–564, 2015.
260. Yang, K., Wan, J., Zhang, S., Tian, B., Zhang, Y., Liu, Z., The influence of surface chemistry and size of nanoscale graphene oxide on photothermal therapy of cancer using ultra-low laser power. *Biomaterials*, 33, 7, 2206–2214, 2012, http://dx.doi.org/10.1016/j.biomaterials.2011.11.064.
261. Katsnelson, M.I., Graphene: Carbon in two dimensions. *Mater. Today*, 10, 1, 20–27, 2007, https://doi.org/10.1016/S1369-7021(06)71788-6.
262. Novoselov, K.S., Geim, A.K., Morozov, S.V., Jiang, D., Katsnelson, M.I., Grigorieva, I.V., Dubonos, S.V., Firsov, A.A., Two-dimensional gas of massless Dirac fermions in graphene. *Nature*, 438, 7065, 197–200, 2005.
263. Harima, Y., Setodoi, S., Imae, I., Komaguchi, K., Ooyama, Y., Ohshita, J., Mizota, H., Yano, J., Electrochemical reduction of graphene oxide in organic solvents. *Electrochim. Acta*, 56, 15, 5363–5368, 2011, http://dx.doi.org/10.1016/j.electacta.2011.03.117.
264. Kauppila, J., Kunnas, P., Damlin, P., Viinikanoja, A., Kvarnström, C., Electrochemical reduction of graphene oxide films in aqueous and organic solutions. *Electrochim. Acta*, 89, 84–89, 2013, http://dx.doi.org/10.1016/j.electacta.2012.10.153.

265. Rameshkumar, P., Praveen, R., Ramaraj, R., Electroanalysis of oxygen reduction and formic acid oxidation using reduced graphene oxide/gold nanostructures modified electrode. *J. Electroanal. Chem.*, 754, 118–124, 2015, http://dx.doi.org/10.1016/j.jelechem.2015.07.011.
266. Liu, Y., Huang, Z., Xie, Q., Sun, L., Gu, T., Li, Z., Bu, L., Yao, S., Tu, X., Luo, X., Luo, S., Electrodeposition of electroreduced graphene oxide-Au nanoparticles composite film at glassy carbon electrode for anodic stripping voltammetric analysis of trace arsenic(III). *Sens. Actuators B: Chem.*, 188, 894–901, 2013, http://dx.doi.org/10.1016/j.snb.2013.07.113.
267. Chen, L., Tang, Y., Wang, K., Liu, C., Luo, S., Direct electrodeposition of reduced graphene oxide on glassy carbon electrode and its electrochemical application. *Electrochem. Commun.*, 13, 2, 133–137, 2011, http://dx.doi.org/10.1016/j.elecom.2010.11.033.
268. Raj, M.A. and John, S.A., Fabrication of Electrochemically Reduced Graphene Oxide Films on Glassy Carbon Electrode by Self-Assembly Method and Their Electrocatalytic Application. *J. Phys. Chem. C*, 117, 8, 4326–4335, 2013.
269. Basirun, W.J., Sookhakian, M., Baradaran, S., Mahmoudian, M.R., Ebadi, M., Solid-phase electrochemical reduction of graphene oxide films in alkaline solution. *Nanoscale Res. Lett.*, 8, 1, 1–9, 2013.
270. Balamurugan, D., Muniyandi, R., Shen-Ming, C., Saraswathi, R., Electrochemically reduced graphene oxide/neodynium hexacyanoferrate modified electrodes for electrochemical detection of paracetamol. *Int. J. Electrochem. Sci.*, 7, 3339–3349, 2012.
271. Zeng, F., Sun, Z., Sang, X., Diamond, D., Lau, K.T., Liu, X., Su, D.S., *In Situ* One-Step Electrochemical Preparation of Graphene Oxide Nanosheet-Modified Electrodes for Biosensors. *ChemSusChem*, 4, 11, 1587–1591, 2011.
272. Keagan, P., Salma, Z., Nazeem, J., Nuralli, M., Prisilla, G.B., Emmanuel, I.I., Electrochemical reduced graphene oxide pencil-graphite *in situ* plated bismuth film electrode for the determination of trace metals by anodic stripping voltammetry. *Int. J. Electrochem. Sci.*, 9, 736–759, 2014.
273. Shan, C., Yang, H., Song, J., Han, D., Ivaska, A., Niu, L., Direct Electrochemistry of Glucose Oxidase and Biosensing for Glucose Based on Graphene. *Anal. Chem.*, 81, 6, 2378–2382, 2009.
274. Unnikrishnan, B., Palanisamy, S., Chen, S.-M., A simple electrochemical approach to fabricate a glucose biosensor based on graphene–glucose oxidase biocomposite. *Biosens. Bioelectron.*, 39, 1, 70–75, 2013, http://dx.doi.org/10.1016/j.bios.2012.06.045.
275. Wu, H., Wang, J., Kang, X., Wang, C., Wang, D., Liu, J., Aksay, I.A., Lin, Y., Glucose biosensor based on immobilization of glucose oxidase in platinum nanoparticles/graphene/chitosan nanocomposite film. *Talanta*, 80, 1, 403–406, 2009, http://dx.doi.org/10.1016/j.talanta.2009.06.054.
276. Liu, Y., Yu, D., Zeng, C., Miao, Z., Dai, L., Biocompatible Graphene Oxide-Based Glucose Biosensors. *Langmuir*, 26, 9, 6158–6160, 2010.

277. Alwarappan, S., Liu, C., Kumar, A., Li, C.-Z., Enzyme-Doped Graphene Nanosheets for Enhanced Glucose Biosensing. *J. Phys. Chem. C*, 114, 30, 12920–12924, 2010.
278. Pumera, M., Scipioni, R., Iwai, H., Ohno, T., Miyahara, Y., Boero, M., A Mechanism of Adsorption of β-Nicotinamide Adenine Dinucleotide on Graphene Sheets: Experiment and Theory. *Chem.—Eur. J.*, 15, 41, 10851–10856, 2009.
279. Dong, X.-C., Xu, H., Wang, X.-W., Huang, Y.-X., Chan-Park, M.B., Zhang, H., Wang, L.-H., Huang, W., Chen, P., 3D Graphene–Cobalt Oxide Electrode for High-Performance Supercapacitor and Enzymeless Glucose Detection. *ACS Nano*, 6, 4, 3206–3213, 2012.
280. Shan, C., Yang, H., Han, D., Zhang, Q., Ivaska, A., Niu, L., Graphene/AuNPs/chitosan nanocomposites film for glucose biosensing. *Biosens. Bioelectron.*, 25, 5, 1070–1074, 2010, http://dx.doi.org/10.1016/j.bios.2009.09.024.
281. Zhang, Y., Liu, S., Wang, L., Qin, X., Tian, J., Lu, W., Chang, G., Sun, X., One-pot green synthesis of Ag nanoparticles-graphene nanocomposites and their applications in SERS, H2O2, and glucose sensing. *RSC Adv.*, 2, 2, 538–545, 2012.
282. Lu, C.-H., Yang, H.-H., Zhu, C.-L., Chen, X., Chen, G.-N., A Graphene Platform for Sensing Biomolecules. *Angew. Chem.*, 121, 26, 4879–4881, 2009.
283. Dong, H., Gao, W., Yan, F., Ji, H., Ju, H., Fluorescence Resonance Energy Transfer between Quantum Dots and Graphene Oxide for Sensing Biomolecules. *Anal. Chem.*, 82, 13, 5511–5517, 2010.
284. Balapanuru, J., Yang, J.-X., Xiao, S., Bao, Q., Jahan, M., Polavarapu, L., Wei, J., Xu, Q.-H., Loh, K.P., A Graphene Oxide–Organic Dye Ionic Complex with DNA-Sensing and Optical-Limiting Properties. *Angew. Chem.*, 122, 37, 6699–6703, 2010.
285. Li, F., Feng, Y., Zhao, C., Li, P., Tang, B., A sensitive graphene oxide-DNA based sensing platform for fluorescence "turn-on" detection of bleomycin. *Chem. Commun.*, 48, 1, 127–129, 2012.
286. Chang, H., Tang, L., Wang, Y., Jiang, J., Li, J., Graphene Fluorescence Resonance Energy Transfer Aptasensor for the Thrombin Detection. *Anal. Chem.*, 82, 6, 2341–2346, 2010.
287. Liu, X., Aizen, R., Freeman, R., Yehezkeli, O., Willner, I., Multiplexed Aptasensors and Amplified DNA Sensors Using Functionalized Graphene Oxide: Application for Logic Gate Operations. *ACS Nano*, 6, 4, 3553–3563, 2012.
288. Zhao, X.-H., Ma, Q.-J., Wu, X.-X., Zhu, X., Graphene oxide-based biosensor for sensitive fluorescence detection of DNA based on exonuclease III-aided signal amplification. *Anal. Chim. Acta*, 727, 67–70, 2012, http://dx.doi.org/10.1016/j.aca.2012.03.044.
289. Liu, B., Sun, Z., Zhang, X., Liu, J., Mechanisms of DNA Sensing on Graphene Oxide. *Anal. Chem.*, 85, 16, 7987–7993, 2013.

290. Zhang, M., Yin, B.-C., Tan, W., Ye, B.-C., A versatile graphene-based fluorescence "on/off" switch for multiplex detection of various targets. *Biosens. Bioelectron.*, 26, 7, 3260–3265, 2011, http://dx.doi.org/10.1016/j.bios.2010.12.037.
291. He, S., Song, B., Li, D., Zhu, C., Qi, W., Wen, Y., Wang, L., Song, S., Fang, H., Fan, C., A Graphene Nanoprobe for Rapid, Sensitive, and Multicolor Fluorescent DNA Analysis. *Adv. Funct. Mater.*, 20, 3, 453–459, 2010.
292. Lin, X., Ni, Y., Kokot, S., A novel electrochemical sensor for the analysis of β-agonists: The poly(acid chrome blue K)/graphene oxide-nafion/glassy carbon electrode. *J. Hazard. Mater.*, 260, 508–517, 2013, http://dx.doi.org/10.1016/j.jhazmat.2013.06.004.
293. Liu, Z., Zhang, A., Guo, Y., Dong, C., Electrochemical sensor for ultrasensitive determination of isoquercitrin and baicalin based on DM-β-cyclodextrin functionalized graphene nanosheets. *Biosens. Bioelectron.*, 58, 242–248, 2014, http://dx.doi.org/10.1016/j.bios.2014.02.051.
294. Ghalkhani, M. and Shahrokhian, S., Adsorptive stripping differential pulse voltammetric determination of mebendazole at a graphene nanosheets and carbon nanospheres/chitosan modified glassy carbon electrode. *Sens. Actuators B: Chem.*, 185, 669–674, 2013, http://dx.doi.org/10.1016/j.snb.2013.05.054.
295. Wang, F., Wu, Y., Lu, K., Ye, B., A sensitive voltammetric sensor for taxifolin based on graphene nanosheets with certain orientation modified glassy carbon electrode. *Sens. Actuators B: Chem.*, 208, 188–194, 2015, http://dx.doi.org/10.1016/j.snb.2014.11.008.
296. Zhang, Y.-Q., Fan, Y.-J., Cheng, L., Fan, L.-L., Wang, Z.-Y., Zhong, J.-P., Wu, L.-N., Shen, X.-C., Shi, Z.-J., A novel glucose biosensor based on the immobilization of glucose oxidase on layer-by-layer assembly film of copper phthalocyanine functionalized graphene. *Electrochim. Acta*, 104, 178–184, 2013, http://dx.doi.org/10.1016/j.electacta.2013.04.099.
297. Zhu, J., Xu, M., Meng, X., Shang, K., Fan, H., Ai, S., Electro-enzymatic degradation of carbofuran with the graphene oxide–Fe3O4–hemoglobin composite in an electrochemical reactor. *Process Biochem.*, 47, 12, 2480–2486, 2012, http://dx.doi.org/10.1016/j.procbio.2012.10.006.
298. Wu, Q., Zhao, G., Feng, C., Wang, C., Wang, Z., Preparation of a graphene-based magnetic nanocomposite for the extraction of carbamate pesticides from environmental water samples. *J. Chromatogr. A*, 1218, 44, 7936–7942, 2011, http://dx.doi.org/10.1016/j.chroma.2011.09.027.

Abbreviation Used for Tables and Figures

AgNps Silver Nanoparticles
AGNRs Armchair graphene nanoribbons
AuNps Gold Nanoparticles

CF	Carboxyfluorescein
CV	Cyclic voltammetry
CVD	Chemical vaporization deposition
FAM	6-Carboxyfluorescein
GCE	Glassy carbon electrode
GO	Graphene oxide
GOD	Glucose oxidase
QDs	Quantum dots
QHE	Quantum hall effect
RGO	Reduced graphene oxide
ROX	Rhodamine-X
SCE	Saturated calomel electrode

9
Germanene

Mohd Imran Ahamed and Naushad Anwar*

Department of Chemistry, Faculty of Science, Aligarh Muslim University, Aligarh, India

Abstract
Over more than two decades, the researchers and scientific community have been extensively studied on two-dimensional (2D) materials such as graphene, phosphorene, silicene, germanene etc. as these materials have enormous properties viz., optical, mechanical, thermal and electronic and applications in various fields such as in optical devices, sensing, catalysis, nano-electronics, semiconductors, supercapacitors, gas adsorption sheets, electrical and mechanical exfoliation etc. In aforementioned 2D materials, germanene is one of them that consist of a number of germanium atoms in their buckled hexagonal honeycomb structures with sp^3 hybridization, show germanene is more flexible over other 2D materials like graphene and thus match easily with the substrate lattices. Germanene exhibits strong spin–orbit coupling, extremely high mobility and massless fermion characteristics that lead to draw enormous interest for high performance devices. It has been reported by researchers that covalent bonding between the two single layer sheets of germanene offers the possibility to open a band gap and intrinsic magnetism that makes germanene attractive for future prospective. On the other hand, the spin–orbit gap in germanene is much larger than the other 2D materials like graphene that make it an ideal prospect to show the Quantum Spin Hall (QSH) effect at experimental temperature that have been studied. In this chapter, we have discussed the structure, properties and applications of germanene that will be helpful for new comers in near future research.

Keywords: Germanene, structure, hybridization, fundamental properties, applications

**Corresponding author*: naushadanwar85@gmail.com

Inamuddin, Rajender Boddula, Mohd Imran Ahamed and Abdullah M. Asiri (eds.)
Monoelements: Properties and Applications, (235–260) © 2020 Scrivener Publishing LLC

9.1 Introduction

The successfully isolation of graphene from graphite since 2004, a series of 2D materials such as silicene, germanene, stannene, phosphorene, antimonene into account from their respective elements of groups 14 and 15 in recent years [1–4]. All the elements of group 14 comprise four electrons in their outermost orbitals as the general valence shell electronic configuration of these group elements is ns^2np^2. The crystal structure of silicon and germanium corresponds to as diamond lattice structure as the carbon atoms are tetrahedrally arranged with four neighbors by a single covalent bond and all the bonds are equivalent leads to sp^3 hybridization (s, p_x, p_y, p_z characters). Graphene consists of a large monolayer sheet of sp^2 hybridized carbon atoms that possesses zero bandgap and large specific surface area [5–9]. Unlike graphene, the energy required to hybridize the three p-orbitals is very low as compared to free standing germanene and also due to the Ge–Ge bond length are much larger than that of C–C bond, the π-bond formation was found to be very difficult as the weak overlapping of p_z orbital in germanene and hence, mixed sp^2–sp^3 hybridization results the buckled structure of germanene. Germanene exhibits very interesting and specific electronic properties with graphene and silicene because it is also a zero bandgap semiconductor that spectacularly prevents its applications on electronic devices, such as field effect transistor [10–14].

With much abstraction with silicon, germanium is also used as a significant semiconductor material as in the conventional semiconductor industry and also possesses high carrier mobility, which provides a potential rewards for their demonstration in high-speed electronic devices. Germanene, the honeycomb structured Ge-atoms have been grown on silicon based nano-electronics would be much favored over graphene and recently a number of work have been performed by researchers for the synthesis, structural arrangements, properties and their potential applications [15–18]. First prediction and confirmation of germanene were performed by Takeda and Shiraishi *et al.* in 1994 that were originated from bulk silicon and germanium that is the 2D counterparts of graphene. They were also imagined that the graphene-like planar structure of germanene may not occur and may act as a poor metal even so, the low buckled form of germanene appears as stable and acts as a semi-metal [19]. Cahangirov *et al.* have studied on the physical properties of low buckled germanene nano-ribbons of the two armchair and zigzag forms and investigated that these ribbons can display interesting electronic and magnetic properties [20].

From these exciting theoretical investigations, Nakano *et al.* have experimentally performed their study which gains the soft synthesis of a monolayer of silicon sheet on a substrate, preserve for the synthesis of honeycomb nano-ribbons of Si and Ge [21]. Pang *et al.* also revealed in their study that these honeycomb ribbons exhibit fascinating electro-magnetic properties [22]. Recently, the hydrogenated germanium with free-standing monolayer, also known as germanane (GeH) has been synthesized using chemical process by Bianco *et al.* and the group have characterized its physical properties experimentally at room temperature and reported that these monolayer 2D materials possess high electron mobility as well as large direct band gap for on/off switching [23]. The further study made by Kawakami *et al.* that reported theoretically the direct bandgap and high mobility transport capability of germanane ~1.5 eV and ~18,000 $cm^2V^{-1}s^{-1}$ respectively, that make it potential applicable in the area of nanoscale digital electronics and opto-electronics [24].

Germanene has similar puckered honeycomb structure like graphene which is shown in Figure 9.1 [25]. Few differences have also occurs between germanene and graphene due to their differences in hexagonal structure. The hexagonal crystal structure in graphene is flat while germanene is buckled. The buckled Ge-atoms results to increase the stability of germanene and amended the carrier transport properties as compared to graphene [19, 20]. Germanene's crystal lattice contains two vertically separated sub-lattices, rather graphene's lattices are confined in a plane. The main difference arises due to the atomic properties of germanene as it possesses highest spin–orbit coupling. It means distinct energy levels in accordance with the spin of electrons. The value of spin–orbit gap in germanene has been reported as 24 meV is very high as compared to graphene to less than 0.05 meV which confirms an idealistic behavior of germanene to exhibit the QSH effect at experimental comprehensible temperature. Several investigations have been carried out by researchers on germanene that reveals some of peculiar properties of germanene such as strain-induced self-doping, electric-field/substrate-induced gaps, high-efficient thermoelectric performances and potential use in battery applications [26–32].

Few study also provides knowledge regarding the germanene nano-sheets which are the Ge counterparts of graphene, have been obtained experimentally consists a chair-like buckled confirmation in their basal plane and these nano-sheets confirms their Dirac-like electronic structures. Due to the discrete coupling of σ and π bonds and also the weak interactions of π–π bonds between the germanium atoms, germanene shows some new physical behavior in addition to graphene. Nevertheless

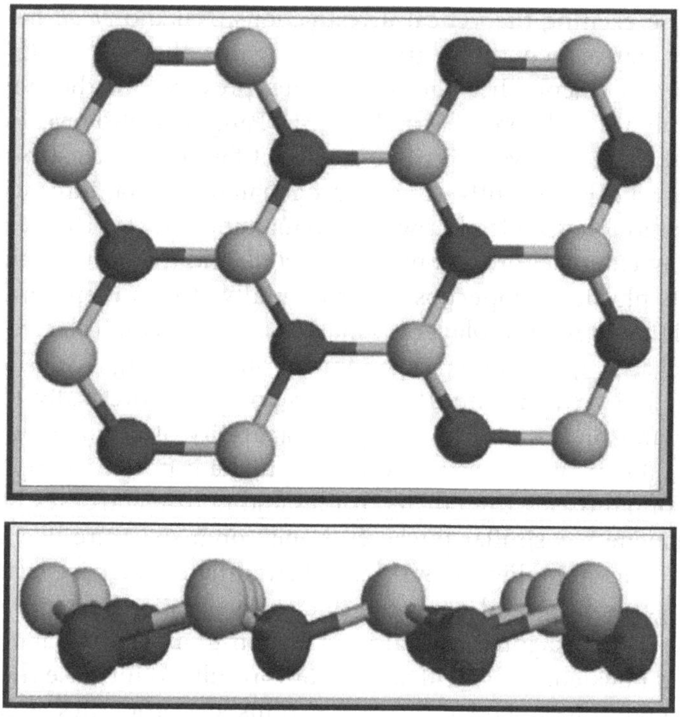

Figure 9.1 Ball and stick model of germanene. The honeycomb lattice is composed of two triangular sub-lattices (blue and brown atoms, respectively). Above panel: top view. Below panel: side view [25].

low band gap and also the less structural stability of germanene, which prevents its in direct applications in logic and high-performance switching devices. However, conjugation with a substance to form a hetero-structure is an effectual process to make it easier to open a bandgap in germanene, e.g., the van der Waals hetero-structure of germanane/germanene has forecasted the opening a band gap of 82–120 meV at the Dirac point because of their intrinsic interface dipole. A number of remarkable and desirable properties such as tunable bandgap, QSH effect, high carrier mobility, surface activity, tremendous mechanical characteristics indicate germanene is a promising substance for future applications in electronic and optical devices [9–13, 33–35].

Germanene shares various interesting and peculiar electronic properties like graphene and silicene. Density functional theory (DFT) calculations have demonstrated that the free-standing germanene is a 2D Dirac fermion system, i.e., the electrons near the K and K′ (Dirac) points of the

Brillouin zone behave as relativistic massless particles that have been confirmed by the Dirac equation, which is the relativistic discrepancy of the Schrodinger equation [2, 35]. On the basis of this study, the Fermi velocity of germanene has been reported around 5×10^5 ms^{-1}, may also explain the high carrier mobility i.e., the carrier interactions with phonons are 25 times lower than that of graphene [36]. Pang *et al.* have also been investigated the nanoribbons of hydrogenated germanene are of the two types such as armchair (AGeNRs) and zigzag (ZGeNRs) with the help of DFT calculations. They have found that there was decrease in the band gap of AGeNRs with increase in the width of ribbon while period of oscillation was thrice. ZGeNRs contain behavior of anti-ferromagnetic (AFM) semiconductor but when it was doped with a single B or N atom, may be changed their behavior to ferromagnetic (FM) which is useful for spintronic applications [22, 37, 38].

The present chapter describes the structure, properties and applications of germanene that have been widely used in many aspects and aimed at drawing the attention of researchers and the scientific community to possibilities offered by the use of germanene-based materials for improving their working characteristics and for replacing rather expensive traditional materials on the basis of energy storage devices.

9.2 Structural Arrangements

9.2.1 Elemental Structures

In light of the growing areas of research on graphene, silicene, germanene is due to their structures, hybridization, doping and functionalization exhibits impressing characteristics to explore potential limits [6, 39, 40]. Mostly in graphene, it can be seen that the sp^2 hybridized C–C bond, the σ_{C-C} bond (in-plane) is one of the strongest bonds and π-bond, the out-of-plane, which contributes to delocalization of electrons responsible for conduction of electron providing weak interactions among the layers of graphene or between graphene and other material. Besides these, in silicene and germanene, the π-bond formation was found to be very difficult as the weak overlapping of p_z orbitals in Si and Ge and hence, mixed sp^2–sp^3 hybridization because of their favorable form of sp^3 hybridization. Figure 9.2 (above) represents the buckled nature in which X atom denotes the group 14 elements (Si, Ge and Sn) which are covalently bond with other three X atoms, results to form a simple hexagonal unit cell [41]. The middle and lower parts of Figure 9.2 represent various forms of lattice arrangement (flat, chair, boat and washboard) are the buckling structures

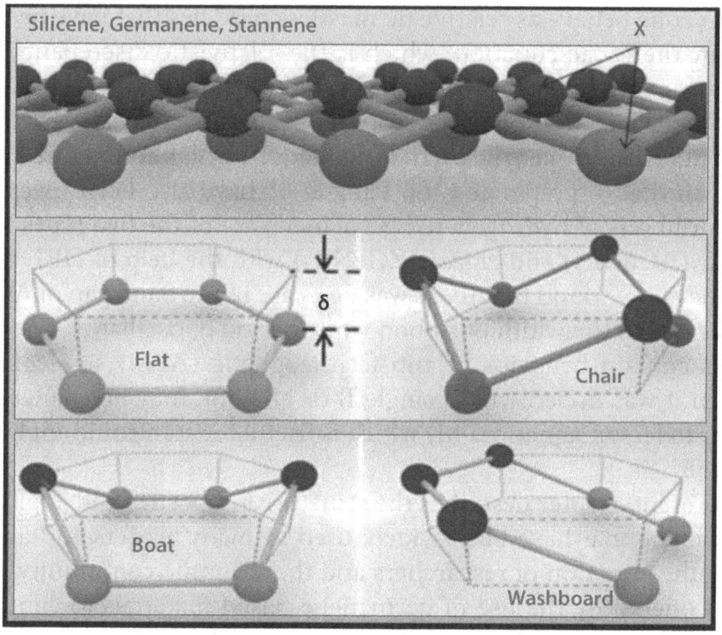

Figure 9.2 Schematic diagrams for the different types of buckled structures: Flat; Chair like; Boat-like, and Washboard [43].

that appear at different points in hexagonal unit cell. As reported by few researchers in their DFT studies that revealed that the chair-like structure is more than the other three structures. Table 9.1 shows the buckling and lattice parameters for the sheets of selected elements. The buckling constant (δ) represents the vertical distance that separates the two atomic planes in all the four structures. Changing in electronic properties could be seen with the variation in the buckling parameters as governed by the electronic state density [41–44].

9.2.2 Decorated Structures

Decorated structure is one of the important structural arrangements that engage foreign elements that are covalently bonds to one another in the direction of cross plane in 2D sheet. Figure 9.3 represents the different geometrical decorated buckle structure of group 14 elements [45]. The stability of the decorated (flat, chair, boat and washboard) structures (Figure 9.3) were more as compared their elemental counterparts (Figure 9.2) assuming their valence band and hence it can be distinguished as free standing sheets.

Table 9.1 Structural and electronic parameters of elemental structures. Electronic quantities for Group 14 elements are derived from hybrid exchange-correlation functional HSE06 calculations with the inclusion of spin orbit coupling [43].

	C	Si	Ge	Sn	P
Lattice constant (a in nm)	0.2468	0.3868	0.4060	0.4673	0.453
Bond length (d in nm)	0.1425	0.2233	0.2344	0.2698	0.336
Buckling parameter (δ in nm)	0	0.045	0.069	0.085	0.250
Effective electron mass M* (m_0)	0	0.001	0.007	0.029	–
Fermi velocity of carriers (10^6 ms^{-1})	1.01	0.65	0.62	0.55	–
Energy gap E_g (meV)	0.02	1.9	33	101	1,000
In-plane stiffness C (Jm^{-2}) Pristine	–	100	53	38	–
Ni doped	–	138	72	63	–

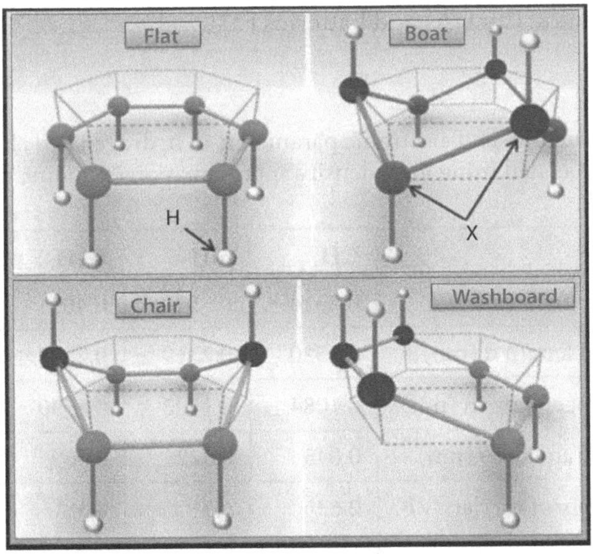

Figure 9.3 Various geometrical configurations of decorated, 2D elemental sheets. In this case, the elemental structures are decorated with hydrogen (X denotes Si, Ge and Sn). The typical decoration sites desired by hydrogen depending on the lattice structure are presented [43].

Justo *et al.* have theoretically reported the hydrogenated and fluorinated elemental planes that provide the stability of the decorated buckled structures upon the group 14 elements i.e., all group 14 (C, SiC, Si, Ge, and Sn) graphene-like sp^3 hybridized configuration show zero bandgap in both flat or buckled structure that opened up with the hydrogenation or fluorination in most materials [18, 41, 46].

Trivedi *et al.* has also been performed their study on hydrogenation of Group 14 elements. They have calculated and reported binding energies of various hydrogenated crystal geometries (Figure 9.3) in order to analyze the change in the density of states and band structure after decoration using hydrogenation method. The hydrogen atoms alternating in stable buckled chair form a symmetrical one up and one down while in unstable boat and washboard conformers have hydrogen atoms alternating on both side of sheets (2up–2down and 3up–3down, respectively). The decorated chair form shows highest binding energy as compared to the other decorated the three ones. The studied parameters of hydrogenated sheets (structural and buckling) of group 14 are listed in Table 9.2 [41, 47]. Sahin *et al.* have also been performed the decorations of Group 14 elemental sheets with s-block elements and first series of transition metals and they have reported the adsorption range of these metal atoms on silicene (SiH) in their theoretical investigations. The analysis of the metal–silicene interactions has a significant importance because of its fundamental relevance to applications in nano-electronics, catalysis and batteries [23].

Table 9.2 Structural and electronic parameters of hydrogenated elemental structures calculated using local density approximation excluding spin–orbit coupling [43].

	CH	SiH	GeH	SnH
Lattice constant a (nm)	0.2514	0.3820	0.4091	0.4719
Metal-metal length d (nm)	0.1520	0.2319	0.2338	0.2846
Metal-hydrogen length (nm)	0.1084	0.1502	0.1530	0.1738
Buckling parameter δ (nm)	0.045	0.072	0.069	0.082
Fermi velocity of carriers VF ($10^6 ms^{-1}$)	0.63	0.51	0.38	–
Energy gap E_g (eV)	4.9, 5.4	2.9, 4.0	2.9, 3.6	–

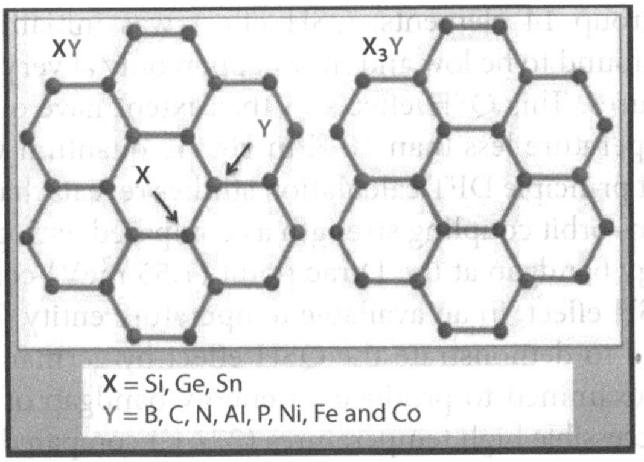

Figure 9.4 Typical configurations of composite structures for silicene, germanene and stannene [43].

9.2.3 Composite Structures

Assuming the Group 14 elements may form classical structures of the types XY and X_3Y where X denotes Si, Ge, Sn whereas Y resembles B, C, N, Al, P, Ni, Fe, and Co and such types of arrangements are shown in Figure 9.4 [41]. The composite of the type XY forms different buckled structures that depend upon the doped materials as portrayed by the unit cell as in Figure 9.2. From the first principle calculation studies by researchers, flat structures as assumed as when trivalent elements like B and Al are doped whereas the buckled structure of chair form have been found with pentavalent atoms (N and P), respectively for X_3Y structures. Such preferences are on the basis of number of valence electrons as the composite structures are assumed to be larger n-doped and the electrons can efforts more tensile strain results in a buckled structures. These behaviors may also be explaining the nature of buckled structure of transition metal complexes [42, 44, 48].

9.3 Fundamental Properties of Germanene

9.3.1 Quantum Spin Hall (QSH) Effect

This phenomenon occurs due to band gap opening at the Dirac point stimulated by spin–orbit coupling that alleviates any two dimensional materials transition from semi-metal behavior to a QSH non-conductor [49].

Among the group 14 elements, QSH effect was initially proposed for graphene and found to be low and may happen only at very low unprocurable temperatures. This QSH effects to that extent have only been investigated at temperature less than 10 K in Hg–Te quantum well structures. Moreover, first principle DFT calculation studies reveals that silicene (SiH) has higher spin–orbit coupling strength as compared with graphene, show a higher energy bandgap at the Dirac point (1.55 meV) consequence in a substantial QSH effect, in an available temperature entity. The same studies also applied to demonstrate the QSH effect by germanene and found that it is also examined to produce an energy bandgap of approximately 23.9 meV at accessible high temperatures (277 K) compared with graphene and silicene induced by spin–orbit coupling, whereas stannene membranes are observed to acquire the highest energy bandgap (nearly 300 meV) over all honeycomb structures [29, 50, 51]. Figure 9.5 described the band structures observed from DFT analysis with comprehension of spin–orbit

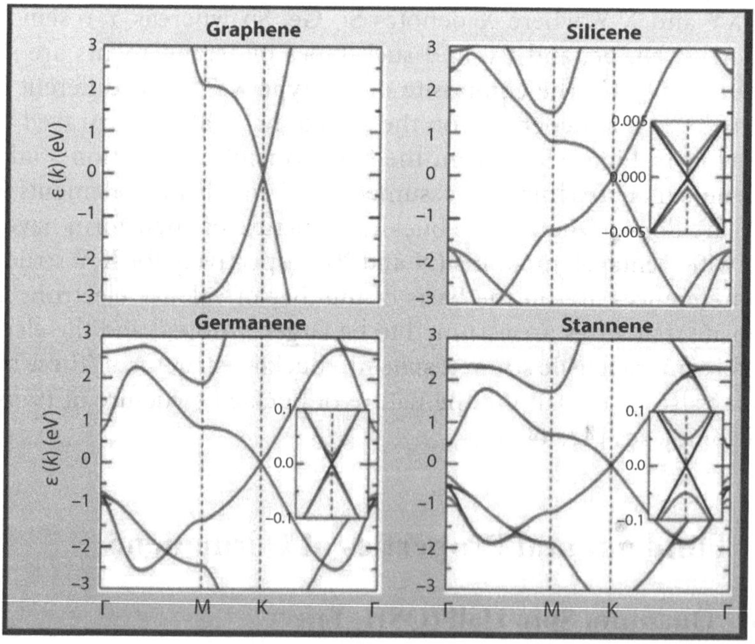

Figure 9.5 Quasi-particle band structures of group 14 elemental structures along high-symmetry lines of the 2D hexagonal Brillouin zone, using the exchange-correlation functional HSE06 including spin–orbit interactions. The inset shows a magnification of the electronic structure around the Dirac point with the solid black lines indicating the band structure obtained without spin–orbit coupling [43].

interactions. Such characteristics of the freestanding structures this group elements develop an opportunity for the reorganization of spin-electronics applications [41, 52].

9.3.2 Mechanical Properties

2D materials possess mechanical characteristics that are important properties for evaluating their utility in nanoelectro-mechanical devices. Few investigations have been conducted by researchers on the mechanical properties of 2D materials via theoretical or experimental method that can be useful to establish a novel pathway to develop functional nanoelectromechanical devices. Manjanath *et al.* have studied on the mechanical properties of Ni-doped group 14 freestanding planes (Si, Ge, and Sn) and pristine structures in different geometries using first principle DFT calculations. The theoretical investigations on that in-plane stiffness of these freestanding planes for both doped and pristine structures, the stiffness is ascertained to low as increase the atomic weight of group 14 elements. As can be concluded from Table 9.1 in both of the cases of doped and pristine structures, the stiffness values have been higher observed for flat structure than that of buckled structures [44]. This is due to the strong sp^2 hybridized arrangement in flat planer structures than that of sp^2–sp^3 hybridized buckled structures in both of the cases. Such hybridized structures make flat planar sheets of germanene mechanically stronger than their buckled counterparts. Doping the elemental sheets (Si, Ge, Sn) with transition metal atoms have larger stiffness value than that of pristine sheets indicates that the doped sheets reinforces their strength. Such a doped behavior is also predicted the other transition metal atoms (Fe, Co) and found the similar behavior as studied in the case of Ni-doping with elemental sheets as a decreasing tends of mechanical stability has been obtained from Si to Sn (Table 9.1; in-plane stiffness) [53].

Wang *et al.* have theoretically investigated two types of stain induced self-doping in germanene, where high tensile strain was assumed to induce doping of p-type while the compressive strain is assumed to induce doping of n-type, respectively. Moreover, the interesting characteristics of germanene can allow the facile control in their engineering properties as compared with self-doping properties of graphene [54]. Based on first principle DFT calculation studies, Kaltsas *et al.* have reported the effect of uniaxial tensile strain on germanane (GeH) of the chair, boat and washboard types. Such theoretical and experimental investigations on the mechanical strength with respect to the excellent electronic capabilities of germanene, render it is a competitive substitutes in modern flexible electronics [55].

9.3.3 Thermal Properties

Seeback, in 1821, first gave the idea to obtain an electric current from steady temperature gradient. This has to be done using a thermoelectric conversion device that was first introduced by him. This device can be used as a cooler in a reverse mode operation by maintaining a steady current in the device (based on Peltier effect). Recently, the pursuit for a highly effective thermoelectric device has attracted enormous attribution due to the considerable potential applications. Following formula also called figure of merit has been used to evaluate the efficiency of thermoelectric conversion

$$ZT = \frac{\sigma S^2 T}{\kappa} \tag{9.1}$$

where σ, S, T are the electric conductance, Seebeck coefficient and absolute temperature, respectively. κ is the total thermal conductance (= $\kappa_e + \kappa_p$, κ_e and κ_p are the electron and phonon contributions) [56–58].

For a good thermoelectric device, the value ZT ≈ 1, while a thermoelectric device approaches the value of ZT ≥3, could effectively engage with conversion methods of conventional energy. Both the theoretical observations and experimental investigations exhibit that low-dimensional or nano-scale devices could demonstrate much larger ZT values as compared to the bulk materials, which shows very excellent outlook of thermoelectric applications. Agosta *et al.* have performed first principle DFT calculation studies to investigate the figure of merit values for 2D germanene and GNRs, respectively and point out the highest ZT values for the hetero-structures as compared to the pure substance as higher temperature. They confirmed that the nano-structuring can improve the overall energy conversion efficiency [59].

9.3.4 Optical Properties

Closely related to the dielectric function, the optical properties mathematically obtained using the following formula

$$\varepsilon(\omega) = \varepsilon_1(\omega) + i\varepsilon_2(\omega) \tag{9.2}$$

where the $\varepsilon_1(\omega)$ and $\varepsilon_2(\omega)$ are the real and imaginary part of the dielectric function have been evaluated from imaginary part using the Kramer–Kronig equation

$$\varepsilon_1(\omega)=1+\frac{2}{\pi}\mathrm{P}\int_0^{\infty}\frac{\omega'\varepsilon_2(\omega')d(\omega')}{\omega'^2-\omega^2} \tag{9.3}$$

$$\varepsilon_2(\omega)=\frac{4\pi^2}{m^2\omega^2}\sum\nolimits_{V,C}\int_{BZ}d^3k\frac{2}{2\pi}\left|eM_{CV}(K)\right|^2\times\delta[E_c(k)-E_V(k)-\hbar\omega] \tag{9.4}$$

where the terms V and C tend for the valence and conduction bands; m and e tend for mass and charge of electrons; ω and p denote the frequency and principal values of incident photons [60–64]. Using the above equations, Ould NE *et al.* have investigated the comparative study on the optical properties on both graphene and germanene. They were reported the optical properties are different in perpendicular and parallel to the direction of electric field, so the graphene and germanene are anisotropic materials. Besides this, the optical properties such as refractive, absorption and reflectivity of germanene is more affordable than graphene structure. Absorption of germanene is higher in visible region than graphene that preserves great potential applications in solar cell [65].

Another investigation for optical properties of germanane have also been performed using diffuse reflectance absorption spectroscopy and reported the bandgap approximately 1.59 eV in visible wavelength range. The calculation of band structure reveals that germanane is a direct bandgap substance both as isolated layers and two layers per unit cell in the crystal structure. Using first principle DFT calculation method by Chen *et al.* to demonstrate the optical properties of hetero-structures of germanene and antimonene simultaneously, and found that, these hetero-structures expose excellent and tunable band gap which can be used to design a high performance nano-electronic device [66]. Tang *et al.* have investigated the optical properties hetero-structure germanene with AlAs under the influence of weak van der Waals interaction, found an indirect band gap approximately 0.494 eV, which can be effectively applied in field effect transistors. So, investigation about various properties of germanene is one of the major research projects in modern condensed matter physics [67].

9.4 Applications of Germanene

9.4.1 Strain-Induced Self-Doping in Germanene

Self-doping is an interesting and peculiar property to define the strain-induced phenomenon. From first principle DFT calculations, Wang *et al.* investigated the changes of band structures of germanene nano-sheets (Ge analogue of graphene) by strains and found that this system exhibits strain-induced self-doping phenomena that are similar with the buckled structure of germanene and can't develop in 2D graphene. In consistent with graphene, the opening of band gap was not occurred under the strains. It can be seen from Figure 9.6, the Dirac point (E_D) of germanene at strain ε = 3.40 eV·A° and $E_D - E_F = 0$ eV, exactly stroked at Fermi level (E_F). Under the tensile strain (ε = −0.05), the Dirac point become less to Fermi level by a factor of 0.181 eV with β = 3:49 eV·A° while at compressive strain of

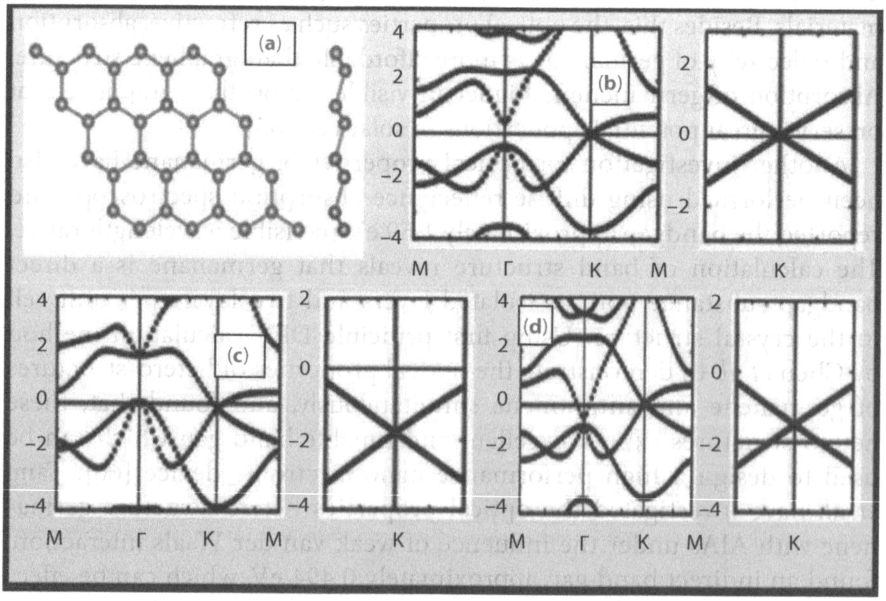

Figure 9.6 (a) The atomic structure of germanene. The band structures of silicene with (b) the zero strain of ε = 0, (c) the compressive strain of ε = −0.05, and (d) the tensile strain of ε = 0.05. The Fermi level E_F is indicated as the line at E = 0 eV [58].

$\varepsilon = 0.05$, the Dirac point become more to Fermi level by a factor of 0.050 eV with $\beta = 3.25$ eV A°. Thus, these strain-effects can induce self-doping in germanene nano-sheets. Large compressive/tensile strain on germanene nano-sheet acts as a p-type doping for its Dirac-like electronic properties. The group also confirmed that long bond lengths and the large buckling heights are necessary things for self-doping in germanene nano-sheets. These studies investigate that germanene has more promising electronic properties compared to graphene in few aspects, and the strain engineering properties could effectively accommodate the electronic structures of Ge-based nano-materials [58].

9.4.2 Battery Applications

Germanene and the functionalized germanene are promising 2D materials which have been widely studied by few researchers in battery applications for energy conversion and storage devices such as in electric vehicles, in portable electronic devices and large scale power grids because of their low cost and high energy density (in Li-ion battery) [2, 20]. Using first-principles DFT calculations by Mortazavi *et al.* to evaluate and analyze the insertion of Li/Na-ion on 2D germanene membrane. They have seen a most stable binding sites and energy of binding after insertion on germanene. The results based on the Bader charge investigation, they were reported the energy density was found to be approximately 369 mAh·g^{-1} for Li/Na ions deposited on separate germanene nano-sheets. Theoretical investigations suggest a good path to develop germanene nano-sheets in Li/Na-ion batteries applications [75].

Similar investigations have also been made by Serino *et al.* They have used the freestanding 2D GeH nano-sheets for Li-ion batteries to control stress, generated by large volume expansion of Ge-based substrates. They have reported the high charge capacity approximately 1,100 mAh·g^{-1} for a GeH–carbon composite as anode for Li-ion battery over 100 charge/discharge cycles. Schwingenschlogl *et al.* studied on germanene on MgX_2 (X = Cl, Br, and I) for Li-ion battery as germanene is a good electrode material for Li-ion batteries due to its low Li diffusion barrier and high Li capacity as silicene. Insertion of MgX_2 between the layers of germanene gives rise to good hybrid substances for Li-ion battery. The first principle DFT study investigates that the Ge/MgX_2/Ge sandwiched structures can accommodate twice the number of Li ions with a separation more than of 9 Å between the two germanene sheets. The energy capacity has been found to be approximately 369 mAh·g^{-1} as in the case of freestanding germane is very low as compared with the separate sheets for silicene (954 mAh·g^{-1}) due to the high atomic mass of Ge [76].

9.4.3 Electronic Devices

Similar to graphene and silicene, germanene also exhibits interesting and peculiar electronic properties. Using first principle DFT calculations, Wang *et al.* have studied the electronic properties of germanene that are adsorbed with various small organic/inorganic molecules such as acetone, ammonia, methanol, ethanol, acetonitrile, etc. The minimum to maximum adsorption energies were found for the studied small organic molecules in the range from −0.459 to −0.231 eV as in the case of germanene than those of silicene and graphene with small organic molecules, confirms that strong interactions between germanene and organic molecules. Their investigations also confirms the tune effectively band gaps are in the range of 3.9–81.9 meV, adsorbed the all small organic molecules by germanene. Besides these, the opening of the band gap is also related to the transfer of charge from studied molecules to host germanene. The transfer of charge provides an internal electric field between the germanene and adsorbed molecules. This field is used to break the sub-lattice symmetry in germanene and band gap opening took place at the Dirac point. From the above studied fact, the group have been concluded that the small molecules adsorbed germanene may be useful to design germanene based devices like field effect transistors in near future. Interestingly, the properties of Dirac cone with an effective mass of small electron are conserved that result in comparatively high mobility of electron. These results are of valuable importance for germanene-based electronic devices [77].

9.4.4 Catalysis

Germanene and its derivative play an important role in photocatalysis that is their potential. applications to convert the solar light into various form of energy such as electrical energy, chemical energy etc. Few researchers first synthesized the 2D materials such as germanene, germanane using various methods and applied their potential applications. Using ion-exchange method, Liu *et al.* have synthesized germanane (GeH). The group have used rhodamine B (RhB) to determine the potential of GeH for degradation, and found that higher the performing tendency of GeH with as compared with N-doped P25 and experimentally evaluated the degradation of RhB ~80% in few minutes (~4 min) whereas ~70% of RhB was degraded when broke down by N-doped P25 for an instance of 30 min. Figure 9.7 describes the photolytic activity for the H_2 evolution by GeH under the visible light region at λ ≥420 nm and found experimentally that the performance of rate of generation of H_2 is achieved about 22 μmol·h^{-1}·g^{-1} while the performance rate were 13 μmol·h^{-1}·g^{-1} and 7 μmol·h^{-1}·g^{-1} for GeH/Pt (1.0 wt%) in H_2O, and H_2O–CH_3OH, respectively. Again under dark condition using GeH/Pt (1.0 wt%) in CH_3OH–H_2O and with GeH powder under visible light conditions, there were no H_2 evolution found which indicates that evolution of H_2 is from the photo-catalytic reaction rather than the decomposition of GeH. To test the stability of GeH, the cyclic experiments were repeated ten times by RhB decomposition under visible light conditions at λ ≥420 nm for 24 h and demonstrated that photocatalytic activity of GeH is high even after repeated decomposition of RhB 10 times with a slight decrease as shown in Figure 9.8. During the photocatalytic reactions, this analysis confirms that the stability of GeH and remarkably show the high photo-reactivity of GeH. This experiment reveals the capability of the GeH as a novel material provides large potential applications for energy systems [67, 68].

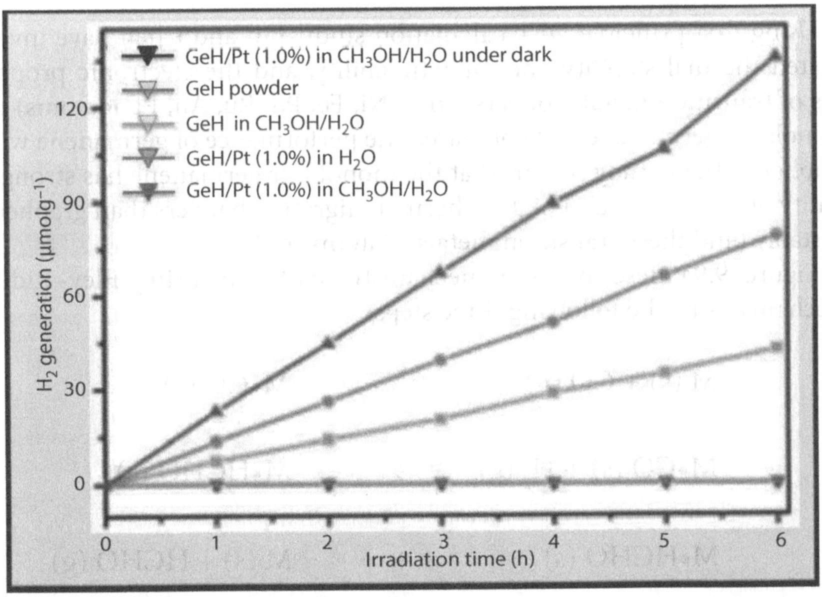

Figure 9.7 Photocatalytic H_2 evolution of GeH under different conditions: GeH/Pt (1.0 wt%) in H_2O–CH_3OH (blue), GeH/Pt (1.0 wt%) in H_2O (yellow), GeH in H_2O–CH_3OH (green), GeH/1.0 wt% Pt in H_2O–CH_3OH under dark conditions (black) and GeH powder only (red) under visible-light (lZ420 nm) irradiation [71].

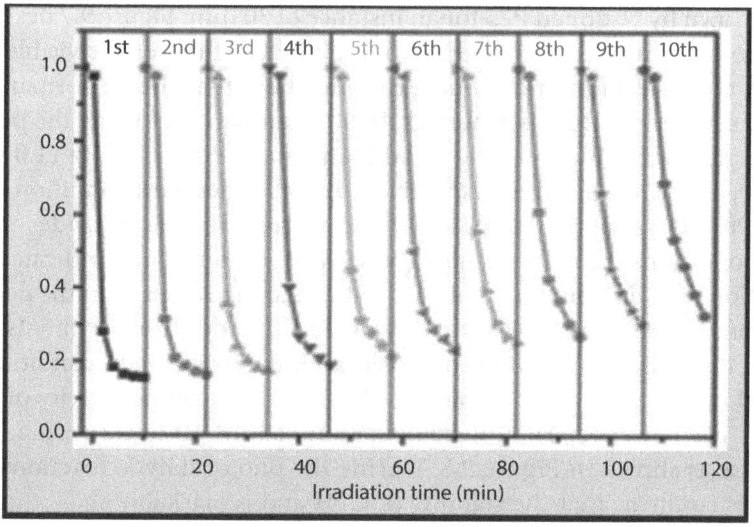

Figure 9.8 Irradiation-time dependence of the relative concentration C/C_0 of the RhB dye and the amount of H_2 evolution in solution over GeH during cycling photocatalytic experiments under visible light [71].

Using first principle DFT calculation study, Lin and Chen have investigated thermal stability, the catalytic ability, and the electronic properties of transition metal atoms (such as Ni, Fe, Pd, Ru, Au, Pt adatoms) on monolayer germanene and compared the performance of germanene with graphene. Their study reveals that the monolayer germanene has stronger ability of adsorption and higher thermal migration barriers than graphene to stably bind these transition metals adatoms [69].

Figure 9.9a describes the mechanistic pathways using Eley–Rideal mechanisms by the following three steps:

$$M\,(s) + CO\,(g) \rightarrow M\bullet CO\,(s)$$

$$M\bullet CO\,(s) + H_2\,(g) \rightarrow M\bullet HCHO\,(s)$$

$$M\bullet HCHO\,(s) \rightarrow M\,(s) + HCHO\,(g)$$

and Figure 9.9b shows the respective energy profile for all these catalytic reaction steps.

For TMs adatoms (Fe, Ni and Ru), their movement on germanene are more adequate for inhibiting the accumulation of the above adatoms.

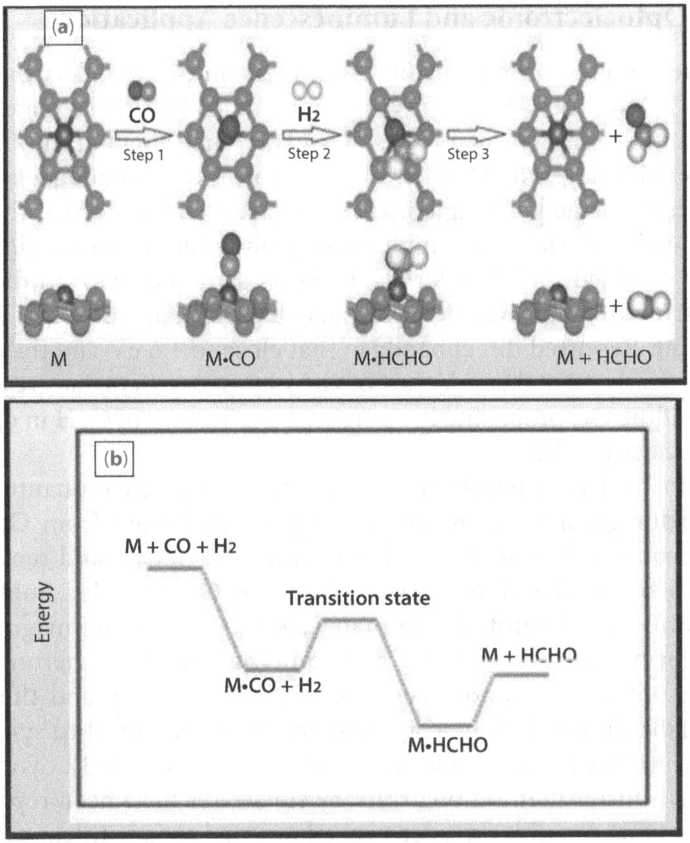

Figure 9.9 The reaction steps (a) and the corresponding energy profile (b) of $CO + H_2 \rightarrow HCHO$ catalyzed by the adatom M on germanene. For (a), both the top and side views are shown [72].

For the accumulation of these adatoms, minimum-energy path calculation was performed to analyze the catalytic transformation of CO contamination in H_2 by these adatoms followed by the reaction path. Also for Fe and Ru adatoms, it has been found that the reaction path was more favorable and thermodynamically and kinetically stable with low potential barriers and stable reaction intermediate but as in the case of Au, Ni, Pd and Pt adatoms, same conditions were not found and therefore, the reactions are difficult to proceed. From the reaction mechanism, they were concluded that with the catalytic ability and high thermal stability, Fe and Ru adatoms on monolayer germanene should be superior catalysts to the removal of CO contamination in hydrogen feed gas of fuel cells and batteries and future catalytic applications [70].

9.4.5 Optoelectronic and Luminescence Applications

Great attention has been made in recent years on the optoelectronic properties of 2D van der Waals (vdWs) materials. When these materials are spread out to mono-layers, mostly they experience alters in the electronic structure. Multilayered 2D materials show low indirect bandgap but when they are transformed to a high direct bandgap, then they were exfoliated to a monolayer material and exhibits strong photoluminescence properties. As Ge is changed to 2D materials, it shows same behavior in their electronic structure substantially. Hydrogenated germanene (GeH) is stable towards air. Reported direct bandgap that close to 1.6 eV and the electron mobility nearly 18,000 $cm^2 \cdot V^{-1} \cdot s^{-1}$ of GeH by researchers, far greater than the crystalline Ge, proposing the potential application of GeH in optoelectronic devices [18, 71, 72].

Jiang *et al.* have investigated the photoluminescence quantum yield (PLQYs) for germanene by substituting the hydrogen from GeH by a methyl group and created a methyl-terminated $GeCH_3$ and reported its PLQYs of band edge fluorescence ~0.2% for $GeCH_3$. This implies that the QYs of the substituted new materials show the same magnitude as monolayer metal chalcogenides, and could be provide numerous advantages for optoelectronic devices, viz., thermal stability and the simple relative fabrication. This new approaches have been used to synthesizes the new family of organic-modified 2D vdWs lattices [73]. Synthesis of GeH and transfer method on arbitrary substrates have been reported by Amamou *et al.* For this transfer procedure, the surface thickness ranges (1–600 nm, approximately 2–1,000 layers) and the surface area >1 cm^2 of GeH taken into account. They were used photoluminescence and X-ray diffraction methods to investigate the optical structures and properties of thick GeH layers that remain unchanged followed by transfer step. It means the GeH may be used on transparent materials for optical studies and also for electronic constituents on non-conducting materials. That is the reason that the 2D GeH materials can be used for future optoelectronic potential applications [74].

To demonstrate the optoelectronic characteristics for germanene 2D materials, ultrathin Ge-flakes have been used by Hu *et al.* using chemical vapour deposition by halide-assisted self-limited growth process, in which thickness of the Ge-flakes are very low (<8.5 nm). The growth mechanisms are confirmed using experimental methods and the theoretical calculations and these studies reveals that the Ge-flakes based phototransistor showed outstanding properties in optoelectronics devices [53].

9.5 Conclusions

In this chapter, we have discussed the various properties of 2D materials based on group 14 elements such as geometrical structure and hybridization, optical electronic and magnetic properties. We have concluded that the elemental germanene are an excellent substitute approach to get the better for the technological barriers that have hindered the current Ge-based electronics. The Ge-based excellent properties of the novel systems have also been discussed that were determined using first principles DFT calculations. Besides these, detail investigations of electronic properties provide a Dirac point and Fermi level in planar and buckled germanene. Germanene-based structures have intense impact in future nanoelectronic systems due to their compatibility with the current semiconductor technology. These studies will also draw the attention of researchers and the scientific community to possibilities offered by the use of germanene based materials for improving their working characteristics and for replacing rather expensive traditional materials on the basis of energy storage devices such as freestanding germanene based Li-ion batteries.

References

1. Jariwala, D., Sangwan, V.K., Lauhon, L.J., Marks, T.J., Hersam, M.C., Emerging Device Applications for semiconducting two-dimensional transition metal dichalcogenides. *ACS Nano*, 8, 1102, 2014.
2. Novoselov, K.S., Geim, A.K., Morozov, S.V., Jiang, D., Zhang, Y., Dubonos, S.V., Grigorieva, I.V., Firsov, A.A., Electric field effect in atomically thin carbon films. *Science*, 306, 666, 2004.
3. Chhowalla, M., Shin, H.S., Eda, G., Li, L.J., Loh, K.P., Zhang, H., The chemistry of two-dimensional layered transition metal dichalcogenide nanosheets. *Nat. Chem.*, 5, 263, 2013.
4. Lalmi, B., Oughaddou, H., Enriquez, H., Kara, A., Vizzini, S., Ealet, B., Aufray, B., Epitaxial growth of a silicene sheet. *Appl. Phys. Lett.*, 97, 223109, 2010.
5. Miro, P., Audiffred, M., Heine, T., An atlas of two-dimensional materials. *Chem. Soc. Rev.*, 43, 6537, 2014.
6. Liu, H., Neal, A.T., Zhu, Z., Luo, Z., Xu, X., Tomanek, D., Ye, P.D., Phosphorene: An unexplored 2D semiconductor with a high hole mobility. *ACS Nano*, 8, 4033, 2014.
7. Jiang, J., Liang, Q., Meng, R., Yang, Q., Tan, C., Sun, X., Chen, X., Exploration of new ferromagnetic, semiconducting and biocompatible Nb_3X_8 (X = Cl,

Br or I) monolayers with considerable visible and infrared light absorption. *Nanoscale*, 9, 2992, 2017.
8. Geim, A.K. and Novoselov, K.S., The rise of graphene. *Nat. Mater.*, 6, 183, 2007.
9. Neto, A.H.C., Guinea, F., Peres, N.M.R., Novoselov, K.S., Geim, A.K., The electronic properties of graphene. *Rev. Mod. Phys.*, 81, 109, 2009.
10. Liu, Y., Weiss, N.O., Duan, X., Cheng, H.C., Huang, Y., Duan, X., Van der Waals heterostructures and devices. *Nat. Rev. Mater.*, 1, 16042, 2016.
11. Prasongkit, J., Amorim, R.G., Chakraborty, S., Ahuja, R., Scheicher, R.H., Amornkitbamrung, V., Highly sensitive and selective gas detection based on silicene. *J. Phys. Chem. C*, 119, 16934, 2015.
12. Li, L., Lu, S.Z., Pan, J., Qin, Z., Wang, Y.Q., Wang, Y., Cao, G.Y., Du, S., Gao, H.J., Buckled germanene formation on Pt(111). *Adv. Mater.*, 26, 4820, 2014.
13. Derivaz, M., Dentel, D., Stephan, R., Hanf, M.C., Mehdaoui, A., Sonnet, P., Pirri, C., Continuous Germanene Layer on Al(111). *Nano Lett.*, 15, 2510, 2015.
14. Davila, M.E., Xian, L., Cahangirov, S., Rubio, A., Lay, G.L., Germanene: A novel two-dimensional germanium allotrope akin to graphene and silicene. *New J. Phys.*, 16, 3579, 2014.
15. Bandaru, P.R. and Pichanusakorn, P., An outline of the synthesis and properties of silicon nanowires. *Semicond. Sci. Technol.*, 25, 024003, 2010.
16. Lebegue, S. and Eriksson, O., Electronic structure of two-dimensional crystals from ab initio theory. *Phys. Rev. B: Condens. Matter Mater. Phys.*, 79, 115409, 2009.
17. Houssa, M., Pourtois, G., Afanas, V.V., Stesmans, A., Electronic properties of two-dimensional hexagonal germanium. *Appl. Phys. Lett.*, 96, 082111, 2010.
18. Garcia, J.C., De Lima, D.B., Assali, L.V.C., Justo, J.F., Group IV graphene- and graphane-like nanosheets. *J. Phys. Chem. C*, 115, 13242, 2011.
19. Takeda, K. and Shiraishi, K., Theoretical possibility of stage corrugation in Si and Ge analogs of graphite. *Phys. Rev. B*, 50, 14916, 1994.
20. Cahangirov, S., Topsakal, M., Ciraci, S., Armchair nanoribbons of silicon and germanium honeycomb structures. *Phys. Rev. B: Condens. Matter Mater. Phys.*, 81, 195120, 2010.
21. Nakano, H., Mitsuoka, T., Harada, M., Horibuchi, K., Nozaki, H., Takahashi, N., Nonaka, T., Seno, Y., Nakamura, H., Soft synthesis of single-crystal silicon monolayer sheets. *Angew. Chem. Int. Ed. Engl.*, 45, 6303, 2006.
22. Pang, Q., Zhang, Y., Zhang, J.M., Ji, V., Xu, K.W., Electronic and magnetic properties of pristine and chemically functionalized germanene nanoribbons. *Nanoscale*, 3, 4330, 2011.
23. Bianco, E., Butler, S., Jiang, S., Restrepo, O.D., Windl, W., Goldberger, J.E., Stability and exfoliation of germanane: A germanium graphane analogue. *ACS Nano*, 7, 4414, 2013.

24. Pinchuk, I.V., Odenthal, P.M., Ahmed, A.S., Amamou, W., Goldberger, J.E., Kawakami, R.K., Epitaxial co-deposition growth of $CaGe_2$ films by molecular beam epitaxy for large area germanane. *J. Mater. Res.*, 29, 410, 2014.
25. Acun, A., Zhang, L., Bampoulis, P., Farmanbar, M., van Houselt, A., Rudenko, A.N., Lingenfelder, M., Brocks, G., Poelsema, B., Katsnelson, M.I., Zandvliet, H.J.W., Germanene: The germanium analogue of graphene. *J. Phys.: Condens. Matter*, 27, 443002, 2015.
26. Guzmán-Verri, G.G. and Lew, Y.V.L.C., Electronic structure of silicon-based nanostructures. *Phys. Rev. B*, 76, 075131, 2007.
27. Katsnelson, M.I. and Fasolino, A., Graphene as a prototype crystalline membrane. *Acc. Chem. Res.*, 46, 97, 2013.
28. Kane, C.L. and Mele, E.J., Z_2 Topological Order and the Quantum Spin Hall Effect. *Phys. Rev. Lett.*, 95, 146802, 2005.
29. Kane, C.L. and Mele, E.J., Quantum Spin Hall Effect in graphene. *Phys. Rev. Lett.*, 95, 226801, 2005.
30. Boettger, J.C. and Trickey, S.B., First-principles calculation of the spin–orbit splitting in graphene. *Phys. Rev. B*, 75, 121402, 2007.
31. Gmitra, M., Konschur, S., Ertler, C., Ambrosch-Draxl, C., Fabian, J., Band-structure topologies of graphene: Spin–orbit coupling effects from first principles. *Phys. Rev. B*, 80, 23, 5431, 2009.
32. Han, W., Kawakami, R.K., Gmitra, M., Fabian, J., Graphene spintronics. *Nat. Nanotechnol.*, 9, 794, 2014.
33. Wang, Y., Ni, Z., Liu, Q., Quhe, R., Zheng, J., Ye, M., Yu, D., Shi, J., Yang, J., Li, J., Lu, J., All-metallic vertical transistors based on stacked dirac materials. *Adv. Funct. Mater.*, 25, 68, 2015.
34. Zhang, R., Zhang, C., Ji, W., Li, F., Ren, M., Li, P., Yuan, M., Wang, P., Tunable electronic properties in the van der Waals heterostructure of germanene/germanane. *Phys. Chem. Chem. Phys.*, 17, 12194, 2015.
35. Novoselov, K.S., Geim, A.K., Morozov, S.V., Jiang, D., Katsnelson, M.I., Grigorieva, I.V., Dubonos, S.V., Firsov, A.A., Two-dimensional gas of massless Dirac fermions in graphene. *Nature*, 438, 197–200, 2005.
36. Roome, N.J. and Carey, J.D., Beyond graphene: Stable elemental monolayers of silicene and germanene. *ACS Appl. Mater. Interfaces*, 6, 7743, 2014.
37. Kan, M., Zhou, J., Li, Y., Sun, Q., Using carbon chains to mediate magnetic coupling in zigzag graphene nanoribbons. *Appl. Phys. Lett.*, 100, 173106, 2012.
38. Kan, M., Zhou, J., Li, Y., Sun, Q., Wang, Q., Kawazoe, Y., Jena, P., Tuning magnetic properties of graphene nanoribbons with topological line defects: From antiferromagnetic to ferromagnetic. *Phys. Rev. B: Condens. Matter Mater. Phys.*, 85, 155450, 2012.
39. Li, L., Yu, Y., Ye, G.J., Ge, Q., Ou, X., Wu, H., Feng, D., Chen, X.H., Zhang, Y., Black phosphorus field-effect transistors. *Nat. Nanotechnol.*, 9, 372, 2014.

40. Kara, A., Enriquez, H., Seitsonen, A.P., Voon, L.C.L.Y., Vizzini, S., Aufray, B., Oughaddou, H., A review on silicene—New candidate for electronics. *Surf. Sci. Rep.*, 67, 1, 2012.
41. Balendhran, S., Walia, S., Nili, H., Sriram, S., Bhaskaran, B., Elemental analogues of graphene: Silicene, germanene, stanene, and phosphorene. *Small*, 11, 640–652, 2014.
42. Ding, Y. and Wang, Y.L., Density functional theory study of the silicene-like SiX and XSi_3 (X = B, C, N, Al, P) honeycomb lattices: The various buckled structures and versatile electronic properties. *J. Phys. Chem. C*, 117, 18266, 2013.
43. Matthes, L., Pulci, O., Bechstedt, F., Massive Dirac quasiparticles in the optical absorbance of graphene, silicene, germanene, and tinene. *J. Phys. Condens. Matter*, 25, 395305, 2013.
44. Manjanath, A., Kumar, V., Singh, A.K., Mechanical and electronic properties of pristine and Ni-doped Si, Ge, and Sn sheets. *Phys. Chem. Chem. Phys.*, 16, 1667, 2014.
45. Tsetseris, L. and Kaltsas, D., Chemical routes to modify, uplift, and detach a silicene layer from a metal substrate. *Chem. Phys. Phys. Chem.*, 16, 5183, 2014.
46. Sahin, H. and Peeters, F.M., Adsorption of alkali, alkaline-earth, and 3d transition metal atoms on silicene. *Phys. Rev. B*, 87, 085423, 2013.
47. Trivedi, S., Srivastava, A., Kurchania, R., Silicene and germanene: A first principle study of electronic structure and effect of hydrogenation-passivation. *J. Comput. Theor. Nanosci.*, 11, 781, 2014.
48. Liu, M., Han, Y., Tang, L., Jia, J.F., Xue, Q.K., Liu, F., Interplay between quantum size effect and strain effect on growth of nanoscale metal thin films. *Phys. Rev. B*, 86, 125427, 2012.
49. Xu, Y., Yan, B., Zhang, H.J., Wang, J., Xu, G., Tang, P., Duan, W., Zhang, S.C., Large-gap quantum spin hall insulators in tin films. *Phys. Rev. Lett.*, 111, 136804, 2013.
50. Liu, C.C., Jiang, H., Yao, Y., Low-energy effective Hamiltonian involving spin–orbit coupling in silicene and two-dimensional germanium and tin. *Phys. Rev. B*, 84, 195430, 2011.
51. Liu, C.C., Feng, W., Yao, Y., Quantum spin hall effect in silicene and two-dimensional germanium. *Phys. Rev. Lett.*, 107, 076802, 2011.
52. Ma, Y., Dai, Y., Guo, M., Niu, C., Huang, B., Intriguing behavior of halogenated two-dimensional tin. *J. Phys. Chem. C*, 116, 12977, 2012.
53. Hu, M., Zhang, X., Poulikakos, D., Anomalous thermal response of silicene to uniaxial stretching. *Phys. Rev. B*, 87, 195417, 2013.
54. Wang, Y. and Ding, Y., Strain-induced self-doping in silicene and germanene from first-principles. *Solid State Commun.*, 155, 6, 2013.
55. Kaltsas, D., Tsatsoulis, T., Ziogos, O.G., Tsetseris, L., Response of silicane and germanane to uni-axial compression: Superstructures, polymorph nanoribbons, and extreme bending. *J. Chem. Phys.*, 139, 124709, 2013.

56. Zhang, X.L., Xie, H., Hu, M., Bao, H., Yue, S.Y., Qin, G.Z., Su, G., Thermal conductivity of silicene calculated using an optimized Stillinger–Weber potential. *Phys. Rev. B*, 89, 054310, 2014.
57. Nolas, G.S., Sharp, J., Goldsmid, H.J., *Thermoelectrics: Basic Principles and New Materials Developments, Springer Series in Material Science*, vol. 45, Springer-Verlag, Berlin, 2001.
58. Goldsmid, H.J., *Springer Series in Material Science*, 1st ed., p. 250, Springer-Verlag, Berlin, 2010.
59. Yang, K., Cahangirov, S., Cantarero, A., Rubio, A., D'Agosta, R., Thermoelectric properties of atomically thin silicene and germanene nanostructures. *Phys. Rev. B*, 89, 125403, 2014.
60. Monkhorst, H.J. and Pack, J.D., Special points for Brillouin-zone integrations. *Phys. Rev. B: Solid State*, 13, 5188, 1976.
61. Andersen, O.K., Linear methods in band theory. *Phys. Rev. B*, 12, 3060, 1975.
62. Singh, D.J., *Plane waves, peudopotential and the LAPW method*, Springer, Boston, Dortrecht, 1994.
63. Peterson, M., Wanger, F., Hufnagel, L., Scheffler, M., Blaha, P., Schwarz, K., Improving the efficiency of FP-LAPW calculations. *Comput. Phys. Commun.*, 126, 294, 2000.
64. Khenata, R., Bouhemadou, A., Sahnoun, M., Ali, H., Reshak, H., Baltache, M., Rabah, Elastic, electronic and optical properties of ZnS, ZnSe and ZnTe under pressure. *Comput. Mater. Sci.*, 38, 29, 2006.
65. Ould NE, M.L., El hachimi, A.G., Boujnah, M., Benyoussef, A., El Kenz, A., Comparative study of electronic and optical properties of graphene and germanene: DFT study. *Optik*, 158, 693, 2018.
66. Chen, X., Yang, Q., Meng, R., Jiang, J., Liang, Q., Tan, C., Sun, X., The electronic and optical properties of novel germanene and antimonene heterostructures. *J. Mater. Chem. C*, 4, 5434, 2016.
67. Tan, C., Yang, Q., Meng, R., Liang, Q., Jiang, J., Sun, X., Ye, H., Chen, X.P., An AlAs/germanene heterostructure with tunable electronic and optical properties *via* external electric field and strain. *J. Mater. Chem. C*, 4, 1871, 2016.
68. Liu, Z., Lou, Z., Li, Z., Wang, G., Wang, Z., Liu, Y., Huang, B., Xia, S., Qin, X., Zhang, X., Dai, Y., GeH: A novel material as a visible-light driven photocatalyst for hydrogen evolution. *Chem. Commun.*, 50, 11046, 2014.
69. Liu, N., Bo, G., Liu, Y., Xu, X., Du, Y., Dou, S.X., Recent progress on germanene and functionalized germanene: Preparation, characterizations, applications, and challenges. *Small*, 20, 1805147, 2019.
70. Lin, Z.Z. and Chen, X., Transition-metal-decorated germanene as promising catalyst for removing CO contamination in H_2. *Mater. Des.*, 107, 82, 2016.
71. Feng, B., Ding, Z., Meng, S., Yao, Y., He, X., Cheng, P., Chen, L., Wu, K., Evidence of Silicene in Honeycomb Structures of Silicon on Ag(111). *Nano Lett.*, 12, 3507, 2012.
72. Ni, Z., Liu, Q., Tang, K., Zheng, J., Zhou, J., Qin, R., Gao, Z., Yu, D., Lu, J., Tunable bandgap in silicene and germanene. *Nano Lett.*, 12, 113, 2012.

73. Jiang, S., Butler, S., Bianco, E., Restrepo, O.D., Windl, W., Goldberger, J.E., Improving the stability and optical properties of germanane *via* one-step covalent methyl-termination. *Nat. Commun.*, 5, 3389, 2014.
74. Amamou, W., Odenthal, P.M., Bushong, O'Hara, D.J., Luo, Y.K., van Baren, J., Pinchuk, I., Wu, Y., Ahmed, A.S., Katoch, J., Bockrath, M.W., Tom, H.W.K., Goldberger, J.E., Kawakami, R.K., Large area epitaxial germanane for electronic devices. *2D Mater.*, 2, 035012, 2015.
75. Mortazavi, B., Dianat, A., Cuniberti, G., Rabczuk, T., Application of silicene, germanene and stanene for Na or Li ion storage: A theoretical investigation. *Electrochim. Acta*, 213, 865, 2016.
76. Serino, A.C., Ko, J.S., Yeung, M.T., Schwartz, J.J., Kang, C.B., Tolbert, S.H., Kaner, R.B., Dunn, B.S., Weiss, P.S., Lithium-ion insertion properties of solution-exfoliated germanane. *ACS Nano*, 11, 7995, 2017.
77. Wang, Y., Ji, W., Zhang, C., Li, S., Li, F., Li, P., Ren, M., Chen, X., Yuan, M., Wang, P., Enhanced band gap opening in germanene by organic molecule adsorption. *Mater. Chem. Phys.*, 173, 379, 2016.

10

2D Graphene Nanostructures for Biomedical Applications

Kiran Rana[1], Rinky Ghosh[2] and Neha Kanwar Rawat[3,4]*

[1]*Department of Textile and Fiber Engineering, Indian Institute of Technology Delhi, Delhi, India*
[2]*Department of Materials Science and Engineering, Indian Institute of Technology Kanpur, Kanpur, India*
[3]*Materials Research Laboratory, Dept. of Chemistry, Jamia Millia Islamia, New Delhi, India*
[4]*Department of Materials Engineering, Indian Institute of Science, Bangalore, India*

Abstract
2-D graphene is an array of interlinked carbon atoms in a single layer of honeycomb structure having a thickness of one atom. Due to its sp^2 hybridization and thin atomic thickness, it exhibits excellent physical and mechanical properties. These materials are having high aspect ratio, excellent electrical conductivity, ease of functionalization and biocompatibility that lead to use in various applications. Among various applications of 2-D graphene, the biomedical applications like tissue engineering, drug delivery, biological imaging, etc. have attracted the more interest due to its unique properties. This review discusses the biomedical applications of two-dimensional graphene nanostructures.

Keywords: 2-D Graphene, biomedical applications, biocompatibility

10.1 Introduction

The fast improvement in the biomedical field is due to the advancement in nanotechnology. The emerging nanotechnology in this field strongly depends upon the materials including its morphology, toxicity, tuneable materials structures, biocompatibility, physiochemical properties and its

Inamuddin, Rajender Boddula, Mohd Imran Ahamed and Abdullah M. Asiri (eds.)
Monoelements: Properties and Applications, (261–284) © 2020 Scrivener Publishing LLC

new techniques [1]. The morphology, size and surface properties of nanomaterials affects the *in vivo* biological performance [2].

The two-dimensional graphene is a new class of nanomaterials having a unique property which makes it popular in biomedical field. In 2010, the noble prize was announced to Andre Geim and Konstantin Novoselvo in physics for a one atom thick crystal graphene [3]. It is an allotrope of carbon and a crystalline material having a high aspect ratio with planer geometry. Graphene is a single layer of carbon atom linked with strong covalent bonds is also known as mono layer graphene [4]. The single layer graphene is a building block of the other carbon-based nanostructures like fullerene, carbon nanotubes (CNTs) and graphite. If these layers are stacked together on each other the 3-D graphite is formed. If the single layer is rotate about one axis, the 1-D carbon nanotube and if it wraps spherically the 0-D fullerene is obtained (Figure 10.1).

These nanosheet materials have strong chemical bonds along the plane but weak van der Waals forces among the different planes [5, 6]. Graphene has many attractive properties having high young modulus (~1 TPa), thermal conductivity (~5,000 $Wm^{-1}K^{-1}$), specific surface area (2,630 m^2g^{-1}), optical transmission (97.7%) and good electrical conductivity [7]. In biomedical applications graphene and its derivatives gain popularity due to its unique properties having biocompatibility, high drug loading capacity and easy cell membrane penetration [8].

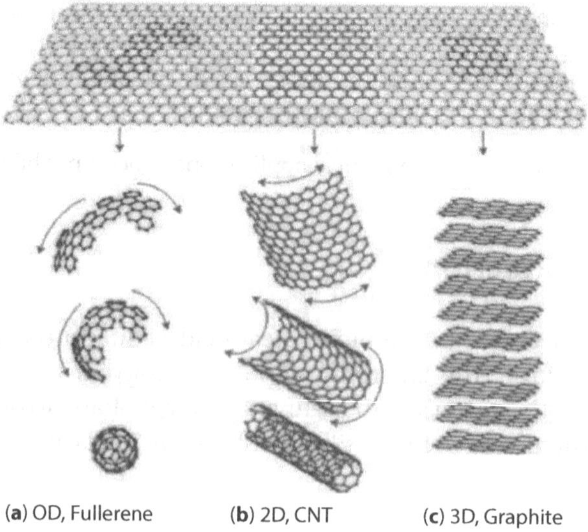

Figure 10.1 The different carbon-based nanostructures are originated from the graphene (a) fullerene (b) CNT (c) Graphite [6].

10.1.1 Synthesis Routes of Graphene

These 2-D materials are synthesized by two methods (Figure 10.2), first is top-down method where the bulk precursor material is required to make the thinner 2-D nanomaterial.

It is a destructive method which produces the products at low yield, the product strongly depends upon the precursor and some methods of this type are highly scalable. Whereas, the second method is constructive bottom-up method. In this method small molecular carbon precursor is used to get the graphene. The template routes, epitaxial growth and substrate free gas phase and chemical vapor deposition (CVD) methods [9] come under bottom-up method.

10.1.2 Graphene and its Derivatives

Other than the single layer graphene, the graphene oxide (GO) and reduced graphene (Figure 10.3) are also attractive nanomaterials in biomedical applications. The graphene oxide is prepared from bulk graphite by introducing the different oxygen functional groups like epoxides, carboxylic, and hydroxyl groups in their basal plane and at the edges through oxidation. This process makes it hydrophilic and with the help of thermal and/or mechanical method it is exfoliated into the monolayers [10].

The three main solution-based methods for the synthesis of graphene oxide are Staudenmaier, Hoffman and Hummers methods by using different concentrated acids which are concentrated nitric acid, concentrated sulfuric acid, potassium chlorate, potassium permanganate [8, 11]. The Hummer method is widely used for the synthesis of graphene oxide. Graphene oxide has good chemical properties including oxygen functionality, biocompatibility, and large surface area which allow it to be used in biosensing, bioimaging biomedical applications [12]. The oxygen functionalization groups of graphene oxide allow to it interact with different biomolecules like enzymes, DNA, and proteins. The different methods are used to convert the graphene into reduced graphene. The different reducing agents are used in the synthesis of reduced graphene oxide. But for biomedical applications the reducing agent should be biocompatible. The ascorbic acid is non-toxic, biocompatible reducing agent for biomedical

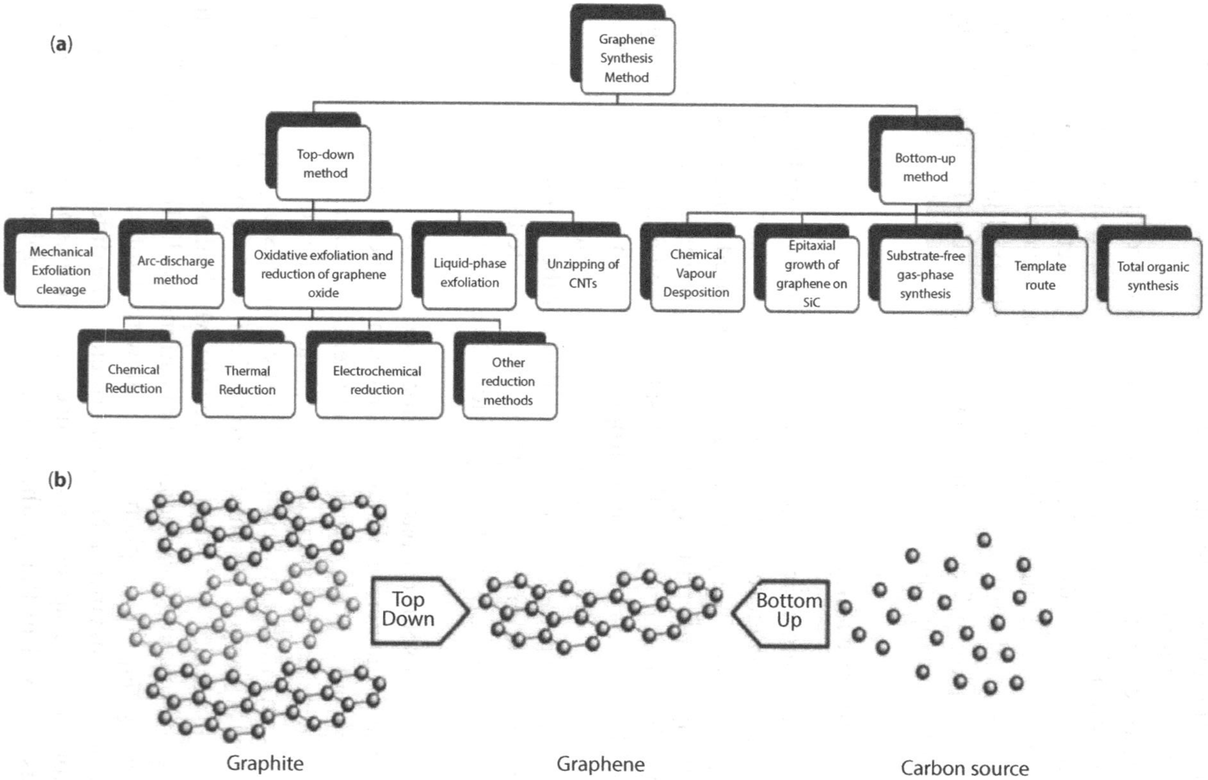

Figure 10.2 (a) The different synthesis methods of graphene and (b) illustration of top-down and bottom-up methods [9].

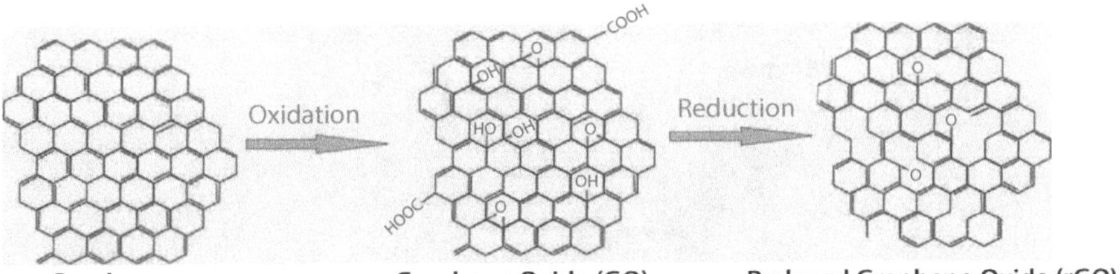

Figure 10.3 The conversion process of graphene oxide and reduced graphene oxide [12].

applications of graphene [13]. This review will give the brief overview of various biomedical applications of graphene and graphene oxide.

10.2 Applications of Graphene in Biomedicine

10.2.1 Tissue Engineering

The human body does not easily regenerate organs and some type of tissues like one after any sudden trauma, disease and inherited defect. Due to these problems, if the tissues are damaged then they cannot be replaced or repaired by the body itself and can affect the surrounding tissues also. In these types of critical areas tissue engineering plays an important role. Tissue engineering is a scientific and multidisciplinary field for the development of biological substitutes like tissues and organs by controlling the parameters of biomechanical and biological. With the help of various engineering principles and with the concepts of life science including cell biology, molecular biology and material engineering, they are used for creating three-dimensional tissue scaffolds for organs and/or repairing the function of tissues [14] like bone repair [15, 16], cardiac tissue [17] and cartilage tissue [18]. Tissue engineering solves the problems associated with organ transplant rejection, non-availability of donor, and surgical reconstruction.

Scaffolds, cells and bioreactors are necessary elements in tissue engineering as shown in Figure 10.4. The combination of all three elements is required for the success of tissue engineering. Scaffold is a structure or a support for the cell growth which mimics the real organs. These structures can be artificial plastics based or built by body proteins. Various types of scaffolds are prepared by different techniques and a variety of biomaterials. But some of the common properties are shared by all types of scaffolds. They should be biocompatible, biodegradable, consistent mechanical properties mostly required in orthopedic applications, porous and interconnected

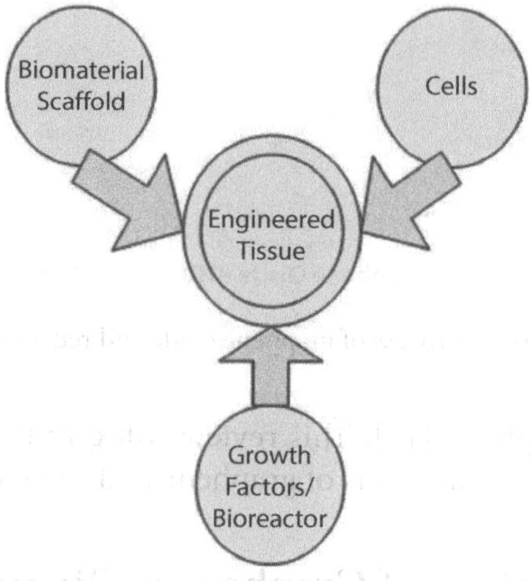

Figure 10.4 The basic components of tissue engineering [19].

structure [19]. The selection of biomaterial for specific application is also an important step.

Cells and growth factor/bioreactor are also part of the triads of tissue engineering. Cells are used to create the tissue. They integrate themselves and convert into tissue. Cells can be used from the patients as a primary cell and conjunction with scaffold to create a tissue for implantation. But due to the invasive nature and difficulty in growth of a specific type of cell, stem cells come into the picture. There are two types of stem cells: embryonic stem cells (pluripotent cells) and adult stem cells (multipotent cells). The embryonic stem cells are limited in use due to controversies of producing teratomas [20]. The cells are required certain biochemical and physical regulatory signals for their survival and in making extracellular matrix which will replace the scaffold [21].

10.2.1.1 Cartilage Tissue Engineering

The articular cartilage is a soft, flexible and firm connective tissue in between the joint of bones like knee (Figure 10.5), elbow, and the ends of the ribs [22]. Articulate cartilage is made by a single type of cell called chondrocytes and it does not contain any blood vessels or nerves. The hyaline tissue is at the surfaces of a joint of a bone called articulate cartilage which provides low friction and has shock absorption property. This extracellular matrix is very hydrated (70–80%) and is made up of collagen type II which provide tensile strength along with proteoglycan aggregates

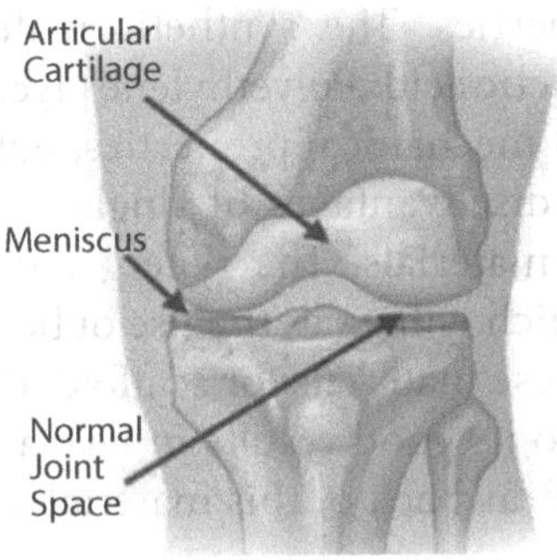

Figure 10.5 Image of a healthy knee [22].

providing compressive strength. The meniscus helps for easy transfer of mechanical load, lubrication, and joint stability.

Generally, the damages in the articulate cartilage are due to aging, trauma, and diseases like osteoarthritis, cancer, etc. [23]. The repairing and healing process of damaged cartilage is too slow because of the absence of blood vessels. So, it is one of the most challenging problems of orthopedic research. Thus, tissue engineering provides a solution for damaged cartilage. Hyaline or hyaline-like tissue is required for the repairing of damaged area. For the development of these types of tissues the specific type of cells is required which produces a chondrogenic phenotype. Chondrocytes, mesenchymal stem cells (MSCs) [23], pluripotent stem cells and fibroblast [24], are the different types of seed cells used in regenerative cartilage engineering. The chondrocytes cell is the important component in articulate cartilage which maintains a cartilaginous matrix. For repairing and rebuilding the cartilage, the autologous strategy is preferable than the allogenic strategy [25]. The standard surgical procedures are autologous transplantation and autologous chondrocyte implantation.

a) Scaffold

Scaffolds are the supporting elements for the cell seeding, proliferation and differentiation which work like an extracellular matrix in natural tissues. Different types of extracellular type natural and synthetic materials are present for cartilage tissue engineering. The natural materials are proteins and polysaccharides like collagen, fibrin, silk, hyaluronic acid, chitosan, cellulose, and alginate. They have good biocompatibility, but weak mechanical properties. The synthetic materials like Poly (lactic-co-glycolic acid), Polylactic acid, Polyethylene glycol and polycaprolactone have good porosity and mechanical properties, but can cause cytotoxicity. So, to overcome these disadvantages of single materials, the researchers focused on composite materials. The synergistic effect is obtained with composite materials which enhances the use of both synthetic and natural materials. Among all these materials for scaffold, the polycaprolactone has been widely used due to its good results *in vivo* and in repairing of damaged cartilage. But the drawback of this material is its hydrophobicity and slower degradation rate.

b) Graphene-Based Scaffolds for Articulate Cartilage

Graphene-based composites improve the mechanical, proliferation and differentiation rates [26]. Liao *et al.* [27] reported that a small amount of graphene oxide (GO) was incorporated in the hydrogel of methoxyl poly(ethyleneglycol)-poly(ε-caprolactone)-acryloyl chloride (MPEG-PCL-AC) and methacrylated chondroitin sulfate (CSMA) which improved the mechanical strength, good biocompatibility and enhanced the specific differentiation into bone cells. This hybrid composite scaffold had a favorable rate of degradability compared with healing rates of damaged cartilage. It provided the high porosity, electrical conductivity and appropriate environment for regeneration. The differentiation rate of cells was enhanced with the porosity of the graphene. The graphene was used as a planar culture platform to increase the cell growth and differentiation [28]. The plastic-based nonbiological materials used in implants behave differently from neighboring tissues under mechanical strain. Some bioactive materials also showed some pathogenic infections. The versatile material graphene is used in implants due to its remarkable properties as they are a non-brittle material, has good mechanical strength, and is easily available. Graphene also enhances the rate of differentiation even in the absence of any cell growth factors like BMP-2. It does not affect the shape of stem cell and hampers the physiological conditions of microenvironment [29]. Zhou *et al.* [30] reported that 3D printed graphene oxide-based gelatin methacrylate–poly (ethylene glycol) diacrylate scaffold promotes the chondrogenic differentiation of MSCs cell. This scaffold showed better mechanical strength and good biocompatibility. The presence of graphene oxide helped with the MSCs cell growth, its adhesion to the scaffold, and differentiation due to its hydrophilic functional groups and good adsorption capacity by π–π interaction and electrostatic interaction. This property of graphene oxide supported the loading of nutritious components on the scaffold which helped cell survival and cell proliferation. Shamekhi *et al.* [31] fabricated the chitosan and graphene oxide scaffold for cartilage tissue engineering. Chitosan and graphene oxide showed antimicrobial, biodegradable, and good mechanical properties. The porosity and permeability of the scaffold was slightly improved with the graphene oxide.

10.2.2 Bone Tissue Engineering

10.2.2.1 Methods of Fracture Repair

The fracture in the bone disturbs the vasculature. Due to this, the mechanical stability is lost as well as the reduction in oxygen level and nutrient supply at the fracture site [32, 33]. The natural process of bone healing is followed by the formation of hemotoma, a blood that comes out from ruptured blood vessels which starts to accumulate in and around the fractured bone. Then, the soft callus is formed by connective tissue cartilage and blood vessels. The fibroblast, chondrocyte and osteoblast make a scaffold between the gaps of fracture. After 6–12 weeks the soft callus is converted into hard callus by enchondral ossification. And last the healed bone will continue to remodel into normal bone [34–36]. In cases where the natural process does not happen such as non-union fracture, infection, traumatic injury, other approaches are considered. These approaches include internal fixation [37], bone grafting such as allografting or autografting and synthetic subsitutes [38]. The goal of internal fixation is to provide stability and helps to osseous healing of broken bone. But there are some complications associates with these internal fixations including bio-incompatibility, allergic reactions, malignancy and osteoporosis. Bone grafting is the gold standard for bone healing fracture repair and spinal fusion [39]. In auto grafting, the natural bone tissues are harvested from the patient. There are limited sites from where the bone tissues are harvested, and it is less effective for defects of irregular shape [40]. The problems with auto grafting are infection, pain and nerve injury. In allograft, the tissues are taken from cadavers or from other patients. The risks with this grafting are transmission of disease and immunologic rejection. It is osteoconductive, but does not produce osteogenic cells [41]. Due to various complications associated with these discussed techniques, researchers are exploring different alternative effective methods.

10.2.2.2 Graphene Used in Bone Tissue Engineering

Tissue engineering is an emerging technique for bone regeneration. This technique requires the substrate or scaffold which helps the osteoprogenitor cell attachment, proliferation and differentiation. Bone scaffold materials are biodegradable, porous, osteoinductive, osteoconductive, osteointegrative and help to give mechanical support. The graphene is successfully used as a scaffold material due its unique properties. Nair and his coworkers reported the 3D scaffold made of graphene oxide nanoflake and gelatin–hydroxyapatite. The incorporation of graphene oxide nanoflakes (0.5 and 1 wt%) in gelatin–hydroxyapatite matrix (GHA) by freeze drying increased its mechanical properties and osteogenic differentiation. The compression strength of the scaffold was higher than the scaffold with graphene oxide flakes. The toxicity to stem cells is reduced on the interaction of graphene oxide with gelatin in the matrix. The osteogenic differentiation of human adipose derived MSCs (ADMSCs) stem cells could be enhanced by graphene oxide without any supplements such as dexamethasone and β-glycerophosphate [42].

The presence of nanostructures at the surface of material which can improve the bioactivities was also studied. The graphene coated collagen scaffold improved its surface structure, cell ingrowth and compressive strength. An appropriate concentration of graphene showed a better result for bone tissue engineering [43].

Nie *et al.* synthesized the porous scaffold with nano-hydroxyapatite (nHA) and reduced graphene oxide (RGO) by *in situ* self-assembly. Some 20% nHA–RGO showed improved cytocompatibility than the RGO scaffold. The *in vivo* subcutaneous implant experiments showed biocompatibility and healing in 6 weeks. According to this study, the scaffold made of suitable concentration of nHA with reduced graphene oxide could be an efficient scaffold for bone repairing [44]. The green fabrication of hybrid scaffold from silk fibroin–graphene oxide (GO) was studied. The silk fibroin (SF) is a natural protein and has potential to be use in biomedical applications due to its unique properties like biocompatibility, biodegradability, and good mechanical strength. The GO/SF hybrid showed the improved and uniform pore size of scaffold. As the concentration of graphene oxide increased (0 to 1 wt%), the pore diameter decreased with improved the uniformity of pore diameter. The presence of GO also improved the resistance to enzyme degradation [45]. For improving the cell attachment and biological properties of poly(ε-caprolactone) (PCL) scaffolds, a low concentration of graphene was used and NaOH surface treatment. The scaffold made by extrusion-based additive manufacturing technique showed a good control over the topography of the scaffold. It also showed good cell proliferation and showed no cytotoxicity [46].

10.2.3 Gene Delivery

Graphene has been considered as a star material, possessing an ultra-high surface area and chemical purity useful for the efficient incorporation of the hydrophobic drug loadings via π–π sp^2 bonded carbon stacking. The favorable interaction between the π–π stacking and therapeutic agents attributed to the presence of delocalized electrons in the π-clouds of graphene. Therefore, intensive efforts have been devoted to recognizing its potential applications in biomedicine for drug/gene delivery and cancer therapy, etc. [47, 48] (Figure 10.6). The novel gene therapy has garnered the utmost importance and is considered as a promising approach to treat genetic disorders, associated with cancer, Parkinson's disease, etc. The gene vector is the important process for gene therapy which acts as the vehicle for delivering of foreign DNA into the recipient cells and further protects the DNA from nuclease degradation and helps in cellular uptake with high transfection efficiency [49, 50]. However, the inefficacy and inadequacy of the gene vector has failed to express its prominent aspects in gene therapy application. The first seminal reports of graphene as therapeutic agents were reported by Dai *et al.* and described the potency of pegylated graphene oxide (PEG-GO) as an efficient nanocarrier for immobilization of various aromatic and insoluble cancer drugs. Recently, graphene and its derivatives are highly explored owing to its ideal functionalization,

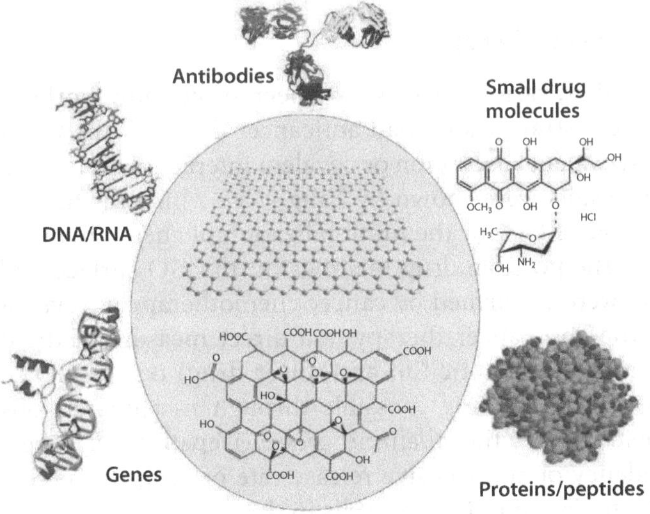

Figure 10.6 Schematic representation of the application of 2D graphene for various drug delivery of therapeutic agents and anticancer drugs [62].

mechanical stiffness with a record-breaking thermal conductivity. The various parameters such as mechanical strength, flexibility, conductivity, electrical and thermal properties are supreme which make them an ideal candidate for tissue engineering applications [51].

The excitement achievements on 2D graphene-based nanomaterial for gene delivery applications have been first reviewed by Liu *et al.* At the foremost the detailed structural idea to compensate the difficulties on carrying/immobilizing double-stranded RNA or DNA was abolished by making the graphene layers thick, ranging for about few nanometers to hundreds of nanometers [52–54]. Generally, the thick layers of graphene oxide (GO) sheets comprised of one to three layers to effectively immobilize single-stranded DNA or RNA over double-stranded by utilizing physisorption or chemical conjugation techniques. Further, approaches reported for the cationic functionalization of polyethyleneimine (PEI) using 2D graphene (PEI-GO) for gene delivery purposes as it exhibited low cytotoxicity, were significantly explored for various biomedical applications. The PEI–GO complex offers electrostatic interaction with the negatively charged phosphates group of DNA/RNA and thereby improves transfection efficiency [55, 56]. The capability of loading drugs arises from the excellent biocompatibility, large specific surface area, physicochemical properties, and the presence of oxygen-containing functional groups, render GO an excellent choice for diverse biological and drug delivery applications.

10.2.4 Cancer Therapy

In recent studies, various groups have been profoundly working on developing different strategies to load anticancer drugs on the GO surface via either non-covalent adsorption or covalent interactions. However the most prominent results were shown by Yang *et al.* which further depicted the significant role of GO for the immobilization of chemotherapeutic cancer drugs [57]. The possible drug loadings on the GO surface and the drug release rate were confirmed by cancer chemotherapy intravenous administration. The anti-cancer therapy is a direct measure of the delivery of an anthracycline antibiotic (an anticancer drug) on the GO surface, the capability of drug loadings possible through π–π interactions with the hydrophobic part of the quainine group. Depan *et al.* have confirmed the dependency of an extensive release rate of drugs from GO following the change in pH around cancer cells [58].

An efficient approach to develop the glioma-targeted drug delivery system based on 2D graphene material has been investigated by Wang *et al.*

Glioma refers to the malignant tumor of the glial tissue of the nervous system [59]. In this novel approach, the highest number of glial cells death, is successfully achieved with the help of Doxorubicin loaded onto chlorotoxin-conjugated graphene oxide (CTX-GO/DOX) via non-covalent interactions [60, 61]. Furthermore, the controlled release of doxorubicin rate can be attained by adjusting the pH values.

The challenges of the precise cancer therapy for an effective target area are to significantly improve the therapeutic efficacy through the homogeneous distribution of therapeutic drugs and have an adequate drug penetration at the targeted tumor [63]. An ultrathin 2D graphene oxide (GO) with ultrasmall particle sizes ameliorating the accumulation of drugs and facilitates the penetration of therapeutic drugs, are potential candidates for drug delivery [64, 65]. Graphene-based materials have gained tremendous attention recently as it precisely and remotely releases drugs and penetrates deep into the tumor site to kill cancer cells, thereby promoting osteogenesis and neurogenesis for cell proliferation and differentiation.

Shang *et al.* have designed and developed hybrid viral molecules based on a rabies virus glycoprotein (RVG) comprised of boron-doped graphene quantum dot (B-GQD) and pH-responsive dendrimers (pH-Den) [66]. These composite hybrid materials facilitate the transports of dual drug penetration to deep tumors which sufficiently improves the distribution of therapeutic agents and controlled the release of drugs to orthotopic tumors by enhancing the tumor retention efficacy [67]. It remarkably impedes the manifestation of tumor growth and exhibited synergistic effects in cancer cell lines, disclose insignificant side effects *in vivo*.

10.2.5 Genotoxicity

Graphene is a highly versatile material and tremendously recognized in real-world medicinal applications. However, its cytotoxicity and safety issues have not been resolved yet. Theranostic applications of nanomaterial require reported studies related to its toxicity. However, authors have reported through *in vitro* analysis of GO on A549 cells revealing no obvious toxicity observed but rather that the cytotoxicity is dose-dependent [68]. In another attempt, Wang *et al.* have reported the biocompatibility of GO when tested against human fibroblast (HDF) cells, confirming GO under a high dose exhibited toxicity while at low dose did not exhibit any obvious toxicity [69]. Many works of literature on graphene family, depicted valuable data have indicated less toxicity and biocompatibility of promising graphene hybrid structures for anti-cancer drug carriers, but the outcomes obtained are so far conflicting. The unique chemical vapor deposition techniques were utilized to synthesis graphene film for human stem cells attachment which allow proliferations are markedly contributing to the biocompatibility without the signs of cytotoxicity. Toxicity is a major concern in using 2D graphene for clinical applications. Recently graphene has been utilized for potential drug carriers in an *in vivo* cancer treatment, and the results were confirmed by extensive studies using cells and animal models. Hondroulis *et al.* have explored the toxicity of graphene-related materials by calculating the trans-endothelial electrical resistance when used against an *in vitro* model of the blood–brain barrier (BBB) [70]. Further outcomes demonstrated the biocompatibility and strong toxicological responses toward individual cell types, thereby, have attracted substantial attention in biomedical applications.

It was found that GO possesses thrombogenicity in response to human platelets which limits its practical applications as therapeutic agents. However, amine-modified graphene showed striking contrasting observation when used intravenously against human platelets, and has proven to be advantageous and exhibited no pulmonary thromboembolism in mice, induced no stimulatory effect on human platelets.

10.2.6 2D Application of Graphene in Biosensing

Graphene and its derivatives have been intensively studied in biosensors owing to its high surface area, excellent conductivity, redox stability, and low Johnson noise [71–75]. Therefore, this graphene-based versatile material evolved as a suitable candidate for sensing and detection of diverse biomolecules such as oligonucleotide, amino acids, dopamine, thrombin, biomolecules, etc.

Wang *et al.* have successfully reported the efficient immobilization of oligonucleotides onto the living cells for *in situ* probing of bio-molecules via graphene-based matrices resulting in a composite electrode [75]. Based on the electronic conductivity and charge transfer capability, these unique features of graphene composites make it a promising material for the early-stage detection of DNA, glucose, and proteins through electrochemical sensing.

Several types of graphene-based biosensors have been constructed due to its zero bandgap semiconductor property rendering it a suitable candidate for biosensing devices. The commonly fabricated biosensors are fluorescence resonance energy transfer (FRET), plasmonic, matrix-assisted laser desorption/ionization time-of-flight mass spectrometer (MALDI-TOF MS), and field-effect transistor (FET) which highly used for the detection of analytes.

Recently Mao *et al.* have investigated a versatile biosensing platform for detection of DNA aptamers and negatively charged bacteria [76]. The developed DNA biosensor exhibited high specificity, ultrasensitivity, and regeneration capability and therefore is a promising tool in disease diagnostics.

10.2.7 Prosthetic Implants

An artificial implantable biomedical device based on carbonaceous materials has offered unique advantages with improved qualities of high interest in terms of mechanical stability, degradation rate, thromboresistant, durability, and strength. The Retinitis Pigmentosa (RP) and Age-related Macula Degeneration (AMD) are such common eye diseases, affecting most of the people around the globe which lead to vision misfortune due to retinal degeneration. Therefore, soft bioelectronics devices have played an important role in the retinal prosthesis and optical sensing. Kim *et al.* have demonstrated the first-ever attempt to successfully achieve the human eye-based soft optoelectronic implantable device using a 2D Ultrathin MoS_2-graphene hemispherically curved image sensor (CurvIS) array [77–79]. This high-density CurvIS array exhibited good quality imaging by stimulating the retinal nerves in response to the optical signals with minimum side effects leading to restoration of vision by stimulating the photoreceptors cells in the retina [80]. Furthermore, the proposed soft materials acquire pixelated optical signals with minimal mechanical distortion results in mimicking the structural features of the human retina.

The abilities of graphene-based materials and its derivatives to combine and functionalize with several other nanomaterial offer several advantages in designing bone plates. Bone injury is a major challenge in the reconstruction of fractures and congenital defects resections, which could be solved using bone plates. However, metals and ceramic-based composites are poor substitutes for bone plates due to stress shielding effect and are inherently brittle and corrosive. In various literature it is claimed that GO on the other hand possesses excellent bioactivity with improved osteogenic potential *in vitro* and *in vivo*, thus is an ideal bone plate material and exhibited similar mechanical characteristics to the bone. The cytotoxicity and biocompatibility of GO were tested by Pang *et al.* [81]. In this *in vivo* study, GO was directly injected into the mouse skeletal muscle to investigate the stimulative effect of the orthopedic implants on the body. It has been illustrated that there has been increased proliferation and adhesion of osteoblast cells on GO surfaces and showed good compatibility to human osteoblasts. Further, it illustrated the potential persuading of the cytotoxicity, resulting from the decrease in the mitochondrial membrane potential.

In recent studies, almost all attention has been given to the GO-based composites bone plates (Figure 10.7). The poly(methyl methacrylate) (PMMA) reinforced with graphene nanoplates, demonstrates its potential properties in orthopedic applications. Additionally, it improved the load-bearing capacity as well as wound healing characteristics, manifested for bone replacement, and significantly, considered as a desirable candidate for an implant [82, 83].

Biomaterials such as hydroxyapatite (HA, $Ca_{10}(PO_4)_6(OH)_2$), is a highly promising material that exhibits exceptionally great biocompatibility and osteoconductivity for use in dental/teeth implants [85, 86]. However, HA showed poor load-bearing capacity for long term use and lower mechanical properties, make them an unsuitable candidate for oral and dental fields. One of the major reasons for its disregard is that it undergoes some chemical reactions between the biological fluids/saliva present in the mouth and the implantable tooth make up of HA. Furthermore, the reaction accelerated by the presence of fermentation sugars and starches, leads to the dissolution of HA in acidic pH condition. Therefore it is another obstacle towards its successful application. However, fluorine substitution to the hydroxyl group of the HA (FHA), has been found to improve the stability against acidic pH and possesses antimicrobial activity towards bacteria causing disintegration of HA. Additionally, graphene and its derivatives as a reinforcing filler to HA have found wide acceptance in orthopedic implants which eventually improve the chemical stability and impart mechanical integrity thus facilitate the stress transfer in the composite. Therefore, this novel bio-composite material based on GO/FHA exhibits long term biostability, high impact resistance, and low dissolution rate in the organic acids, making them future prospects for dental implants (Figure 10.8).

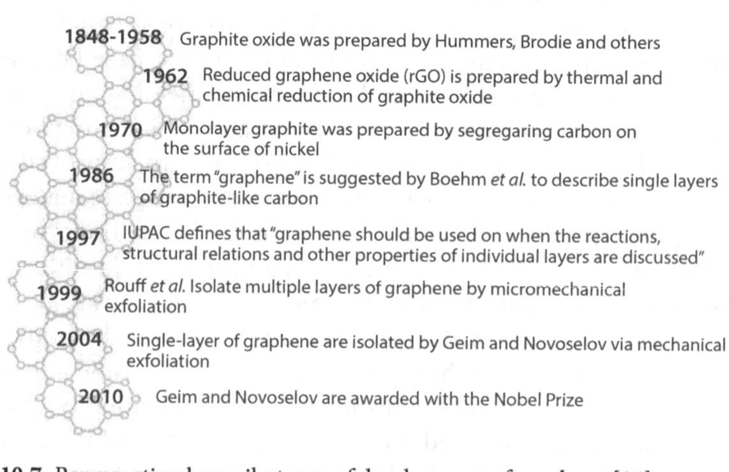

Figure 10.7 Representing key milestones of development of graphene [84].

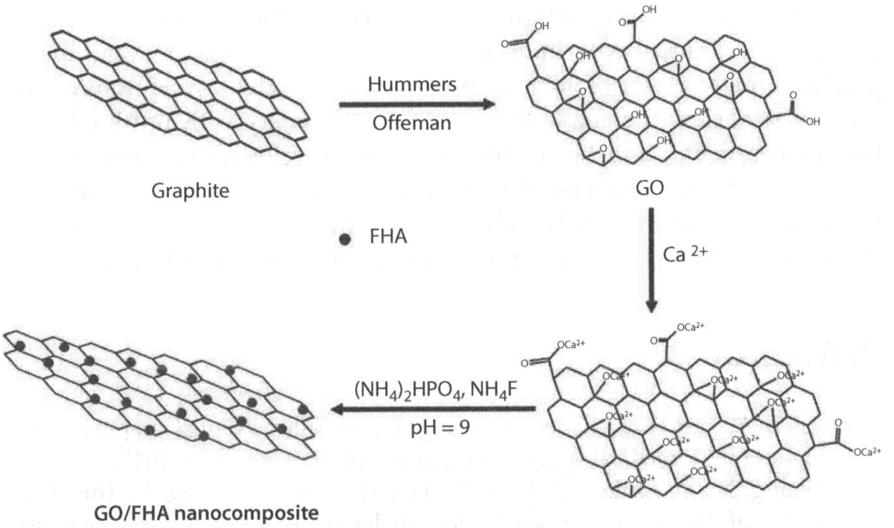

Figure 10.8 Showing a synthesis method of GO/FHA nanocomposite [87].

10.3 Conclusion

During the last few decades, two-dimensional graphene has been widely exploited in biomedical fields. Recently, these materials have created chaos among biomedical scientists as evidenced from the significant increase in the trend of literature studies related to graphene potential applications in biomedicine. 2D graphene and its derivatives are carbonaceous based materials that offer several advantages owing to its ultrahigh surface area, excellent conductivity and mechanical integrity. Additionally, numerous pre-clinical test results reported the utmost importance of graphene in drug delivery/cancer therapy, owing to its fine-tuned physicochemical property and bioactivity. Functionalization of graphene provides an opportunity to tailoring its property for future prospects in nanomedicine. Graphene family have been extensively exploited for gene delivery, tissue engineering, prosthetic implants, cancer therapy, and biosensing field. However the long term toxicity concern of using graphitic materials has adverse health impacts and must be considered in the designing and development of graphene for biomedical applications. Therefore, proper understanding of the potential toxicity results from the *in vivo* degradation of the graphene in response to external stimuli is of fundamental importance for future development in biomedicine. Although many reports have suggested the toxicity studies of graphene and the resultant outcomes depicted to be safer when used at a certain doses. Therefore, the complete characterizations and animal experiments are essential for the safer designing of graphene materials for biological applications. Thereby, tremendous efforts are still needed to persuade the long term biosafety with minimal risks for environmental and health for future use in clinical purpose. Future research shall also focus on determining the safety aspects of graphene-based materials by efficiently collaborated with various branches of science and engineering in conditions relevant to their intended clinical use.

References

1. Chen, Y., Tan, C., Zhang, H., Wang, L., Two-dimensional graphene analogues for biomedical application. *Chem. Soc. Rev.*, 44, 2681, 2015.
2. Huang, X., Li, L., Liu, T., Hao, N., Liu, H., Chen, D., Tang, F., The Shape Effect of Mesoporous Silica Nanoparticles on Biodistribution, Clearance, and Biocompatibility *in Vivo*. *ACS Nano*, 5, 5390–5399, 2011.
3. It's still all about graphene. *Nat. Mater.*, 10, 1, 2011.
4. Tadyszak, K., Wychowaniec, J.K., Litowczenko, J., Biomedical Applications of Graphene-Based Structures. *Nanomaterials*, 8, 944, 2018.
5. Rafiei-Sarmazdeh, Z., Zahedi-Dizaji, S.M., Kang, A.K., *Two-Dimensional Nanomaterials*, IntechOpen, 2019, http://dx.doi.org/10.5772/intechopen.85263
6. Mas-Balleste, R., Gomez Navarro, C., Gomez Herrero, J., Zamora, F., 2D Materials: To graphene and beyond. *Nanoscale*, 3, 20, 2011.
7. Alam Bhuyan, Md. S., Uddin, Md. N., Islam, Md. M., Alam Bipasha, F., Hossain, S.S., Synthesis of graphene. *Int. Nano. Lett.*, 6, 65–83, 2016.
8. Dideikin, A.T. and Vul, A.Y., Graphene Oxide and Derivatives: The Place in Graphene Family. *Front. Phys.*, 6, 149, 2019.
9. Leea, X.J., Hiewa, B.Y.Z., Lai, K.C., Lee, L.Y., Gana, S., Thangalazhy-Gopakumar, S., Rigby, S., Review on graphene and its derivatives: Synthesis methods and potential industrial implementation. *J. Taiwan Inst. Chem. E.*, 98, 163–180, 2018.
10. Bojana, V., Properties of two-dimensional graphene-like materials. *Nanomater. Energy*, 4, 18–29, 2014.
11. Poh, H.L., Sanek, F., Ambrosi, A., Zhao, G., Soferb, Z., Pumera, M., Graphenes prepared by Staudenmaier, Hofmann and Hummers methods with consequent thermal exfoliation exhibit very different electrochemical properties. *Nanoscale*, 4, 3515, 2012.
12. Priyadarsini, S., Mohanty, S., Mukherjee, S., Basu, S., Mishra, M., Graphene and graphene oxide as nanomaterials for medicine and biology application. *J. Nanostructure Chem.*, 8, 123–137, 2018.

13. Shin, S.R., Li, Y.-C., Jang, H.L., Khoshakhlagh, P., Amir Nasajpour, M.A., Zhang, Y.S., Tamayol, A., Khademhosseini, A., Graphene-based materials for tissue engineering. *Adv. Drug Deliv. Rev.*, 105, 255–274, 2016.
14. Langer, R. and Vacanti, J.P., Tissue Engineering. *Science*, 260, 920–926, 1993.
15. Shadjou, N. and Hasanzadeh, M., Graphene and its nanostructure derivatives for use in bone tissue engineering: Recent advances. *Biomed. Mater. Res.*, 104, 1250-1275, 2016.
16. Dubey, N.K., Bentini, R., Islam, I., Cao, T., Castro Neto, A.H., Rosa, V., Graphene: A Versatile Carbon-Based Material for Bone Tissue Engineering. *Stem Cells Int.*, 2015, 804213, 1–12, 2015.
17. Shin, S.R., Zihlmann, C., Akbari, M., Assawes, P., Cheung, L., Zhang, K., Manoharan, V., Zhang, Y.S., Yüksekkaya, M., Wan, K.-t., Nikkhah, M., Dokmeci, M.R., Tang, X., Khademhossein, A., Reduced Graphene Oxide-GelMA Hybrid Hydrogels as Scaffolds for Cardiac Tissue Engineering. *Small*, 12, 3677–3689, 2016.
18. Pathmanapan, S., Sadagopan, A., Kumar, S., Graphene and Composite based Biomaterials in Cartilage Tissue Engineering Application. *Am. J. Pharmacol.*, 1, 2018.
19. O'Brien, F.J., Biomaterials & scaffolds for tissue engineering. *Mater. Today*, 14, 88–95, 2011.
20. Castells-Sala, C., Alemany-Ribes, M., Fernández Muiños, T., Recha-Sancho, L., López-Chicón, P., Aloy-Reverté, C., Caballero-Camino, J., Márquez-Gil, A., Semino, C.E., Current Applications of Tissue Engineering in Biomedicine. *J. Bioeng. Bioelectron.*, 2013.
21. Mhanna, R. and Hasan, A., Introduction to Tissue Engineering, in: *Tissue Engineering for Artificial Organs*, pp. 1–34, 2017.
22. Ponemone, V., Gupta, S., Suthar, M., Emerging potential of cell-based therapies for articular cartilage repair and regeneration. *Adv. Tissue Eng. Regen. Med.*, 3, 2, 342–353, 2017.
23. Tuli, R., Li, W.-J., Tuan, R.S., Current state of cartilage tissue engineering. *Arthritis Res. Ther.*, 5, 235–238, 2003.
24. Huang, H., Xu, H., Zhang, J., *Current Tissue Engineering Approaches for Cartilage Regeneration*, Intechopen, 84429, 2018, http://dx.doi.org/10.5772/intechopen.84429
25. Baugé, C. and Boumédiene, K., Use of Adult Stem Cells for Cartilage Tissue Engineering: Current Status and Future Developments. *Stem Cells Int.*, 438026, 14, 2015.
26. Pathmanapan, S. and Sadagopan, S.K.A., Graphene and Composite based Biomaterials in Cartilage Tissue Engineering Application. *Am. J. Pharmacol.*, 1, 1006, 2018.
27. Liao, J.F., Qu, Y., Chu, B.Y., Zhang, X.N., Qian, Z.Y., Biodegradable CSMA/PECA/Graphene Porous Hybrid Scaffold for Cartilage Tissue Engineering. *Sci. Rep.*, 5, 1, 9879, 2015, https://doi.org/10.1038/srep09879

28. Lee, W.C., Lim, C.H., Kenry, S.C., Loh, K.P., Lim, C.T., Cell-Assembled Graphene Biocomposite for Enhanced Chondrogenic Differentiation. *Small*, 11, 8, 963–969, 2014.
29. Nayak, T.R., Andersen, H., Makam, V.S., Khaw, C., Bae, S., Xu, X., Ee, P.-L.R., Ahn, J.-H., Hong, B.H., Pastorin, G., Ozyilmaz, B., Graphene for Controlled and Accelerated Osteogenic Differentiation of Human Mesenchymal Stem Cells. *ACS Nano*, 5, 6, 4670–4678, 2011.
30. Zhou, X., Nowicki, M., Cui, H., Zhu, W., Fang, X., Miao, S., Lee, S.-j., Keidar, M., Zhang, L.G., 3D biopritined graphene oxide—Incorporated matrix for promoting chondrogenic differentiation of human bone marrow mesenchymal stem cells. *Carbon*, 116, 615–624, 2017.
31. Shamekhi, M.A., Mirzadeh, H., Mahdavi, H., Rabiee, A., Mohebbi-Kalhori, D., Eslaminejad, M.B., Graphene oxide containing chitosan scaffolds for cartilage tissue engineering. *Int. J. Biol. Macromol.*, 127, 396–405, 2019.
32. Bahney, C.S., Zondervan, R.L., Allison, P., Theologis, A., Ashley, J.W., Ahn, J., Miclau, T., Marcucio, R.S., Hankenson, K.D., Cellular Biology of Fracture Healing. *J. Orthop. Res.*, 37, 35–50, 2019.
33. Beamer, B., Hettrich, C., Lane, J., Vascular Endothelial Growth Factor: An Essential Component of Angiogenesis and Fracture Healing. *HSS J.*, 6, 85–94, 2010.
34. Sathyendra, V. and Darowish, M., Basic Science of Bone Healing. *Hand Clin.*, 29, 473–481, 2013.
35. Ghiasi, M.S., Chen, J., Vaziri, A., Rodriguez, E.K., Nazarian, A., Bone fracture healing in mechanobiological modeling: A review of principles and methods. *Bone Rep.*, 6, 87–100, 2017.
36. Holmes, D., Closing the gap. *Nature*, 550, S194–S195, 2017.
37. Perez, J.R., Kouroupis, D., Li, D.J., Best, T.M., Kaplan, L., Corre, D., Tissue engineering and cell-based therapies for fracture and bone defect. *Front. Bioeng. Biotechnol.*, 6, 105, 2018.
38. Ansari, M., Bone tissue regeneration: Biology, strategies and interface studies. *Prog. Biomater.*, 8, 223–237, 2019.
39. Flierl, M.A., Smith, W.R., Mauffrey, C., Irgit, K., Williams, A.E., Ross, E., Peacher, G., Hak, D.J., Stahel, P.F., Outcomes and complication rates of different bone grafting modalities in long bone fracture nonunions: A retrospective cohort study in 182 patients. *J. Orthop. Surg. Res.*, 8, 33, 2013.
40. Mistry, A.S. and Mikos, A.G., Tissue Engineering Strategies for Bone Regeneration. *Adv. Biochem. Eng./Biotechnol.*, 94, 1–22, 2005.
41. Baldwin, P., Li, D.J., Auston, D.A., Mir, H.S., Yoon, R.S., Koval, K.J., Autograft, allograft, and bone graft substitutes: Clinical evidence and indications for use in the setting of orthopaedic trauma surgery. *J. Orthop. Trauma*, 33, 203–213, 2019.
42. Nair, M., Nancy, D., Krishnan, A.G., Anjusree, G.S., Vadukumpully, S., Nair, S.V., Graphene oxide nanoflakes incorporated gelatin–hydroxyapatite scaffolds enhance osteogenic differentiation of human mesenchymal stem cells. *Nanotechnology*, 26, 161001, 2015.

43. Nishida, E., Miyaji, H., Takita, H., Kanayama, I., Tsuji, M., Akasaka, T., Sugaya, T., Sakagami, R., Kawanam, M., Graphene oxide coating facilitates the bioactivity of scaffold material for tissue engineering. *Jpn. J. Appl. Phys.*, 53, 06JD04, 2014, http://dx.doi.org/10.7567/JJAP.53.06JD04
44. Nie, W., Peng, C., Zhou, X., Chen, L., Wang, W., Zhang, Y., Ma, P.X., He, C., Three-dimensional porous scaffold by self-assembly of reduced graphene oxide and nano-hydroxyapatite composites for bone tissue engineering. *Carbon*, 116, 325–337, 2017.
45. Wanga, L., Lua, C., Lia, Y., Wuc, F., Zhaoc, B., Dong, X., Green fabrication of porous silk fibroin/graphene oxide hybrid scaffolds for bone tissue engineering. *RSC Adv.*, 5, 96, 78660–78668, 2013.
46. Wang, W., Caetano, G., Ambler, W.S., Blaker, J.J., Frade, M.A., Mandal, P., Diver, C., Bártolo, P., Enhancing the Hydrophilicity and Cell Attachment of 3D Printed PCL/Graphene Scaffolds for Bone Tissue Engineering. *Materials*, 9, 992, 2016.
47. Liu, Z., Robinson, J.T., Sun, X., Dai, H., PEGylated nanographene oxide for delivery of water-insoluble cancer drugs. *J. Am. Chem. Soc.*, 130, 10876–10877, 2008.
48. Zhang, L., Lu, Z., Zhao, Q., Huang, J., Shen, H., Zhang, Z., Enhanced chemotherapy efficacy by sequential delivery of siRNA and anticancer drugs using PEI-grafted graphene oxide. *Small*, 7, 460–464, 2011.
49. Shen, H., Zhang, L., Liu, M., Zhang, Z., Biomedical applications of graphene. *Theranostics*, 2, 283–294, 2012.
50. Yang, Z.R., Wang, H.F., Zhao, J., Peng, Y.Y., Wang, J., Guinn, B.A., Huang, L.Q., Recent developments in the use of adenoviruses and immunotoxins in cancer gene therapy. *Cancer Gene Ther.*, 14, 599–615, 2007.
51. Geim, A.K., Graphene: Status and prospects. *Science*, 324, 1530–1534, 2009.
52. Liu, Z., Robinson, J.T., Sun, X., Dai, H., PEGylated nano-graphene oxide for delivery of water insoluble cancer drugs. *J. Am. Chem. Soc.*, 130, 33, 10876–10877, 2008.
53. Sun, X., Liu, Z., Welsher, K., Robinson, J.T., Goodwin, A., Zaric, Dai, H., Nano-Graphene Oxide for Cellular Imaging and Drug Delivery. *Nano Res.*, 1, 3, 203–212, 2008.
54. Loh, K.P., Bao, Q.L., Eda, G., Chhowalla, M., Graphene oxide as a chemically tunable platform for optical applications. *Nat. Chem.*, 2, 1015–1024, 2010.
55. Feng, L., Zhang, S., Liu, Z., Graphene based gene transfection. *Nanoscale*, 3, 1252–1257, 2011.
56. Chen, B., Liu, M., Zhang, L., Huang, J., Yao, J., Zhang, Z., Polyethylenimine-functionalized graphene oxide as an efficient gene delivery vector. *J. Mater. Chem.*, 21, 7736–7741, 2011.
57. Yang, X.Y., Zhang, X.Y., Liu, Z.F., Ma, Y.F., Huang, Y., Chen, Y., High-efficiency loading and controlled release of doxorubicin hydrochloride on graphene oxide. *J. Phys. Chem. C*, 112, 17554–17558, 2008.

58. Depan, D., Shah, J., Misra, R.D.K., Controlled release of drug from folate-decorated and graphene mediated drug delivery system: Synthesis, loading efficiency, and drug release response. *Mat. Sci. Eng. C Bio. S.*, 31, 1305–1312, 2011.
59. Wang, H., Gu, W., Xiao, N., Ye, L., Xu, Q., Chlorotoxin-conjugated graphene oxide for targeted delivery of an anticancer drug. *Int. J. Nanomed.*, 9, 1433–1442, 2014, preference.
60. Wang, H., Gu, W., Xiao, N., Ye, L., Xu, Q., Chlorotoxin-conjugated graphene oxide for targeted delivery of an anticancer drug. *Int. J. Nanomed.*, 9, 1433–1442, 2014.
61. Pattnaik, S., Swain, K., Lin, Z., Graphene and graphene-based nanocomposites: Biomedical applications and biosafety. *J. Mater. Chem. B*, 4, 7813–7831, 2016.
62. Liu, J., Cui, L., Losic, D., Graphene and graphene oxide as new nanocarriers for drug delivery applications. *Acta Biomater.*, 9, 9243–9257, 2013.
63. Li, H.J., Du, J.Z., Liu, J., Du, X.J., Shen, S., Zhu, Y.H., Wang, X., Ye, X., Nie, S., Wang, J., Smart superstructures with ultrahigh ph-sensitivity for targeting acidic tumor microenvironment: Instantaneous size switching and improved tumor penetration. *ACS Nano*, 10, 6753–6761, 2016.
64. Li, Y., Yuan, H., von dem Bussche, A., Creighton, M., Hurt, R.H., Kane, A.B., Gao, H., Graphene microsheets enter cells through spontaneous membrane penetration at edge asperities and corner sites. *Proc. Natl. Acad. Sci. U.S.A.*, 110, 12295–12300, 2013.
65. Shen, H., Zhang, L., Liu, M., Zhang, Z., Biomedical applications of graphene. *Theranostics*, 2, 283–294, 2012.
66. Su, Y.L., Kuo, L.W., Hsu, C.H., Chiang, C.S., Lu, Y.J., Chang, S.J., Hu, S.H., Rabies virus glycoprotein-amplified hierarchical targeted hybrids capable of magneto-electric penetration delivery to orthotopic brain tumor. *J. Control. Release*, 321, 159–173, 2020.
67. Su, Y.L., Yu, T.W., Chiang, W.H., Chiu, H.C., Chang, C.H., Chiang, C.S., Hu, S.H., Hierarchically targeted and penetrated delivery of drugs to tumors by size-changeable graphene quantum dot nanoaircrafts for photolytic therapy. *Adv. Funct. Mater.*, 27, 1700056, 2017.
68. Chang, Y., Yang, S.T., Liu, J.H., Dong, E., Wang, Y., Cao, A., Liu, Y., Wang, H., *In vitro* toxicity evaluation of graphene oxide on A549 cells. *Toxicol. Lett.*, 200, 201–210, 2011.
69. Wang, K., Ruan, J., Song, H., Zhang, J., Wo, Y., Guo, S., Cui, D., Biocompatibility of Graphene Oxide. *Nanoscale Res. Lett.*, 6, 8, 2011.
70. Hondroulis, E., Zhang, Z.Q., Chen, C.Y., Li, C.Z., Impedance based nanotoxicity assessment of graphene nanomaterials at the cellular and tissue level. *Anal. Lett.*, 45, 272–282, 2011.
71. Ohno, Y., Maehashi, K., Yamashiro, Y., Matsumoto, K., Electrolyte gated graphene field-effect transistors for detecting pH and protein adsorption. *Nano Lett.*, 9, 3318–3322, 2009.

72. Mohanty, N. and Berry, V., Graphene-based single-bacterium resolution biodevice and DNA transistor: Interfacing graphene derivatives with nanoscale and microscale biocomponents. *Nano Lett.*, 8, 4469–4476, 2008.
73. Zhou, M., Zhai, Y.M., Dong, S.J., Electrochemical sensing and biosensing platform based on chemically reduced graphene oxide. *Anal. Chem.*, 81, 5603–5613, 2009.
74. Song, Y.J., Qu, K.G., Zhao, C., Ren, J.S., Qu, X.G., Graphene oxide: Intrinsic peroxidase catalytic activity and its application to glucose detection. *Adv. Mater.*, 22, 2206–2210, 2010.
75. Wang, Y., Li, Z.H., Hu, D.H., Lin, C.T., Li, J.H., Lin, Y.H., Aptamer/graphene oxide nanocomplex for *in situ* molecular probing in living cells. *J. Am. Chem. Soc.*, 132, 9274–9276, 2010.
76. Mao, S., Yu, K., Chang, J., Steeber, D.A., Ocola, L.E., Chen, J., Direct Growth of Vertically-oriented Graphene for Field-Effect Transistor Biosensor. *Sci. Rep.*, 3, 1696, 2013.
77. Akinwande, D., Petrone, N., Hone, J., Two-dimensional flexible nanoelectronics. *Nat. Commun.*, 5, 5678, 2014.
78. Lee, H. *et al.*, A graphene-based electrochemical device with thermoresponsive microneedles for diabetes monitoring and therapy. *Nat. Nanotechnol.*, 11, 566–572, 2016.
79. Choi, C., Choi, M.K., Liu, S., Kim, M.S., Park, O.K., Im, C., Kim, J., Qin, X., Lee, G.J., Cho, K.W., Kim, M., Joh, E., Lee, J., Son, D., Kwon, S.H., Jeon, N.L., Song, Y.M., Lu, N., Kim, D.H., Human eye-inspired soft optoelectronic device using high-density MoS2-graphene curved image sensor array. *Nat. Commun.*, 8, 2017, https://doi.org/10.1038/s41467-017-01824-6.
80. Meyer, J.-U., Retinal Implant—A bioMEMS Challenge. ELSEVIER *Sens. Actuators A*, 97–98, 1–9, 2002.
81. Pang, L., Dai, C., Bi, L., Guo, Z., Fan, J., Biosafety and antibacterial ability of graphene and graphene oxide *in vitro* and *in vivo*. *Nanoscale Res. Lett.*, 12, 1, 564, 2017.
82. Vaishya, R., Chauhan, M., Vaish, A., Bone cement. *J. Clin. Orthop. Trauma*, 4, 157–63, 2013.
83. Puska, M., Moritz, N., Aho, A.J., Vallittu, P.K., Morphological and mechanical characterization of composite bone cement containing polymethylmethacrylate matrix functionalized with trimethoxysilyl and bioactive glass. *JMBM*, 59, 11–20, 2016.
84. Xie, H., Cao, T., Rodríguez-Lozano, F.J., Luong-Van, E.K., Rosa, V., Graphene for the development of the next-generation of biocomposites for dental and medical applications. *Dent. Mater.*, 33, 765–774, 2017.
85. Tan, Q., Zhang, K., Gu, S., Ren, J., Mineralization of surfactant functionalized multi-walled carbon nanotubes (MWNTs) to prepare hydroxyapatite/MWNTs nanohybrid. *Appl. Surf. Sci.*, 255, 7036–7039, 2009.

86. Kealley, C., Elcombe, M., Riessen, A.V., Ben-Nissan, B., Development of carbon nanotube-reinforced hydroxyapatite bioceramics. *Physica B*, 385–386, 496–498, 2006.
87. Bai, Y., Bai, Y., Gao, J., Ma, W., Su, J., Jia, R., Preparation and characterization of reduced graphene oxide/fluorhydroxyapatite composites for medical implants. *J. Alloys Compd.*, 688, 657–667, 2016.

11

Graphene and Graphene-Integrated Materials for Energy Device Applications

Santhosh, G.[1]* **and Bhatt, Aarti S.**[2]

[1]*Department of Mechanical Engineering, NMAM Institute of Technology, Nitte Karkala, India*
[2]*Department of Chemistry, NMAM Institute of Technology, Nitte Karkala, India*

Abstract
The demand for the development of high-performance energy devices has increased significantly in recent years. A huge interest in graphene has been raised due to its crystal structure and properties that provide unique opportunities to address the increasing demands of energy devices. Since its discovery, graphene has been considered as a "wonder material". Its presence causes an improvement in battery performance, especially lithium ion battery (LIB). Moreover, graphene has the flexibility and compatibility to blend with metal and metallic compounds; thereby forming efficient electroactive composite materials. However, the demand for commercialization of graphene-based batteries is not so high. In this chapter, by citing some of the recent works, we will discuss the common drawbacks of graphene-based materials and also measures being taken to overcome the same. This will provide an opportunity to the scientific community to effectively employ graphene-based materials for high performance energy devices.

Keywords: Energy devices, graphene, lithium ion battery, electroactive composite

11.1 Introduction

Graphene is a two-dimensional (2D) material, comprising of sp^2 hexagonal planar allotrope of carbon with strong covalent bonds. Nowadays, worldwide energy consumption is rising sharply due to escalating demand

**Corresponding author*: drsanthug@gmail.com

Inamuddin, Rajender Boddula, Mohd Imran Ahamed and Abdullah M. Asiri (eds.)
Monoelements: Properties and Applications, (285–328) © 2020 Scrivener Publishing LLC

for energy. As a result, the storage of energy devices such as rechargeable batteries and supercapacitors are extensively researched around the globe [1, 2]. Although considerable improvement in energy storage devices has been seen in the last few decades, still it is very hard to meet the demands to create and fabricate the efficient storage devices. With the increasing demand for flexible, light weight and high energy density storage devices, greater efforts are being made to prepare and fabricate suitable electrodes to suit the desired performance. The last decade witnessed the fabrication of graphene-based materials for energy storage applications exhibiting very unique properties compared to its three-dimensional (3D) bulk graphite [3, 4].

The researchers have attempted the use of graphene-based composites in energy storage devices. However, assembling these composites by the traditional approach exhibits poor properties restricting the efficiency of devices [5–8]. Graphene is considered as one of the most ideal materials for energy device applications because of its broad range of properties including surface area, mechanical strength, chemical, electrical and thermal stability [9]. Graphene offers many morphological benefits such as high accessible surface with exposed active sites and fast reaction kinetics which helps in electrochemical energy storage, resulting in high power and energy densities [10]. The various applications of graphene in energy devices are presented in Figure 11.1 [11].

Graphene can be synthesized by various methods, such as exfoliation [12], chemical vapor deposition (CVD) (Figure 11.2), arc discharge (AD) method and by reducing graphene oxide [12–15].

Mechanical exfoliation of a graphite crystal is one of the distinguished processes to isolate the graphene sheet, but the process can't be scaled up beyond small flakes. The large-scale synthesis includes thermal deposition and growth by chemical vapor deposition [16]. The graphene deposited on silicon carbide substrate with sp^2 bonding is achieved by heating under ultrahigh vacuum at very high temperatures from 1,000 to 1,500 °C. Hence, chemical vapor deposition has gained a lot of consideration to synthesize graphene. The graphene synthesized by CVD is found to be of high quality with large surface area. Further, graphene platelets and multi-layered graphene are obtained by gas phase synthesis technique and arc discharge which are also investigated. The graphite oxide (GO) layers that play an important role in improving the device efficiency are exfoliated in solvents which offer large production with effectively low cost [15].

The recent studies of GO having phenol hydroxyl and epoxide groups have provided significant breakthroughs in the energy device applications [18]. The use of new functionalities can be improved by increasing the interlayer spacing of oxidized graphite by disrupting graphene lattice

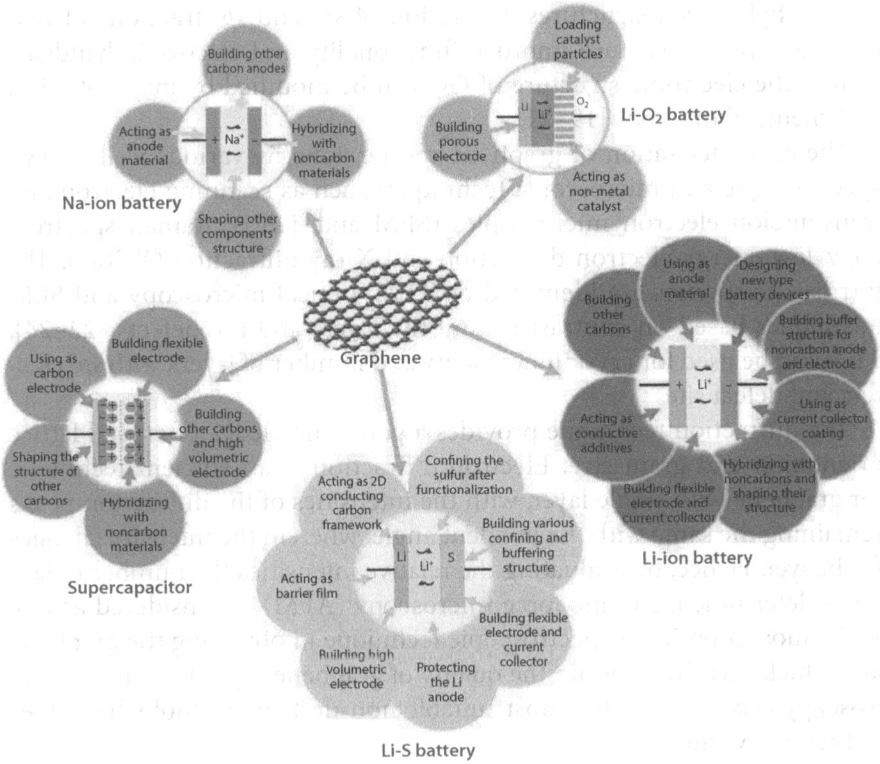

Figure 11.1 Graphene as the major constituent in various energy devices [11].

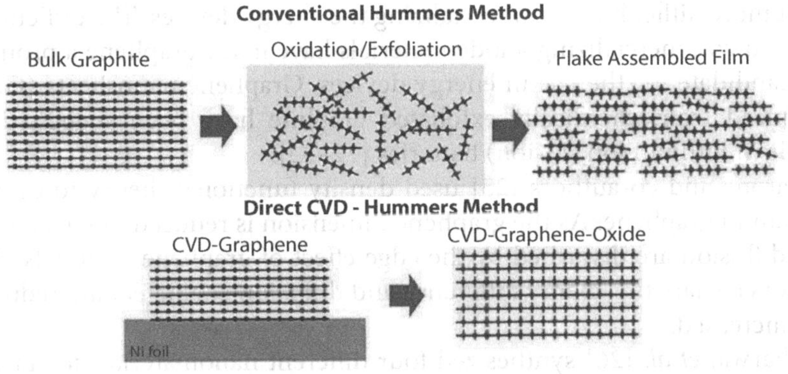

Figure 11.2 Graphene and graphene oxide derived from graphite from conventional Hummer's and CVD methods [17].

and sp^2 hybridized structures, the ration of sp^2 and sp^3 fractions of oxidized graphene structure improve functionality and electronic bandgap. Hence, the electronic structure of GO can be modified by many physical and chemical methods [19–21].

The characterization of graphene can be done by various well-known spectroscopic and microscopic techniques such as scanning electron and transmission electron microscopies (SEM and TEM), Raman spectroscopy, low energy electron diffraction and X-ray diffraction (X-Ray). The intrinsic properties are identified by TEM, optical microscopy and SEM techniques based on contrast to find the layers and the defects [22–24]. TEM can be used for crystalline quality and number of layers with sp^2 clusters and defect sites.

The diffraction technique provides a good knowledge of crystal lattice orientation and grain size. Electron diffraction is an efficient technique for graphene with single layer, with the intensities of the diffraction peaks remaining the same with the incident angle, wherein the intensity changes for bilayer. Hence, by evaluating the relative intensities the number of layers is determined. Atomic force microscopy (AFM) is considered as one of the most prominent spectroscopic technique in obtaining the graphene layer thickness. To ascertain the quality of graphene crystals, Raman spectroscopy is considered the most suitable non-destructive tool which does not require vacuum.

11.1.1 Anode Materials for Electrodes

The boundless efforts are being made to produce various graphene-based anode materials with very high energy density (Ed). However, the change in volume due to charge and discharge cycles reduces the cycle life making it more difficult in commercializing the energy devices. The difficulties faced in commercializing anode materials have made graphene a promising candidate for the use in energy devices. Graphene nanosheets (GNs) produced from chemically exfoliated graphite have been suggested as anode in Lithium ion (Li-ion) batteries [27–32].

Barone and co-authors [25] used density functional theory to diffuse lithium in graphene. As the graphene dimension is reduced, the reactivity and diffusion are disturbed by the edge effect of graphene materials. The authors declare the energy difference and diffusion coefficient are reduced and increased.

Bharwaj *et al.* [26] synthesized four different nanomaterials for Li-ion batteries such as graphite, multilayer carbon nanotubes (MlCNTs), oxidized and non-oxidized graphene nanoribbons (OGNr and NoGNr).

The electrochemical measurements suggest the potential of reduced-graphene nanoribbons (rGNr) is higher (1,400 and 820 $mAhg^{-1}$) than that of MlCNTs (200–370 $mAhg^{-1}$) with an efficiency of 53%. Uthaisar and co-authors [27] investigated the properties of reduced graphene oxide (rGO) synthesized by various thermal processes. The authors found that the GO synthesized by Hummer's method has high oxygen content (33 mol%) with 1.2 carbon/oxygen ratio (C/O ratio). Further, the C/O ratio is found to increase from 4.2 to 11.6 with 7–17 mol% of the oxygen. The electrochemical studies of GO reveal that the charge/discharge capacity values are 1,000 and 500 $mAhg^{-1}$ and that of the reduced graphene oxide, the capacity increased from 500 to 1,000 $mAhg^{-1}$. The behavior is attributed to the higher Li intake capacity due to the defect oxygen groups.

Lee *et al.* [28] studied the charge capacity of GO prepared by using different oxidation degrees. The authors exhibited that synthesized GO has strong stability with one of the materials having maximum capacity of 2,000 $mAhg^{-1}$. Chen and co-authors [29] presented a simple technique to produce porous-graphene nanosheets (PGNs). The electrochemical results presented describe the poor stacking of PGNs delivering highest storage capacity at high discharge rate. The delivered capacity approaches 480 $mAhg^{-1}$ after 100 cycles and 320 $mAhg^{-1}$ after 300 cycles at 2 Ag^{-1}. Fan *et al.* [30] also fabricated PGNs by CVD method with MgO templates. The results indicate the nano porous structures with reversible capacity of 1,723 $mAgh^{-1}$ with excellent cycling stability. Hu *et al.* [31] synthesized the graphene paper and the investigations reveal that storage capacity is highly dependent on paper thickness. The electrochemical tests are done for graphene papers with 1.5 and 3 μm thickness. For graphene paper with 1.5 μm the reversible capacity is ~200 $mAhg^{-1}$ and ~140 $mAhg^{-1}$ for 3 μm thick paper. The results show that the capacity decreases as the paper thickness increases due to the restacking of GNs. Yang *et al.* [32] studied the behavior of carbon coated graphene sheets. The thin layer of carbon enriches the rate capacity and cycle life at different current rates. These results authorize the rate capability of composite anodes and other anodes based on graphene.

11.1.2 Cathode Materials for Electrodes

Graphene-integrated materials used as cathodes in energy devices are researched predominantly to enrich the overall performance of the devices.

He *et al.* [33] used spray-drying system to synthesize graphene-coated $Li[Li_{0.2}Mn_{0.5}4Ni_{0.13}Co_{0.13}]O_2$. The result demonstrated by the authors validates that graphene coating could increase the efficiency and rate performance of the device. Sun *et al.* [34] prepared rGO-based composite films with free-standing $VO_{2.07}$ by a hydrothermal process. The results exhibit high reversible capacity of 160 $mAhg^{-1}$ with good capacity retention of 83% after 200 cycles. Ding *et al.* [35] synthesized graphene based nano $LiMn_1/3Ni_{1/3}$ $Co1/3O_2$ particles by coprecipitation technique. Prepared nanostructures about 50 nm were coated uniformly with graphene. The electrochemical studies of the coated nanostructures display better discharge with capacity retention of 95.5%. Prabakar *et al.* [36] prepared graphene-based sandwiched composite structures. The material demonstrated an efficient conducting network supressing the formation of a solid–electrolyte interface with $LiNi_{0.5}Mn_{1.5}O_4$. The rate capability of prepared composite structure with 2.5% graphene showed better electrochemical results. The exhibited capacity retention of $LiNi_{0.5}Mn_{1.5}O_4$ is attributed to the formation of sufficient conducting channels by graphene by suppressing the electrolyte decomposition. Wang and co-authors [37] showcased a carbon coated $LiFePO_4$ altered with rGO. The cathode material showed very high specific capacity with just 5 wt% rGO. The rate performance of the cathode material is considered to be dominating with discharge capacities of 160.4 $mAhg^{-1}$ and 115.0 $mAhg^{-1}$ at 0.2 and 20 °C. The materials also proved to have superior cycling stability with 10% decay at 10 °C after 1,000 cycles.

11.2 Graphene-Integrated Electrodes for Lithium-Ion Batteries (LIBs)

Battery technology has engulfed the whole world. Batteries have made a profound impact in our modern lives by making devices mobile and long lasting. Currently, the most dominant and widely deployed battery is the Lithium Ion Battery (LIB). The preference for LIBs over other batteries lies in the unique performance attributes of LIBs viz., (i) high gravimetric and volumetric energy density, (ii) high coulombic efficiency and energy efficiency, (iii) long cycle life, (iv) nil memory effect, and (v) flexibility in design. No wonder, LIBs have become an integral part of many modern devices and appliances such as smart phones, laptops and other portable electric devices. The high expectations from LIBs has led to a growing trend of research in developing next-generation LIBs for electric vehicles (EVs), grid energy storage applications and aerospace applications.

11.2.1 The Working of LIBs

The early 1970s saw the advent of the first non-rechargeable lithium battery. The inherent instability of Li metal made it impossible to successfully commercialize a rechargeable lithium battery. The focus shifted to non-metallic solution using lithium ions and in the early 1980s, Goodenough and his team invented the first rechargeable LIB using lithium cobalt oxide (LCO) as cathode [38].

A typical LCO battery (Figure 11.3) has LCO as cathode, graphite as anode, a separator and electrolyte made of lithium salt suspended in an organic solvent. During discharge, Li^+ flows from anode to cathode through the electrolyte. Simultaneously, the electrons flow through external open circuit in the reverse direction. The LCO structure comprises of an octahedral array of edge sharing cobalt atoms (Co^{3+}) sandwiched between two layers of oxygen atoms (O^{2-}). Such a lattice structure provides a two-dimensional migration channel (repeating for every three layers of CoO_2) for easy feasibility of Li^+, leading to high energy density. On charging, the reactions reverse. Here, the lithium ions are sandwiched between hexagonal graphite layers (C_6 i.e. for every six carbon atoms only one lithium ion can intercalate) in an alternate manner.

$$\textit{Anodic half-reaction: } Li_xC_6 \leftrightarrows C_6 + xLi^+ + xe^-$$

$$\textit{Cathodic half-reaction: } Li_{1-x}CoO_2 + xLi^+ + xe^- \leftrightarrows LiCoO_2$$

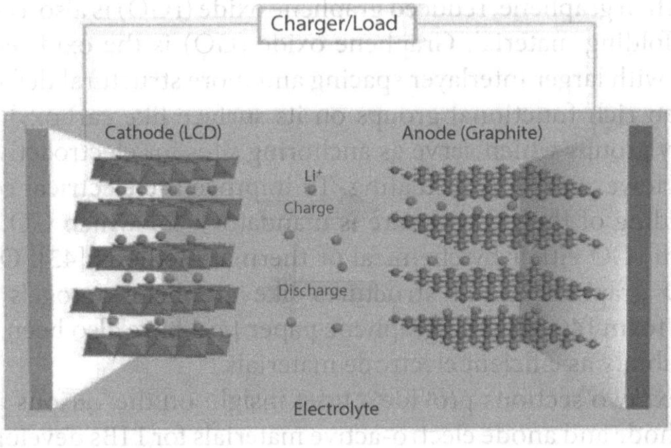

Figure 11.3 Schematic diagram explaining working principle of LCO-graphite battery [41].

$$\textit{Overall reaction: } Li_xC_6 + Li_{1-x}CoO_2 \leftrightarrows LiCoO_2 + C_6$$

The back and forth movement of lithium ions occurs by oxidation (at anode) and reduction (at cathode) of Co^{3+} to Co^{4+} and Co^{4+} to Co^{3+} respectively. However, the reversibility is possible only for $x < 0.5$ [39]. Since the electrochemistry involves insertion (intercalation) and extraction (deintercalation) of Li^+ to and fro across the electrodes, LIBs are often referred to as "rocking chair" batteries or "swing" batteries [40].

The electrode materials are the most significant components of LIBs which determine the battery characteristics. Several electrode materials have been and are being investigated for superior electrochemical performances so as to apply them in future technologies. Among them, graphene-based electrodes are the most promising. The sp^2 structure of graphene provides an impressive lithium storing capacity. Moreover, graphene is also preferred in terms of excellent electrical conductivity, good thermal conductivity and chemical and mechanical stability. In addition, the flexibility of graphene to be shaped into different structures have made them suitable matrix candidates for composites. Graphene can be shaped to spherical form encapsulating the electroactive material within it. It can even be mechanically mixed with active materials during the electrode preparation process or can be wrapped around the host materials. The most commonly reported is the anchored structure wherein the active materials are attached on the graphene surface. The 2D graphene layers have also been used as building block materials for sandwich-like and layered structures with electroactive materials.

Other than graphene, reduced graphene oxide (rGO) is also a commonly used scaffolding material. Graphene oxide (GO) is the oxidized form of graphene with larger interlayer spacing and more structural defects. It also has oxygen rich functional groups on its surface like carboxyl, hydroxyl and epoxy groups which serve as anchoring sites for electroactive materials [42]. However, GO is insulating. To improve the electrical conductivity, rebuilding of the sp^2 structure is mandatory, for which GO is further reduced to rGO either by chemical or thermal methods [43]. Other than these, 3D graphene-based structures like graphene aerogels [44, 45], graphene foam [46–48] and graphene paper [49] have also been tested for their suitability as efficient electrode materials.

The next two sections provide a brief insight on the various graphene-based cathode and anode electro-active materials for LIBs developed in the recent times. More focus will be on the structural design and electrochemical performances as well as limitations of the graphene-based electrodes for commercial applications of LIBs.

11.2.2 Graphene-Integrated Cathodes

The three commonly employed metal oxide structures for cathode materials are (i) the olivine $LiFePO_4$ structure which is considered to be a safe alternative with impressive thermal stability and practical capacity (170 $mAhg^{-1}$) but has a major drawback of poor ionic and lithium diffusion rate; (ii) the spinel structured $LiMn_2O_4$ though has the advantage of good ion diffusion rate and structural stability, is unable to provide high practical capacity (<120 $mAhg^{-1}$); (iii) the layered structure like $LiMO_2$ [M=Co, Mn, Ni] which produces practically the highest capacity (~180 $mAhg^{-1}$) but lacks in chemical and/or structural stability with variation in lithium content during cycling. The practical capacity, electrical conductivity, chemical and structural stability of the metal oxides can be enhanced by the synergistic effect of graphene.

11.2.2.1 Graphene/$LiFePO_4$ as Cathode

The current research on flexible cathode materials is highly concentrated on $LiFePO_4$ (LFP). LFP structure has a high stability due to the strong P–O bonds. Phosphorus occupies the tetrahedral sites whereas the octahedral sites are taken up by Li^+ and Fe^{2+}. Such kind of arrangement gives rise to one-dimension diffusion channel wherein the Li^+ diffusion is limited to the edge shared LiO_6 units. The diffusion gets further blocked by impurities or defects present along the 1D diffusion path length. The low Li^+ diffusion rate (~10^{-14} cm^2s^{-1}) with poor electronic conductivity (10^{-9} Scm^{-1}) does not permit the growth in capacity. These two drawbacks can be compensated by using graphene which provides a conductive network.

Pioneered by Ding in 2010, LFP/graphite electrode has been exploited to its full length as a flexible LIB cathode material. Ding's group could achieve a capacity of 160 $mAhg^{-1}$ at 0.2 °C for LFP/rGO electrode while the maximum capacity reported by bare LFP is just 113 $mAhg^{-1}$ [50]. Encouraged by the results, Zhou *et al.* [51] considered the benefits of nanochemistry to prepare LFP/graphene composites. By employing a simplistic spray-drying method, the team could successfully disperse LFP nanoparticles on the 3D graphene nanostructures. This led to easy diffusion of Li^+ with a high capacity of 70 $mAhg^{-1}$ at 60 °C. Similarly, LFP/rGO with carbon coating were developed by *in situ* synthesis of LFP nanoparticles [52]. Polypyrrole was polymerized on graphene oxide which provides a carbon coating on the substrate surface; thereby restricting the size and agglomeration of the dispersed LFP nanoparticles. The hybrid material provided a capacity of 72 $mAhg^{-1}$ at a very high rate of 60 °C. The high capacity was attributed to the formation of a highly conducting network of polypyrrole and rGO with least electron resistance. In addition, LFP/graphite composites with excellent cycling performance of over 5,000 cycles have also been developed [53]. A recent study [54] investigates the advantages of using graphite foil prepared by chemical vapor deposition method as an efficient current collector for LFP electrode. The graphite foil has the flexibility to be rolled, bent or even twisted; all the time retaining its excellent tensile strength (45 MPa) and high electrical conductivity (4,149 Scm^{-1}). In presence of these graphite foils, the cathode material could provide reversible capacities of about 138, 123 and 112 $mAhg^{-1}$ at 0.2, 0.5 and 1 °C respectively.

11.2.2.2 Graphene/$LiMn_2O_4$ as Cathode

$LiMn_2O_4$ (LMO) has a cubic close packed arrangement of oxygen in spinel structure with Li^+ occupying tetrahedral sites and Mn ions occupying the octahedral sites. Lithium diffusion occurs through a 3D channel. However, at the operating voltage, LMO suffers from phase transition (cubic to tetragonal) due to Jahn–Teller effect of high spin $Mn^{3+}:3d^4(t_{2g}^3e_g^1)$ leading to capacity fading. The metal oxide also suffers from dissolution of Mn^{2+} into the electrolyte during cycling; restricting its reversible capacity to a meagre 100 $mAhg^{-1}$. Many alternative solutions for these practical problems have been developed out of which the use of graphene stands out. The conductive network of graphene plays a pivotal role in improving the lithium diffusion during intercalation and de-intercalation of Li^+ from/into the electrode.

Prabakar and his team developed a sandwiched structure of graphene with Ni doped LMO [55]. The graphene/$LiNi_{0.5}Mn_{1.5}O_4$ structure achieved a high coulombic efficiency because of the prevention of electrolytic decomposition on the electrode surface by graphene. As a result, the capacity and cyclic retention of the electrode improved (107 $mAhg^{-1}$ at 50 °C) when compared to the undoped electrode [56]. Chen and co-workers have developed a graphene/LMO cathode with conformal coating of graphene on LMO NPs. Such a coating leads to efficient charge transfer because of which high cycling efficiency can be retained even at high rate and low temperature. The electrode under study could retain 96% capacity at 0.2 °C and freezing temperature of −20 °C [57]. Composites with more than one type of metal oxides are recently gaining attention as a promising alternative for LIB cathode materials. Y_2O_3/$LiMn_2O_4$ wrapped graphene has shown to exhibit a capacity of 129.3 $mAhg^{-1}$ (at 1 °C) with a high capacity retention of 89.3% at the end of 500 cycles [58]. Similar conclusions were drawn for graphene/ZnO/LMO which showed excellent electrochemical performance and cyclic stability [59].

Also, graphene can be employed to modify the cathode surface so as to isolate the cathode from the electrolyte to prevent metal dissolution. Recent studies have shown the influence of graphene on the interfacial reactions between the electrolyte and LMO cathode surface [60]. Graphene/LMO cathode has more uniform interfacial layer with the electrolyte and can significantly store more metal ions after cycling when compared to bare LMO electrode. Using computational analysis, it was shown that graphene induces Mn dissolution suppression by (i) acting as selective ionic filter which allows movement of Li^+ but not manganese ions, and (ii) conversion of dissolution likely Mn^{3+} species to more stable Mn^{4+} via the defect sites.

11.2.2.3 Graphene-Layered Cathode Material

The layered lithium metal oxides ($LiMO_2$) that are commonly employed as cathode materials in LIBs generally have transition metals like Co, Mn or Ni. Though the layered structures provide high capacity, the structural and chemical stability of these oxides are questionable. During cycling, the metal ions from the metal octahedral sites change place to the lithium octahedral sites via a neighboring tetrahedral site [61]. This kind of structural instability is more prominent if the octahedral-site stabilization energy (OSSE) is low. The OSSE values of metal ions follow the order Co > Ni > Mn. This is the reason that Mn^{3+} undergoes easy migration and gets converted from layered to spinel structure. Co^{3+} gives the most stable metal oxide structure; however, it suffers from chemical instability on extracting more than 50% lithium from $LiCoO_2$. Furthermore, Co is expensive and toxic when compared to Mn; the toxicity of Ni lies intermediate of Co and Mn. Nevertheless, $LiNiO_2$ is not a suitable candidate for LIBs as Ni^{3+} too suffers from chemical instability with variation in lithium concentration. Moreover, the synthesis of stoichiometric $LiNiO_2$ is challenging as Ni^{3+} decomposes at temperature above 250 °C. Owing to such critical disadvantages, layered structures with single transition metal oxide have failed to deliver a commercially applicable electrode material. Instead, since the past decade, researchers have shown interest in layered structures with mixed transition metal oxides $Li(Ni_xMn_yCo_{1-x-y})O_2$ ($0 \leq x,y \leq 1$ and $0 \leq x + y \leq 1$), also known as NMC electrodes. The NMC cathode shows high charge capacity like $LiNiO_2$, structural stability due to Mn^{3+} and appreciable rate capability like $LiCoO_2$ [62]. Moreover, NMC cathodes are cheaper and less toxic than $LiCoO_2$.

Various synthetic methods have been adopted to arrest the synergism of graphene and NMC for developing cathode material with high electrochemical performance. For instance, NMC particles wrapped with rGO have been prepared by spray-drying method followed by thermal treatment. The capacity of NMC enhanced from 6 to 88.5 $mAhg^{-1}$ (at 5 °C) on wrapping with rGO [63]. A facile way of introducing rGO as active material is by substituting the carbon black slurry with rGO during the electrode preparation. When half of the slurry was substituted with rGO, the composite electrode achieved a capacity of 110 and 55 $mAhg^{-1}$ at 6 and 20 °C respectively. Surprisingly, in the absence of rGO, the pristine NCM cathode material reported nil capacity till 6 °C [64].

Other than the above classes of materials, there are several other metal compounds such as V_2O_5 and $Li_3V_2(PO_4)_3$ [65, 66] that are commonly combined with graphene for use as cathode materials in LIBs. In recent times, use of sulfides as cathode materials has also been put forward. Though MoS_2 is well-known for its suitability as anode material for Li–S batteries, MoS_x particles with $x > 2$ have shown to operate at higher voltage, which makes them a suitable cathode material [67]. With MoS_3 as host material and rGO as anchoring material, MoS_3/rGO electrode rendered a superior capacity of 900 $mAhg^{-1}$ at 5 °C for 2,500 cycles. With silicon-carbon as anode, the full cell battery exhibited much higher energy density (1,725 $Whkg^{-1}$/7,100 WhL^{-1}) than the present LIBs. Many a times, because of the defects present in the basal planes of graphene, attempts have been made to employ graphene itself as a cathode active material without any additives [68].

11.2.3 Graphene-Integrated Anodes

For a high-performance LIB, the cathode should have a high voltage while the voltage should be low for anode. At the same time, both the electrodes must produce high capacity and high energy density. Graphene has several advantages of mechanical stability, large surface area, high electrical conductivity and good charge carrier mobility. However, in practice, LiC_6 phase is not stable as the binding energy of Li to C is quite low. Also, the coulombic repulsion between lithium atoms situated in the opposite directions of the graphene layer result in poor Li storage capacity. The major two challenges faced in fabricating anode with the desired characteristics are (i) development of solid-electrolyte interface (SEI) layer which prevents the interaction between electrode and electrolyte, (ii) dendrite growth due to prolonged cycling leading to low cycle efficiency. SEI formation is strongly correlated with the electrode surface area. As the reactivity of Li is high, during SEI formation, some of the Li ions get trapped in the SEI layer and are no more available during cycling. An electrode with higher surface area consumes more Li during SEI formation. Because of the irreversible Li insertion in the SEI layer, the initial irreversible capacity increases. An ideal SEI layer must be uniformly thin but dense, ion-conducting but electrically insulating. In practice, the SEI films are brittle and when subjected to continuous charge/discharge cycles, they may eventually crack down aggravating dendrite growth. Graphene, which is the major anode material for LIBs, suffers from the above two drawbacks. An alternative for this is to composite the organic framework with host materials by (i) insertion (e.g. lithium titanate), (ii) alloying (e.g. Si, Sn, Ge) and (iii) conversion (e.g. metal oxides, nitrides, sulfides).

11.2.3.1 Graphene/$Li_4Ti_5O_{12}$ as Anode

$Li_4Ti_5O_{12}$ (LTO) is a promising active anode material for LIBs due to its zero strain insertion reaction, which implies Li insertion/extraction without any structural change, leading to a high cycle life. It also has a comparatively high operating voltage (1.55 V *vs* Li/Li^+) preventing lithium dendrite formation and making it more stable and safer than other electrode materials. However, it faces some intrinsic disadvantages of poor electronic conductivity (<10^{-13}Scm^{-1}) and low Li^+ diffusion (10^{-9} to 10^{-16} cm^{2}s^{-1}). These issues are generally addressed by combining LTO with carbon materials like graphene.

Sol–gel method is a well-known synthetic method by which LTO particles with uniform dispersion can be obtained [69]. The presence of graphene controls the size and agglomeration of LTO particles. To state an example, using sol-gel method LTO particles of 200–500 nm size have been synthesized to form graphene/LTO nanocomposite with improved rate capability. The reversible capacity of the electrode increased from 44.1 to 108.4 mAhg^{-1} at 10 °C in the 1,000 charge–discharge cycles [70]. Many such efforts on developing graphene/LTO composite with superior rate capability and high capacity have been reported [71]. Taking the research one step ahead, Jeong *et al.* have used N-doped rGO to disperse LTO particles. The role of nitrogen is to improve electrical conductivity and avoid restacking of graphene layers. It is observed that at 2.3 wt% nitrogen, LTO/N-doped rGO delivered the maximum electrical conductivity of 1.60 Scm^{-1} [72].

11.2.3.2 Graphene/Si or Ge as Anode

Silicon (Si), with high theoretical capacity (4,200 $mAgh^{-1}$), is a promising anodic material for LIBs. However, it experiences rapid capacity decay because of large volume expansion (400%) during lithiation/delithiation. Germanium (Ge) is another metal which, like Si, has gained popularity for its practical applications. The electronic conductivity of Ge is 100 folds higher and diffusivity is 400 folds higher compared to Si. However, it has a comparatively low theoretical capacity of 1,600 $mAhg^{-1}$. Furthermore, the large volume change (~370%) during lithiation/delithiation is a matter of concern. Si and Ge both react with Li by alloying/dealloying mechanism (Equation 11.1). Graphene, as a supporting material, can provide voids for minimizing volume expansion and also contribute in improving conductivity.

$$\text{Si (or Ge)} + x\text{Li} \rightarrow \text{Li}_x\text{Si (or Ge)} \tag{11.1}$$

An effective Si/graphene composite must have uniform dispersion of Si particles over the carbon framework promoting good contact between Si and graphene. In the absence of such arrangement, excessive cycling will lead to fracture. Several graphene-based Si electrodes synthesized by different methods have shown higher capacity and cycling performance [73–76]. A sandwich model of Si/graphene composites, with Si occupying the intermediate space between the two graphene layers have been recently reported. The highlight of this work is the achievement of capacity (2,515 $mAhg^{-1}$ at 0.4 °C for 200 cycles) that is six times greater than that of free-standing graphene. The credit is given to the highly conducting graphene layers which most probably prevent the damage of SEI during lithiation/delithiation [77]. Recently, Huang *et al.* have synthesized spherical graphite/Si/GO/carbon composite electrode which displayed a high discharge capacity of 1,212 $mAhg^{-1}$ at 200 mAg^{-1} and a capacity retention rate of 81.7% even after 100 cycles [78].

A nanocomposite of Ge/graphene exhibits better cycling performance than pure Ge. Graphene layers are used as supporting materials for carbon coated Ge resulting in the formation of sandwich-like structure. Such a composite with high electrochemical performance makes it suitable to be used as anodic material in LIBs [79]. Medvedev *et al.* studied charge capacity of Ge/GO as a function of processing temperature. Such kind of studies is beneficial for next-generation LIBs for extended applications. The authors report an increase in capacity up to a temperature of 650 °C; with 250 $mAhg^{-1}$ capacity retention after 50 cycles. The Ge/GO composite processed at 650 °C attained a capacity of 740 $mAhg^{-1}$ at 2,000 mAg^{-1} which is credited to the strong interaction between Ge and GO [80]. In a more advanced work, Ge NPs were anchored on rGO interwoven with carbon nano tubes (CNTs). Graphene checks on the volume expansion while CNT provides mechanical stability to the conductive network. The synergistic effects of graphene and CNT leads to better cycling performance with an excellent energy capacity of 863.8 $mAhg^{-1}$ at 100 mAg^{-1} at the end of 100 cycles [81].

11.2.3.3 Graphene/Metal Oxides as Anodes

11.2.3.3.1 Transition Metal Oxides

Transition metal oxides interact with Li by "conversion" mechanism. During conversion, the active material undergoes complete reduction to form metal. Other than transition metal oxides, even sulfides, phosphides and nitrides also follow the same mechanism. The lithiation involves multiple electron transfer resulting in high capacity. However, the metal oxides suffer from poor electrical conductivity. Also, the formation of amorphous lithium oxide during conversion, results in volume expansion.

$$nLi + MO \rightarrow Li_nO + M \tag{11.2}$$

Co_3O_4 NPs undergo a highly reversible electrochemical reactions making them an anticipated electrode active material. On the other hand, graphene plays its part well in improving conductivity and its porous nature prevents any kind of major volume change in Co_3O_4 occurring during conversion [82]. Exploiting the advantages of graphene, sandwich-structured 3D Co_3O_4/graphene [83, 84] and graphene-coated Co_3O_4 hybrids [85] with superior electrochemical performances have been investigated as anode materials for advanced LIBs. In a recent attempt, Zhu *et al.* designed the synthesis of nanostructured Co_3O_4 based anode for next generation LIBs. They adopted a two-step calcination method to fabricate hollow Co_3O_4 NPs embedded in nitrogen and sulphur co-doped rGO framework. Because of the porous structure of metal oxide and the presence of N and S atoms in rGO, the electrode could achieve a high capacity value of 1,590 $mAhg^{-1}$ at 1 mAg^{-1} after 600 cycles [86].

Core-void-shell structure is an alternative way to achieve superior electrochemical characteristics. The shell is the hollow graphene which adds to the electrochemical conductivity of the metal oxide NPs occupying the core. Graphene also prevents the agglomeration of the NPs. The intermediate void accommodates the large volume changes, thereby providing electrode stability during charge-discharge cycling. Moreover, the large surface area creates more reactive sites on the electrode surface thereby enhancing the rate capability. Jiang *et al.* have encapsulated Fe_3O_4 NPs in hollow graphene shell and the composite electrode reported a high capacity of 1,236.6 $mAhg^{-1}$ which is better than bare Fe_3O_4 or a simple physical mixture of graphene and Fe_3O_4 [87]. The magnetite NPs have also been designed in a hierarchical structure with graphene foam as the supporting material without the use of binder additives. Such a hierarchical structure leads to enhancement in electronic and ionic diffusion leading to improved capacity. As a result, the Fe_3O_4/graphene reported a capacity of 1,198 $mAhg^{-1}$ at the end of 400 cycles which is far superior to the capacity of pristine Fe_3O_4 (926 $mAhg^{-1}$) and pure graphene (300 $mAhg^{-1}$) [46]. Also, Fe_3O_4 intertwined with CNTs have been sandwiched between rGO layers to form Fe_3O_4/CNT/rGO composite electrode. The electrode, due to the presence of CNTs along with rGO, performed fairly well even on increasing current density to 10 Ag^{-1} and delivered a reversible capacity of 540 $mAhg^{-1}$ [88].

Among the different forms of graphene, holey graphene (hG) is recently gaining attraction for building high-performance electrode materials for battery applications. hG is a graphene framework with nanoscale holes on its surface. The porous configuration further improves ion diffusion and conductivity of the composite. Moreover, hG has lesser surface groups compared to rGO [89]. Sun *et al.* prepared hG by etching GO with H_2O_2 which formed the conductive framework for Nb_2O_5 NPs. The NPs can be successfully dispersed in hG by annealing and self-assembly methodology. The authors observed that the specific capacity of Nb_2O_5/graphene system (35 $mAhg^{-1}$ at 100 °C) doubled to 75 $mAhg^{-1}$ when coupled with hG [90]. Yan and his group further stated that performance and cycle durability can be improved by using oxygen deficient Nb_2O_5 (Nb_2O_{5-x}) [91]. The synthetic approach adopted by Sun and co-workers, has been further simplified by Zhu *et al.* who developed a facile *ex situ* method wherein Fe_3O_4 NPs were dispersed in hydrosols followed by uniformly anchoring them on 3D hG and evaluated its anodic performance. Based on the ultra-high discharge capacity and cyclic stability witnessed, the authors claim to have achieved the best results for any of the Fe_3O_4 based anodes reported till date. The interesting feature of the work is the versatile methodology which can be easily extended to fabricate various hG composite materials with metal oxides such as SnO_2, MnO and MoS_2 [92].

Fe_2O_3 with a high theoretical capacity (1,006 $mAhg^{-1}$) suffers from poor electrical conductivity and large volume changes (74.7%). To address these issues, Fe_2O_3 is generally combined with rGO. Xia *et al.* have used the electrochemical approach to create a robust 3D conductive graphene foam electrodes with multilayers of Fe_2O_3 alternately arranged with rGO spacers. The porous rGO spacers aid in electron diffusion and volume change control. The hierarchical electrode can be constructed with up to 7 vertical layers with a discharge capacity of 1,175 $\mu Ahcm^{-2}$ with a notable capacity retention of 84% after 1,000 cycles [47]. In an attempt to increase the carbon content, α-Fe_2O_3 NPs were dispersed in graphene layer which are alternately placed with CNT to form a hierarchical structure. The carbon layers relieve the stress caused by Fe_2O_3 and also prevent the aggregation of the NPs resulting in a high retention capacity [93].

MnO_2 is also an exciting metal oxide which has characteristics suitable for battery applications. A MnO/graphene hybrid is constructed by arranging MnO_2 nanosheets (50–100 nm) between layers of graphene sheets with micropores (0.4–0.8 nm). The dense stacked packing with microporous structure resulted in a high capacity of 1,000 $mAhg^{-1}$ at 0.1 Ag^{-1} whereas the rate capability at 2 Ag^{-1} is 604 $mAhg^{-1}$. The volumetric capacity also increased up to five times than pure graphene. However, the electrode suffered from initial capacity loss and displayed a low value of initial coulombic efficiency of 65%. The poor display was credited to SEI layer formation and decomposition of electrolyte during charge/discharge [94]. Xiao *et al.* could improve this performance by employing electrode active material in core–shell structure; MnO_2 nanowires forming the core which are surrounded by graphene. The electrode not only had excellent initial coulombic efficiency of 83% but also had high cycle stability of 500 cycles at 1 Ag^{-1} [95].

11.2.3.3.2 Tin Oxides

SnO_2 can function as an effective anode material owing to its high theoretical capacity (782 $mAhg^{-1}$). However, when subjected to high volume change, lowering in electrochemical reversibility and conductivity is observed. The SnO_2–Li interaction proceeds by a two-step "conversion-alloying" mechanism. Similar to transition metal oxides, in the first step, the metal oxide on interaction with lithium gets reduced to metallic form. The metallic Sn form is reversible only for SnO_2 NPs and not for SnO_2 in bulk form. The second step involves the alloy formation between metallic Sn and Li with Li_4Sn as the most stable composition.

$$SnO_2 + 4Li^+ + 4e^- \rightarrow Sn + 2Li_2O \tag{11.3}$$

$$Sn + xLi^+ + xe^- \rightarrow Li_xSn\ (0 \leq x \leq 4) \tag{11.4}$$

A graphene based carbon coated SnO_2 has been synthesized using glucose as a soft template agent. Glucose can control the growth of SnO_2 particles and porous carbon (PC) coating on SnO_2 prevents the particles from aggregation. At 0.1 °C, the rGO/PC/SnO_2 anode delivers a capacity of 2,238.2 $mAhg^{-1}$ with a high retention of 1,467.8 $mAhg^{-1}$ at the end of 150 cycles [96]. Arnaiza *et al.* developed a SnO_2/rGO nanocomposite electrode which when coupled with activated carbon as cathode delivered excellent energy density of 60 $Whkg^{-1}$ at 2.9 $kWkg^{-1}$ [97]. A recent study also reports the durability of SnO_2/S-doped rGO electrode wherein the presence of sulphur introduced defects on graphene surface resulting in easy Li diffusion. The capacity reported was 897 $mAhg^{-1}$ at 100 mAg^{-1} and a good cycle stability of 500 cycles with 88% retention at 500 mAg^{-1} [98]. Similarly, flexible graphene based electrodes with fluorine doped SnO_2 [99] and mixed metal oxides [100] have reported high specific capacity, initial discharge capacity and cycling stability.

11.2.3.3.3 Titanium Oxides

TiO_2, like LTO, follow the mechanism of insertion/extraction of lithium ions and electrons without affecting the lattice dimensions. However, like other metal oxides, TiO_2 suffers from low conductivity which can be improved by the presence of graphene materials.

A hierarchical structure of TiO_2 with N-doped graphene have been evaluated for anodic material in LIBs. The large surface area and porous structure leads to high electron/Li^+ transport. Also, the high-level N content (7.34%) improved the conductivity and electrochemical activity of the hybrid electrode resulting in high cyclic retention of 165 $mAhg^{-1}$ at 1 °C for 200 cycles and rate capability of 96 $mAhg^{-1}$ at 20 °C [101]. In a recent work, Wang and his group have investigated the influence of graphene content on the electrode performance. It was observed that composites with 18.8 wt% organic content displayed the best specific capacity of 264 $mAhg^{-1}$ at 1 °C and capacity retention value of 171 $mAhg^{-1}$ at 5 °C for 100 charge/discharge cycles. A drop in lithium storage capacity was observed when the graphene content exceeded 27.7 wt% [102].

11.2.3.4 Graphene/Sulfides as Anodes

Metal sulfides are another category of materials which have great demand in battery technology because of their high theoretical capacity. Sulfides generally function by alloying or conversion mechanism. During cycling, at the initial stage itself, the metal sulfide layers exfoliate into small pieces because of weak S–S bond. As a result, sulfides possess a lower volume expansion (103% for MoS_2) and shorter Li^+ diffusion length when compared to other metal oxides. However, like oxides, sulfides too are prone to low conductivity and shorter cycle life.

There are several reports highlighting the superior electrochemical performance of MoS_2 with graphene materials [103–105]. MoS_2 NPs with graphene aerogel as scaffolding material was prepared without any polymeric binders or conductive additives. The free-standing electrode showed remarkable retention of 101.1% after 700 cycles [45]. Chao *et al.* have fabricated a free standing flexible MoS_2/rGO porous electrode by self-assembly process. They observed that the composite with 75 wt% of MoS_2 delivered the best capacity of 800 $mAhg^{-1}$ at 100 mAg^{-1}. Interestingly, the electrode does not show capacity fading over 500 cycles even when the current density is increased up to 400 mAg^{-1} [105].

The graphene network provides homogeneous dispersion of SnS_2 particles without aggregation, large Li storage capacity and high conductivity. SnS_2 nanoflakes anchored 3D graphene foam electrode displayed a remarkable capacity of 1,386.7 $mAhg^{-1}$ at 0.1 Ag^{-1} with a long cycling life (818.4 $mAhg^{-1}$ at 1 mAg^{-1} for 500 cycles). The outstanding performance of the electrode is attributed to higher conductivity and reduced strain during lithiation/delithaiation [48]. A free standing SnS_2/graphene paper composite, prepared by two phase hydrothermal process, has shown a reversible capacity of 593 $mAhg^{-1}$ at 100 mAg^{-1} for 200 cycles. Here SnS_2 nanospacers between the graphene layers help in electrolyte penetration and prevention of graphene restacking [49].

11.3 Graphene-Integrated Nanocomposites for Supercapacitors (SCs)

Supercapacitors (SCs) are electrochemical energy storage devices which work on the basic principle of storing and releasing electrical energy on demand by reversible electrochemical processes. The mass interest in development of advanced supercapacitor is due to the rapid depletion of conventional energy sources. SCs are low-cost and environment-friendly energy storage and conversion devices which lie intermediate between capacitors (high power density but low energy density) and batteries (high energy density but low power density). SCs are presently being used as a power source in portable devices like mobile phones and laptops. However, in the coming days, SCs may also be utilized to supply electrical energy to remote places without the use of any grids or wiring required. Also, SCs with high power density are being considered for hybrid electric vehicles to achieve rapid charging, high mileage and controlling energy fluctuation during braking. Thus, the future applications of SCs demand a high energy density with a high-power output. As a result, the present research is focused on chemical and/or physical modifications of the SC components to develop a high-performance supercapacitor.

11.3.1 Working Mechanism of Supercapacitors

11.3.1.1 Electrochemical Double Layer Capacitors (EDLC)

EDLC is an energy storage device which works on the principle of electrostatic interactions. Here, capacitance is produced due to electrical double layer that is developed at the electrode/electrolyte interface during absorption/desorption of electrolyte ions subject to charge/discharge. Since the specific capacitance of EDLC is largely dependent on surface area and pore size of the active material, carbon-based materials are generally preferred. Activated carbon (AC) and carbon nanotubes (CNT) have good stability but are unable to achieve high capacitance because of the presence of micropores and slow interfacial performance respectively. Graphene, on the other hand, is garnering attention because of its high EDL capacitance (550 Fg^{-1}). Moreover, the hexagonal structure of sp^2 carbon atoms imparts excellent physical and mechanical properties.

11.3.1.2 Pseudo-Capacitors

Pseudo-capacitors follow faradic reactions i.e., charge is stored due to rapid and reversible redox reactions occurring within the bulk material as well as on its surface. This leads to production of higher energy density than EDLCs. Simultaneously, it becomes unavoidable to prevent volumetric expansion and structural modifications during charge/discharge, which results in lower cycling stability. Pseudocapacitive materials generally include metal oxides, hydroxides, sulfides, nitrides and conductive polymers.

11.3.1.3 Hybrid Supercapacitors

The requirement for energy storage devices with high energy density and good cyclic stability has resulted in the inception of hybrid supercapacitors which involve both faradic and non-faradic processes as discussed above.

Table 11.1 Comparison of the different types of capacitors [107].

Content	Electrochemical double layer capacitor (EDLC)	Pseudo-capacitor	Hybrid capacitor
Electrode material	Carbon	Metal oxides and Polymers	Carbon and Metal oxides/ Polymers
Mechanism	Charge storage	Through the redox reactions	By faradaic and non-faradaic processes.
Properties	Low energy density, good rate capability, good cyclic stability, low capacitance	High capacitance, energy density and power density	High energy and power density, good cyclability

This is done by reinforcement of pseudocapacitive materials on carbon structures. Such hybrid structures deliver high energy density and are stable for several charge/discharge cycles.

Table 11.1 gives the salient features of the three types of capacitors [106]. The supercapacitors can also be further classified as symmetric and asymmetric supercapacitors (ASC) based on similar and dissimilar electrodes respectively. In practice, ASCs have higher energy density and working voltage.

11.3.2 Graphene-Integrated Supercapacitors (GSCs)

Graphene has become the most sought after electro-active material for flexible supercapacitors owing to its advantages of high conductivity, large surface area and excellent mechanical strength. Graphene also enjoys the benefit of being highly flexible to provide the desired 0D, 1D, 2D and 3D structures, with various possible structural conformations like graphene foam, graphene films and graphene paper. A major challenge in successful employment of graphene as electrode material is the restacking of graphene layers which causes hindrance in effective ion exchange between electrode and electrolyte. A remedial solution to this is intercalation of nanomaterials in the interlayer spaces. Also, the graphene storage capacity is highly influenced by the presence of defects on its surface. As a matter of fact, capacitive performance of graphene can be enhanced by defect engineering, i.e., introducing an optimum amount of defect sites on host surface to shorten the electron/ion diffusion path length. Chemical doping of graphene structures with heteroatoms (N, B, P and S) can induce local lattice changes. This can impact the pore size and generate several defect sites on the materials surface. Accordingly, the electrolyte interactions and faradic reactions may improve thereby increasing the specific capacitance. As an illustration, boron with three valence electrons can give rise to a hole charge carrier on substitution with carbon. The result is an increase in charge density at the Fermi level leading to increase in conductivity and EDL capacitance [108]. Chen *et al.* employed urea as a source of N-containing functional groups to obtain N-doped porous rGO fibers. Compared to pristine porous rGO fibers, the N-doped composite electrodes displayed ultrahigh specific surface area (389 m^2g^{-1} *vs.* 204 m^2g^{-1}) and specific capacitance (1,132 $mFcm^{-2}$ *vs.* 415 $mFcm^{-2}$) [109].

In the following sections, a brief review on the different types of hybrid/modified graphene-based electrode materials employed in supercapacitors have been given. The contribution of researchers, in the past five years, to improve the electrochemical performance of SCs has also been highlighted.

11.3.2.1 Graphene/Organic Material Nanocomposites

Addition of graphene to carbonaceous materials can lead to increase in electrochemical performance. This has been demonstrated by Huang and his group by coating rGO with N-doped hydrothermal carbon [110]. The composite electrode could deliver a specific capacitance of 340 Fg^{-1} (at 0.1 Ag^{-1}) in 6M KOH solution as against pure rGO in an inorganic solvent (150–230 Fg^{-1}) [111]. Improvements were also observed for rate capability (203 Fg^{-1} at 50 Ag^{-1}) and cyclic stability (up to 2,000 cycles).

In another work, graphene-cellulose nanofiber (CNF) hybrid aerogel has been employed as electrode material for flexible SCs. The rGO/CNF composite possessed a web-like structure with a rough and wrinkled surface. CNF played a major role in avoiding restacking of graphene nanosheets and upholding hydrophilicity for easy access to active sites and smooth inward/outward flow of electrolyte. The benefit of CNF's presence is explicit in the marked electrical conductivity (100 Sm^{-1}) and high EDLC capacitance value of 207 Fg^{-1} at 5 mVs^{-1} (as against 188 Fg^{-1} for graphene film) with a capacity retention of 99.1% on completion of 5,000 cycles. The measurements were conducted using identical electrodes in PVA/H_2SO_4 electrolyte solution [112].

Carbon nanotubes (CNTs) are high conducting organic materials with hexagonally placed sp^2 carbon atoms in a rolled form. The striking feature of CNT is the huge mean free path available for charge carriers without giving any room for Joule heating. To increase the capacitance, rGO/amino functionalized CNT composites were prepared by hydrothermal method and investigated in aqueous KOH. The outer and inner diameters of the CNT-NH_2 material were 8–10 and 2–3 nm respectively. The specific capacitance achieved by the nanocomposite was 76.2 Fg^{-1} [113].

Similar to CNTs, activated carbon (AC) is another class of organic materials used as electrode material due to its porous nature. However, AC suffers from low conductivity. In a eco-friendly synthetic strategy, agricultural waste (corncob) was used to derive AC which is oxidized (OAC) and later blended with rGO by reduction method. The rGO sheets served as a carrier for OAC particles and simultaneously controlled the pore distribution and active surface area, which resulted in high rate capability. The lithium ion capacitor, built by using OAC/rGO as cathode and Si/C as anode, delivered a high energy density of 141 $Whkg^{-1}$ at 1,391 Wkg^{-1}, with 78.98% capacity retention at 0.4 Ag^{-1} for 1,000 cycles [114].

Recently, chemically modified graphene structures are gaining attention for their applications in energy storage devices. The basic idea of modifying the graphene surface is to introduce electro-active groups so as to improve the conductivity and capacitance of the material. In this regard, introduction of amino group on graphene surface has shown to improve energy storage capacity [115, 116]. Recently, a successful attempt has been made by Kang and his group to investigate novel diaminopyrene modified rGO electrode for supercapacitor applications. Prepared by a simple solvothermal method, the modified electrode displayed better capacitance (397.63 Fg^{-1} at 0.5 Ag^{-1}) when compared to pristine rGO (80.29 Fg^{-1}). On conversion to symmetric supercapacitor, it excelled with remarkable characteristics (capacitance of 82.70 Fg^{-1} at 0.5 Ag^{-1}; energy density of 25.84 $Whkg^{-1}$ at 375 Wkg^{-1}; power density of 7,500 Wkg^{-1} at 10 Ag^{-1}) with electrochemical stability up to 20,000 cycles [117]. Likewise, there are several reports on combining electrochemically active quinone molecules with graphene for high energy density SC devices [115–121].

11.3.2.2 Graphene/Conducting Polymer Nanocomposites

Conducting polymers (CPs) are electrically conducting materials with a conjugated backbone. CPs are basically psuedocapacitive materials with high theoretical capacitance and charge mobility. Additionally, they are easy to synthesize and are eco-friendly. The lacking feature is the cyclic stability which diminishes due to volume change. Excessive cycling can result in swelling and thickening of the electrode material, which may prove dangerous. Integrating graphene materials with CPs are reported to improve the stability and rate capability of the latter.

CPs have inherent properties of large specific conductance and several surface active sites which need to be explored for the application of pseudocapacitors. Polyaniline (PANI), due to its easy synthesis and good capacitive characteristics, has a high preference as an electro-active material, provided its rate capability and cycle life can be improved. Grafting PANI with carbonaceous material is one of the alternatives [122]. As an example, the synergistic effects of high pseudocapacitance of PANI and large surface area of rGO led to a high specific capacitance of 1,019 Fg^{-1} at 1 mVs^{-1} scan rate with 80% retention after 100 cycles [123]. In another study, rGO/PANI composites with three dimensional structures were prepared by steamed water regulation method. The porous structure of rGO ensures uniform dispersion of PANI fibres and efficient electrolyte infiltration. The electrode achieved a specific capacitance of 1,182 Fg^{-1} at 1 Ag^{-1} in 1M H_2SO_4. When assembled to a symmetric SC device, a capacitance of 808 Fg^{-1} (at 1 Ag^{-1}) and high energy density of 28.06 $Whkg^{-1}$ (at 0.25 $kWkg^{-1}$) was obtained [124]. Later, Wu and co-workers reported a similar observation for 3D rGO/PANI gel structure synthesized by self-assembly method. They confirmed that insertion of PANI on GO nanosheets before reduction favours easy electrolyte diffusion, which results in high electrochemical performance. The authors also studied the effect of PANI concentration and electrolyte chosen on the working mechanism. A high concentration of PANI resulted in a thick coat of PANI on a curled surface of rGO nanosheets, having a negative impact on ion transfer. Use of a solvent like DMSO, which has low solubility for PANI, gave rise to aggregated PANI chains on the host surface, deteriorating the capacitance to a large extent [125]. Wang *et al.* fabricated PANI-rGO/MWCNT ternary nanocomposite for high performance supercapacitor applications. The presence of MWCNT improves the capacitive behavior while PANI dispersed in rGO imparts a 3D structure which aids in electron/ion transfer [126].

Like PANI, several attempts have been made to fabricate graphene based polypyrrole (PPy) nanocomposites [127, 128]. To understand the significance of graphene backbone, the surface microstructure of pure PPy and PPy/rGO nanocomposites have been compared by varying the graphene content. It was observed that PPy exhibited a "cauliflower" like morphology with agglomeration which on hybridization with rGO, gradually got converted to sheet-like structures. Moreover, a noteworthy leap in specific surface area (5 times more than pure PPy) was observed for PPy/rGO-10 (126 m^2g^{-1}). The same electrode reported a high specific conductance (290.1 Fg^{-1} at 0.2 Ag^{-1}) in 3M KCl with 97.5% retention after 20,000 cycles (at 2 Ag^{-1}) [129]. Composites with core-shell structures have also been attempted. PPy nanotube with graphene nanoflake coating to form core-shell like tubular structure has been fabricated by Qi *et al.* The presence of graphene based coating throughout PPy nanotube surface amplified the electronic conductivity of the latter. A solid-state flexible SC built using this nanocomposite reported a capacitance of 128 $mFcm^{-2}$ at 1.8 $mAcm^{-2}$. The energy density furnished was 11.4 $\mu Whcm^{-2}$ at 720 $\mu Wh\ cm^{-2}$ power density. The capacitance observed after 5,000 cycles was 80% [130]. Recently holey graphene (hG) (graphene framework with high porosity) frameworks have been used as scaffolding materials for PPy. The advantage of hG lies in its contribution towards large surface area and mechanical strength. A 3 dimensional hG/PPy aerogel structure with 75 wt% PPy imparted an appreciable specific capacitance of 418 Fg^{-1} at 0.5 Ag^{-1} with good rate capability (80%) and a reasonable cycling stability (74% after 2,000 cycles) [131].

Poly(3,4-ethylenedioxythiophene) (PEDOT) is also considered as an electro-active material due to its reasonably good electrical conductivity and pseudocapacitance. However, because of its low solubility, it is generally blended with poly(styrene sulfonic acid) (PSS). PEDOT can achieve a specific capacitance of 70–130 Fg^{-1} depending on the polymerization technique utilized [132]. The synthetic strategy, choice of electrolyte and processing parameters also play crucial role in their capacitive performance [133]. When used with sodium dodecyl sulphate as electrolyte, the PEDOT/GO composite attains large surface area and enhanced electron/ion diffusion rate. The positive outcome is due to SDS which produces negative charges on the GO surface and also plays a crucial role in dispersion. The capacitance delivered was 79.6 $mFcm^{-2}$ at 10 mVs^{-1} with 12.5 $Whcm^{-2}$ energy density liberated [134]. The preparation of PEDOT/GO composite in to a hydrogel structure not only improves the conductivity but also mechanical strength. In one of the studies, the compressive fracture stress for PEDOT/GO hydrogel assembly was evaluated to be 29.6 MPa with 2.1 MPa modulus at 10 $rads^{-1}$. The electrode attained electrical conductivity of 0.73 Scm^{-1} with a specific capacitance of 174.4 nFg^{-1} [135].

In modern research, ternary electrodes employing metal oxides with polymer-graphene composites are being reported. The advantages of metal oxides in electrode materials have been given in detail in the next section. However, the main intention of metal oxide addition is enhancement in specific capacitance. In this direction, a series of conductive polymers (PANI, PPy, PEDOT and polythiophene-PTh) were polymerized on ZnS/rGO composite to form a ZnS/rGO/CP ternary electrode. It was observed that ZnS/rGO/PANI displayed the best capacitance (722.0 Fg^{-1} at 1 Ag^{-1})

and cycle stability (76.1% after 1,000 cycles). The electrode also exhibited the best energy density (349.7 $Whkg^{-1}$) and power density (18 $kWkg^{-1}$) among all the other polymer coated ZnS/rGO electrodes. The ZnS/rGO/CP system with ~1.6 V discharge velocity makes them suitable for supercapacitor applications [136].

11.3.2.3 Graphene/Metal Oxide Nanocomposites

Among metal oxides, transition metal oxides of Ru, Mn, Co, Ni and Fe are more favored for electrochemical applications due to their multiple valence states possible. Metal oxide, in itself, can provide high energy density and capacitance but suffers from mechanical and electrochemical stability in the long run. Materials with metal oxide nanoparticles (NPs) in collaboration with graphene-like structures can bring together the best of the individual components. The organic component will serve as the nucleation site for the growth of metal oxides and provide a porous but strong backbone for agglomeration free NPs. Metal oxides on the other hand, can hinder the restacking of graphene sheets and contribute in the overall specific capacitance of the hybrid material.

Ruthenium oxide is the most sought-after metal oxides because of its outstanding thermal stability and capacitance of ~720 Fg^{-1} (for amorphous RuO_2) [137]. At low electrolyte pH, the redox reaction occurs rapidly with $Ru^{+4} \leftrightarrow Ru^{+2}$. Laser treated graphene/RuO_2 composite as a positive electrode for SC and light-emitting applications have been recently studied. Here, the NPs were found uniformly distributed on graphene surface. This resulted in unconstrained electron/ion transport pathway. Pairing it with activated carbon as negative electrode, an ASC device was designed which reported an outstanding specific capacitance of 1,139 Fg^{-1} at a high current density of 300 Ag^{-1}, besides a large energy density of 55 $Whkg^{-1}$ at a power density of 12 $kWkg^{-1}$ [138].

Manganese oxides (MnO, MnO_2, Mn_2O_3, Mn_3O_4) are another group of oxides which are known for their high pseudocapacitive values (~1,400 Fg^{-1} for MnO_2). The redox reaction generally proceeds by the following mechanism

$$MnO_2 + H^+ + e^- \leftrightarrow MnOOH$$

Patil and his group studied the electrochemical performance of 3D MnO_2 grown *in situ* over graphene nanosheet surface. The presence of graphene improves the electron transfer and cycling efficiency along with enhancement in faradic reaction rate. The authors report a specific capacitance of 683 Fg^{-1} at 2.2 Ag^{-1} with an outstanding cycling efficiency capacitance retention of 96.9% after 5,000 cycles. Even at a high current density of ~10.86 Ag^{-1}, a remarkable energy density of ~7.99 $kWhkg^{-1}$ and a high-power density of ~6.235 $kWkg^{-1}$ is achieved [139]. In another work, MnO_2 nanoflowers have been deposited on graphene paper by a cost-effective and environment friendly electrochemical deposition method. The graphene paper/MnO_2 nanocomposite, obtained after 10 cycles of deposition, delivered ultrahigh capacitance of 385.2 Fg^{-1} at 1 mVs^{-1} with excellent retention after 5,000 cycles. When fabricated into an ASC with pure graphene paper as negative electrode, the storage system exhibited outstanding energy density of 6.14 $mWhcm^{-2}$ and power density of 36 $mWcm^{-2}$ [140]. A similar method has been employed by Cui et al. to successfully deposit MnO_2 NPs on rGO/lignin based porous carbon (PC) composite. The MnO_2/PC/rGO ternary electrode exhibited improved specific capacitance of 1,136 $mFcm^{-2}$ at 1 $mAcm^{-2}$ because of the conductive network provided by the carbonaceous material and the large surface area of MnO_2 NPs. When assembled into a symmetric supercapacitor, energy density of 0.253 $mWhcm^{-3}$ at 0.5 $mAcm^{-2}$ and a power density of 0.018 Wcm^{-3} at 5 $mAcm^{-2}$ was obtained. In addition, the supercapacitor also displayed a good cycling stability (~85.2% capacitance retention for 5,000 cycles).

NiO is a metal oxide which too is stable at more than two oxidation states. Its theoretical capacitance is reported to be ~2,584 Fg^{-1} (for NiO_2). But due to limited electrical conductivity, oxides of nickel can achieve only 50–1,776 Fg^{-1} of capacitance. NiO in an alkaline pH follows the following redox reaction [141].

$$NiO + OH^- \leftrightarrow NiOOH + e^-$$

NiO/rGO composite has a better capacitive performance and cyclic stability than the individual components. The superior performance is credited to the porous nature of the 3D graphene framework providing higher electro-active sites and smaller diffusion pathways [142]. Additionally, even core–shell structure for graphene-based nickel oxide nanocomposites, comprising of inorganic core and organic shell, with desirable electrochemical characteristics have been reported [143, 144]. A high specific capacitance has also been reported by combining $Ni(OH)_2$ with graphene [145] and holey graphene [146]. In the latter case, the highly porous nature of hG added to the storage capacity and rate capability values.

Cobalt oxide, with its high theoretical capacitance (~3,560 Fg^{-1}) but low conductivity, has also been employed as supercapacitor electrode material. To exemplify, vertically aligned graphene nanosheets were taken on a carbon cloth as a supporting structure. Co_3O_4 NPs were deposited. This strategy resulted in achievement of maximum capacitance of 3,482 Fg^{-1} at a low scan rate of 1 mVs^{-1}, which is very close to the theoretical value of Co_3O_4. Further, a flexible solid-state SC device manifested a superior capacitance (580 Fg^{-1}), extremely high cycling capacity (86.2%) for 20,000 cycles, and huge energy density (80 $Whkg^{-1}$) and power density (20 $kWkg^{-1}$) values [147].

Other than the above mentioned oxides, much work has been reported on other transition metal oxides too, viz., Fe_3O_4 [148], CuO [149], ZnO [150] and V_2O_5 [151]. Similar to the above reported metal oxides, these oxides also act as spacer molecules preventing the restacking of graphene sheets and/or as components contributing in faster ion/electron transport which result in enhancement in specific capacitance.

Attention has also been drawn towards lanthanide oxides, which are well-known for their utilization in fuel cells, catalysis and electronics. The redox characteristics arising due to multiple *4f* electron transitions and availability of voids in the crystal structure have prompted many researchers to apply the rare earth oxides for energy storage applications [152, 153]. As an illustration, Eu_2O_3 nanorods, synthesized by hydrothermal method, have been anchored on rGO by self-assembly process. The Eu_2O_3/rGO electrode displayed a capacity of 345.9 Fg^{-1} at 2 Ag^{-1} current density. The high charge mobility of Eu_2O_3 nanorods together with flexibility of rGO has been credited for the superior electrochemical performance [154].

Parallel to metal oxides, bimetallic oxides also possess substantial pseudocapacitive behavior. Many a times, the bimetallic oxides surpass the performance of the individual components [155]. Makkar and Ghosh employed a facile "one-pot" hydrothermal method to prepare $MnFe_2O_4$/rGO nanocomposite as a cathodic material for ASC. The synthetic procedure was optimized to attain the highest specific capacitance of 768 Fg^{-1} at 8 Ag^{-1} for the hybrid with 80% $MnFe_2O_4$ and 20% rGO. When coupled with rGO, the all solid-state flexible ASC device exhibited an energy density of 27.7 $Whkg^{-1}$ at power density of 750 Wkg^{-1} [156]. In a recent work, $MnFe_2O_4$/rGO was developed by decorating $MnFe_2O_4$ NPs on 3D crumpled graphene structure, by a single step aerosol-assisted capillary compression process. The composite was applied not only as an electrochemical supercapacitor but also as hydrogen peroxide sensor [157]. Mn–Co oxide/rGO hybrid nanostructures have been designed for enhanced energy storage performances. The presence of graphene modifies Mn-Co oxide structure. The oxide structure enabled high collection of free electrons at Fermi level leading to a large specific capacitance of 2,749 Fg^{-1} (381.8 $mAhg^{-1}$) with a remarkable 95.7% retention at 10 Ag^{-1} for 8,000 cycles. On fabrication to

ASC, a high energy density of 35.5 $Whkg^{-1}$ at 1008.2 Wkg^{-1} was obtained [158]. In another work, $NiCoO_2$/rGO/Ni foam, synthesized by hydrothermal method, exhibited an appreciable capacitance retention of 89% for 2,500 charge/discharge cycles. The authors credit the availability of more number of surface-active sites for the high cycling stability [159].

Besides metal oxides, metal sulfides and nitrides are also known for their pseudocapacitive nature. MoS_2, in particular, has been extensively studied for energy storage applications. MoS_2 is known to have graphene-like structure with high surface area and active edge sites promoting faster redox reactions. Composites of graphene and MoS_2 are basically fabricated for high capacitive energy storage applications [160, 161]. Also, several reports are available on metal nitride/graphene composites displaying admirable electrochemical characteristics as supercapacitor electrodes [162, 163].

Another facile approach is the employment of metallic nanoparticles or metal alloys for SC applications. Metallic nanoparticles like Ag and Ni were immobilized on rGO to form $(Ag_xNi_{(1-x)})_y/rGO_{(100-y)}$ nanocomposite with hierarchical structure. The combined effect of the two metals improved the electrochemical output with $(Ag_{0.5}Ni_{0.5})_{90}/rGO_{10}$ hybrid manifesting the best capacitance of 897 Fg^{-1} at 1 Ag^{-1} and energy density of 80 $Whkg^{-1}$ at power density of 400 Wkg^{-1}. The initial capacitance resistance was ~77% at the end of 4,000 cycles [164]. In a different work, the high electrical conductivity, enhanced faradic redox reactions and superior electrochemical performance of CoNi alloy has been exploited to fabricate a hybrid electrode material with rGO for high-performance supercapacitors. The authors claimed a better performance by snow-flake dendritic CoNi alloy compared to spherical CoNi alloy. $(CoNi)_{60}/rGO_{40}$ composite delivered a high specific capacitance (501 Fg^{-1} at 6 Ag^{-1}), large energy density (21.08 $Whkg^{-1}$ at 1,650 Wkg^{-1}) and good cycling efficiency (~85% retention after 4,000 cycles), manifesting the synergistic benefits of the CoNi/rGO hierarchical system [165].

In the present times, ternary electrodes are gaining wider attention. These electrodes have the advantages of all the three individual components while at the same time, the disadvantages of each component is taken care by the remaining two. In other words, the amalgam of three electro-active materials contributes constructively in the overall electrochemical performance. As a first of its kind, mixed metal sulfides in a carbon rich environment have been tested for their suitability in ASC applications by Karimi and co-workers. The authors employed PPy, Fe–Co sulfide and graphene foam to construct a ternary binder-free composite on Ni foam (NF) electrode. The nanocomposite delivered a high specific capacitance of 3178 Fg^{-1} at 2 Ag^{-1}. FeCoS can exist in various valence states giving rise to faradic reversible reactions and high charge storage capacity. The presence of the conducting polymer PPy further enhances pseudocapacitance by reducing the electron transport path length and providing more active sites. Apart from increasing electronic conductivity, rGO played a crucial role in developing an effective interfacial contact between PPy and FeCoS resulting in efficient utilization of the active material. The ternary electrode was used as a positive electrode and combined with rGO/NF acting as negative electrode, to create a hybrid ASC which exhibited a high specific capacitance of 94 Fg^{-1} at 1 Ag^{-1} and a high energy density of 28.3 $Whkg^{-1}$ at 810 Wkg^{-1} power density. Also, a remarkable cycling efficiency with capacitance retention of 97.5% for 2,500 cycles was achieved [166]. In another example, the capacitance of boron nitride (BN) has been shown to enhance on doping with $Ni(OH)_2$ and scaffolding with rGO. A specific capacitance value of 349 Fg^{-1} was delivered by rGO/BN/$Ni(OH)_2$ composite in 0.5 M KOH solution [167].

11.4 Conclusion

In past one decade, the upsurge in publications relating to novel cathodic and anodic materials for LIBs has scaled new heights. Graphene, as a scaffolding material, has contributed extremely in enhancing the capacity, rate capability and cycling efficiency of the active material. Graphene fits well in the electrode design by imparting high conductivity (for faster electron/ion transfer), defect sites (for active lithiation/delithiation) and porous structure (for maximizing Li storage capacity and diminishing large volume and structural changes during cycling). Other than these inherent properties, graphene also benefits from ability for uniform dispersion of the active material and resistance towards particle aggregation during excessive cycling. Furthermore, graphene also provides the mechanical robustness which is very much essential for practical applications of a free-standing electrode. The flexibility of graphene lies in the various structural form it can easily undertake, out of which hierarchical, core-shell and sandwich-like structures are garnering more attention. Though graphene-based electrodes look promising for future lithium ion battery technology, there is still scope for development. The composite electrodes must be designed to attain high surface area and conductivity. In fact, much of the discharge capacity and coulombic efficiency wastage can be minimized if the attention is focussed on enhancing power conversion efficiency of the active materials. Another major loophole in the research of graphene-based electrodes is the lack of knowledge on electrode-electrolyte interfaces. Much of the reported composite electrodes have suffered from initial capacity loss. A deeper insight on the mechanism of interactions occurring at these interfaces will definitely be helpful in selection of active materials and electrolytes for high performance batteries. The future perspectives of graphene-based electrodes should be concerned with developing simple and affordable methodology to design composite electrodes for large-scale production. Efforts must be centred to design next-generation batteries with high voltage cathode, low voltage anode and a compatible electrolyte. It will also be interesting to include flame retardants, which can put off the risk of fire and thereby improve battery safety.

The significant role played by SCs in energy storage devices is well-known. The presence of graphene as one of the integrated component for SC electrodes is high beneficial in terms of electrochemical and mechanical stability. In the present chapter, the contribution and achievements of SC devices fabricated by combining graphene with organic materials, conducting polymers and metal oxides are exemplified. Nevertheless, graphene-based nanocomposites have several future challenges to be addressed. A low-cost, eco-friendly, facile synthetic method for large scale production is onc amongst thcm. For applications requiring high energy density and power density, a careful selection of the electrode constituents and electrolyte material becomes essential. Also, there is still room for evolvement in performance of the electrode materials. This is possible by understanding the mechanism involved in electrode/electrolyte interfacial reactions and thereby designing strategies to control these reactions. Another barrier is the control of pore size. A high porosity is essential for large energy storage. But at the same time, during charge/discharge cycling, several compatibility issues between the porous surface and the electrolyte ions crop up. The requisites for a highly flexible supercapacitor are high capacitance with large retention, good conductivity and ultrahigh energy density and power density with minimal volumetric capacitance loss. Electrode with have high mechanical and thermal stability with long lifespan would be an added advantage. Only by consideration of the above mentioned issues, it will be possible to achieve the required traits of next-generation SCs for energy storage and energy harvesting applications.

References

1. Cong, H.P., Chen, J.F., Yu, S.H., Graphene-based macroscopic assemblies and architectures: An emerging material system. *Chem. Soc. Rev.*, 43, 7295–7325, 2014.
2. Han, C., Zhang, N., Xu, Y.J., Structural diversity of graphene materials and their multifarious roles in heterogeneous photocatalysis. *Nano Today*, 11, 351–372, 2016.
3. Goodenough, J.B., Energy storage materials: A perspective. *Energy Storage Mater.*, 1, 158–161, 2015.
4. Geim, A.K. and Novoselov, K.S., The rise of graphene. *Nat. Mater.*, 6, 183–191, 2007.
5. Wang, X. and Shi, G., Flexible graphene devices related to energy conversion and storage. *Energy Environ. Sci.*, 8, 790, 2015.
6. Novoselov, K.S., Falko, V.I., Colombo, L., Gellert, P.R., Schwab, M.G., Kim, K., A roadmap for graphene. *Nature*, 490, 192–200, 2012.
7. Mao, M., Hu, J., Liu, H., Graphene-based materials for flexible electrochemical energy storage. *Int. J. Energy Res.*, 39, 727–740, 2015.
8. Hu, K., Kulkarni, D.D., Choi, I., Tsukruk, V.V., Graphene–polymer nanocomposites for structural and functional applications. *Prog. Polym. Sci.*, 39, 1934–1972, 2014.
9. Bonaccorso, F., Colombo, L., Yu, G.H., Stoller, M., Tozzini, V., Ferrari, A.C., Ruoff, R.S., Pellegrini, V., Graphene, related two-dimensional crystals, and hybrid systems for energy conversion and storage. *Science*, 347, 124651, 2015.
10. Raccichini, R., Varzi, A., Passerini, S., Scrosati, B., The role of graphene for electrochemical energy storage. *Nat. Mater.*, 14, 271–279, 2015.
11. Wei, L., Zhengjie, L., Yaqian, D., Quan-Hong, Y., Feiyu, K., Graphene-based Materials for Electrochemical Energy Storage Devices: Opportunities and Challenges. *Energy Storage Mater.*, 2, 107–138, 2016, http://dx.doi.org/10.1016/j.ensm.2015.10.002.
12. Novoselov, K.S. *et al.*, Electric field in atomically thin carbon films. *Science*, 306, 5696, 666–669, 2004.
13. Obraztsov, A.N. *et al.*, Chemical vapor deposition of thin graphite films of nanometer thickness. *Carbon*, 45, 10, 2017–2021, 2007.
14. Gattia, D.M., Vittori Antisari, M., Marazzi, R., AC arc discharge synthesis of single-walled nanohorns and highly convoluted graphene sheets. *Nanotechnology*, 18, 25, 255604, 2007.
15. Hummers, W.S., Jr. and Offeman, R.E., Preparation of graphitic oxide. *J. Am. Chem. Soc.*, 80, 6, 1339, 1958.
16. Van Bommel, A.J., Crombeen, J.E., Van Tooren, A., LEED and Auger electron observations of the SiC(0001) surface. *Surf. Sci.*, 48, 2, 463–472, 1975.
17. Lockett, M., Sarmiento, V., Balingit, M., Oropeza-Guzman, M.T., Vazquez-Mena, O., Direct chemical conversion of continuous CVD graphene/graphite films to graphene oxide without exfoliation. *Carbon*, 158, 1, 202–209, 2019.

18. Wang, Z. *et al.*, Direct electrochemical reduction of single-layer graphene oxide and subsequent functionalization with glucose oxidase. *J. Phys. Chem. C*, 113, 32, 14071–14075, 2009.
19. Choi, K. *et al.*, Terahertz and optical study of monolayer graphene processed by plasma oxidation. *Appl. Phys. Lett.*, 102, 13, 131901, 2013.
20. Rani, J.R. *et al.*, Controlling the luminescence emission from palladium grafted graphene oxide thin films *via* reduction. *Nanoscale*, 5, 12, 5620–5627, 2013.
21. Lim, J. *et al.*, Terahertz, optical, and Raman signatures of monolayer graphene behavior in thermally reduced graphene oxide films. *J. Appl. Phys.*, 113, 183502–183505, 2013.
22. Nair, R.R. *et al.*, Fine structure constant defines visual transparency of graphene. *Science*, 320, 5881, 1308, 2008.
23. Meyer, J.C. *et al.*, The structure of suspended graphene sheets. *Nature*, 446, 7131, 60–63, 2007.
24. Tung, V.C. *et al.*, High-throughput solution processing of large-scale graphene. *Nat. Nanotechnol.*, 4, 1, 25–29, 2009.
25. Uthaisar, C. and Barone, V., Edge effects on the characteristics of Li diffusion in graphene. *Nano Lett.*, 10, 2838–2842, 2010.
26. Bharwaj, T., Antic, A., Paven, B., Barone, V., Fahlman, B.D., Enhanced electrochemical lithium storage by graphene nanoribbons. *J. Am. Chem. Soc.*, 132, 12556–12558, 2010.
27. Uthaisar, C., Barone, V., Fahlman, B.D., On the chemical nature of thermally reduced graphene oxide and its electrochemical Li intake capacity. *Carbon*, 61, 558–567, 2013.
28. Lee, W., Suzuki, S., Miyayama, M., Lithium storage properties of graphene sheets derived from graphite oxides with different oxidation degree. *Ceram. Int.*, 39, S753–S756, 2013.
29. Kuo, S.L., Liu, W.R., Wu, H.C., Lithium storage behavior of graphene nanosheets-based materials. *J. Chin. Chem. Soc.*, 59, 10, 1220–1225, 2012.
30. Liu, Y., Fan, Q., Tang, N., Wan, X., Liu, L., Lv, L., Du, Y., Study of electronic and magnetic properties of nitrogen doped graphene oxide. *Carbon*, 60, 538–561, 2013.
31. Hu, Y., Li, X., Geng, D., Cai, M., Li, R., Sun, X., Influence of paper thickness on the electrochemical performances of graphene papers as an anode for lithium ion batteries. *Electrochem. Acta*, 91, 227–233, 2013.
32. Yang, Y.Q., Wu, K., Pang, R.Q., Zhou, X.J., Zhang, Y., Wu, X.C., Wu, C.G., Wu, H.X., Guo, S.W., Graphene sheets coated with a thin layer of nitrogen-enriched carbon as a high-performance anode for lithium-ion batteries. *RSC Adv.*, 3, 14016–14020, 2013.
33. He, Z., Wang, Z., Guo, H., Li, X., Xianwen, W., Yue, P., Wang, J., A simple method of preparing graphenecoated $Li[Li_{0.2}Mn_{0.5}4Ni_{0.13}Co_{0.13}]O_2$ for lithium-ion batteries. *Mater. Lett.*, 91, 261–264, 2013.
34. Sun, Y., Yang, S.B., Lv, L.P., Lieberwirth, I., Zhang, L.C., Ding, C.X., Chen, C.H., A composite film of reduced graphene oxide modified vanadium oxide

nanoribbons as a free standing cathode material for rechargeable lithium batteries. *J. Power Sources*, 241, 168–172, 2013.
35. Ding, Y.H., Ren, H.M., Huang, Y.Y., Chang, F.H., He, X., Fen, J.Q., Zhang, P., Co-precipitation synthesis and electrochemical properties of graphene supported $LiMn_1/3Ni_1/3Co_1/3O_2$ cathode materials for lithium-ion batteries. *Nanotechnology*, 24, 375401–375408, 2013.
36. Prabakar, S.J.R., Hwang, Y.H., Lee, B., Sohn, K.S., Pyo, M., Graphene-sandwiched $LiNi_{0.5}Mn_{1.5}O_4$ cathode composites for enhanced high voltage performance in Li ion batteries. *J. Electrochem. Soc.*, 160, 6, A832–A837, 2013.
37. Wang, B., Wang, D., Wang, Q., Liu, T., Guo, C., Zhao, X., Improvement of the electrochemical performance of carbon-coated $LiFePO_4$ modified with reduced graphene oxide. *J. Mater. Chem. A*, 1, 135–144, 2013.
38. Mizushima, K., Jones, P.C.C., Wiseman, P.J.J. *et al.*, $Li_xCoO_2(0 < x<-1)$: A new cathode materials for batteries of high energy density. *Mater. Res. Bull.*, 15, 783–789, 1980.
39. Gold Peak Industries (Taiwan) Ltd., GPI International Limited, Taipei 10647, TAIWAN R. O. C, *Lithium ion technical handbook*, vol. 1, pp. 1–13, 2003.
40. Megahed, S. and Scrosati, B., Lithium-ion rechargeable batteries. *J. Power Sources*, 51, 79–104, 1994.
41. Cai, X., Lai, L., Shen, Z., Lin, J., Graphene and graphene-based composites as Li-ion battery electrode materials and their application in full cells. *J. Mater. Chem. A*, 5, 15423, 2017.
42. Compton, O.C. and Nguyen, S.T., Graphene oxide, highly reduced graphene oxide and graphene: Versatile building blocks for carbon-based materials. *Small*, 6, 711–723, 2010.
43. Raccichini, R., Varzi, A., Passerini, S. *et al.*, The role of graphene for electrochemical energy storage. *Nat. Mater.*, 14, 271–279, 2014.
44. Wang, B., Al Abdulla, W., Wang, D. *et al.*, A three dimensional porous $LiFePO_4$ cathode material modified with a nitrogen-doped graphene aerogel for high-power lithium ion batteries. *Energy Environ. Sci.*, 8, 869–875, 2015.
45. Lee, W.S.V., Peng, E., Loh, T.A.J. *et al.*, Few-layer MoS_2-anchored graphene aerogel paper for self-supported electrode materials. *Nanoscale*, 8, 8042–8047, 2016.
46. Zhang, N., Yan, X., Huang, Y. *et al.*, Electrostatically assembled magnetite nanoparticles/graphene foam as a binder-free anode for lithium ion battery. *Langmuir*, 33, 8899–8905, 2017.
47. Xia, Z.Y., Christian, M., Arbizzani, C. *et al.*, A robust, modular approach to produce graphene-MO_x multilayer foams as electrodes for Li-ion batteries. *Nanoscale*, 11, 5265–5273, 2019.
48. Ren, Y., Lv, W., Wen, F. *et al.*, Microwave synthesis of SnS_2 nanoflakes anchored graphene foam for flexible lithium-ion battery anodes with long cycling life. *Mater. Lett.*, 174, 24–27, 2016.

49. Wen, H., Kang, W., Liu, X. *et al.*, Two-phase interface hydrothermal synthesis of binder-free SnS_2/graphene flexible paper electrodes for high-performance Li-ion batteries. *RSC Adv.*, 9, 23607–23613, 2019.
50. Ding, Y., Jiang, Y., Xu, F. *et al.*, Preparation of nano-structured $LiFePO_4$/graphene composites by co-precipitation method. *Electrochem. Commun.*, 12, 10–13, 2010.
51. Zhou, X., Wang, F., Zhu, Y. *et al.*, Graphene modified $LiFePO_4$ cathode materials for high power lithium ion batteries. *J. Mater. Chem.*, 21, 3353–3358, 2011.
52. Ha, S.H. and Lee, Y.J., Core-shell $LiFePO_4$/carbon-coated reduced graphene oxide hybrids high-power lithium-ion battery cathodes. *Chem.— Eur. J.*, 21, 2132–2138, 2015.
53. Abe, Y. and Kumagai, S., Effect of negative/positive capacity ratio on the rate and cycling performances of $LiFePO_4$/graphite lithium-ion batteries. *J. Energy Storage*, 19, 96–102, 2018.
54. Cui, L.F., Wang, X.P., Chen, N. *et al.*, A versatile graphene foil. *J. Mater. Chem. A*, 5, 14508–14513, 2017.
55. Prabakar, S.J.R., Hwang, Y.H., Lee, B. *et al.*, Graphene-sandwiched $LiNi_{0.5}Mn_{1.5}O_4$ cathode composites for enhanced high voltage performance in Li ion batteries. *J. Electrochem. Soc.*, 160, A832–A837, 2013.
56. Sreelakshmi, K.V., Sasi, S., Balakrishnan, A. *et al.*, Hybrid composites of $LiMn_2O_4$–Graphene as rechargeable electrodes in energy storage devices. *Energy Technol.*, 2, 257–262, 2014.
57. Chen, K.S., Xu, R., Luu, N.S. *et al.*, Comprehensive enhancement of nanostructured lithium-ion battery cathode materials *via* conformal graphene dispersion. *Nano Lett.*, 17, 2539–2546, 2017.
58. Ju, B., Wang, X., Wu, C. *et al.*, Electrochemical performance of the graphene/Y_2O_3/$LiMn_2O_4$ hybrid as cathode for lithium-ion battery. *J. Alloys Compd.*, 584, 454–460, 2014.
59. Aziz, S., Zhao, J., Cain, C. *et al.*, Nanostructured $LiMn_2O_4$/graphene/ZnO composites as electrodes for lithium ion batteries. *J. Mater. Sci. Technol.*, 30, 427–433, 2014.
60. Jaber-Ansari, L., Puntambekar, K.P., Kim, S. *et al.*, Suppressing manganese dissolution from lithium manganese oxide spinel cathodes with single-layer graphene. *Adv. Energy Mater.*, 5, 1500646, 2015.
61. Chebiam, R.V., Prado, F., Manthiram, A., Structural instability of delithiated $Li_{1-x}Ni_{1-y}Co_yO_2$ cathodes. *J. Electrochem. Soc.*, 148, A49–A53, 2001.
62. Choi, J. and Manthiram, A., Role of chemical and structural stabilities on the electrochemical properties of layered $LiNi_{1/3}Mn_{1/3}Co_{1/3}O_2$ cathodes. *J. Electrochem. Soc.*, 152, A1714, 2005.
63. He, J.R., Chen, Y.F., Li, P.J. *et al.*, Synthesis and electrochemical properties of graphene modified $LiCo_{1/3}Ni_{1/3}Mn_{1/3}O_2$ cathodes for lithium ion batteries. *RSC Adv.*, 4, 2568–2572, 2014.

64. Jiang, K.C., Xin, S., Lee, J.S. *et al.*, Improved kinetics of $LiNi_{1/3}Mn_{1/3}Co_{1/3}O_2$ cathode material through reduced graphene oxide networks. *Phys. Chem. Chem. Phys.*, 14, 2934–2939, 2012.
65. Aliahmad, N., Liu, Y., Xie, J. *et al.*, V_2O_5/grapheme hybrid supported on paper current collectors for flexible ultrahigh capacity electrodes for lithium-ion batteries. *ACS Appl. Mater. Interfaces*, 10, 16490–16499, 2018.
66. Cheng, B., Zhang, X.D., Ma, X.H. *et al.*, Nano-$Li_3V_2(PO_4)_3$ enwrapped into reduced graphene oxide sheets for lithium-ion batteries. *J. Power Sources*, 265, 104–109, 2014.
67. Chang, U., Lee, J.T., Yun, J.M. *et al.*, In situ self-formed nanosheet MoS_3/reduced graphene oxide material showing superior performance as a lithium-ion battery cathode. *ACS Nano*, 13, 1490–1498, 2019.
68. Wang, D.W., Sun, C., Zhou, G. *et al.*, The examination of graphene oxide for rechargeable lithium storage as a novel cathode material. *J. Mater. Chem. A*, 1, 3607–3612, 2013.
69. Guo, Q.J., Wang, Q., Chen, G. *et al.*, Enhancing the electrochemical performance of $Li_4Ti_5O_{12}$ for Li-ion battery anodes by a sol-gel assisted molten salt method and graphene modification. *RSC Adv.*, 6, 110032–110039, 2016.
70. Liu, H.P., Wen, G.W., Bi, S.F. *et al.*, High rate cycling performance of nanosized $Li_4Ti_5O_{12}$/graphene composites for lithium ion batteries. *Electrochim. Acta*, 192, 38–44, 2016.
71. Cao, N., Wen, L.N., Song, Z.H. *et al.*, $Li_4Ti_5O_{12}$/reduced graphene oxide composite as a high-rate anode material for lithium ion batteries. *Electrochim. Acta*, 209, 235–243, 2016.
72. Jeong, J.H., Kim, M.S., Kim, Y.H. *et al.*, High-rate $Li_4Ti_5O_{12}$/N-doped reduced graphene oxide composite using cyanamide both as nanospacer and a nitrogen doping source. *J. Power Sources*, 336, 376–384, 2016.
73. Jiang, H., Zhou, X., Liu, G. *et al.*, Free-standing Si/graphene paper using Si nanoparticles synthesized by acid-etching Al–Si alloy powder for high-stability Li-ion battery anodes. *Electrochim. Acta*, 188, 777–784, 2016.
74. Yi, R., Zai, J., Dai, F. *al.*, Dual conductive network-enabled graphene/Si–C composite anode with high areal capacity for lithium-ion batteries. *Nano Energy*, 6, 211–218, 2014.
75. Ko, M., Chae, S., Jeong, S. *et al.*, Elastica-silicon nanoparticle backboned graphene hybrid as a self-compacting anode for high-rate lithium ion batteries. *ACS Nano*, 8, 8591–8599, 2014.
76. Chang, J., Huang, X., Zhou, G. *et al.*, Multilayered silicon nanoparticle/reduced graphene oxide hybrid as a high-performance lithium-ion battery anode. *Adv. Mater.*, 26, 758–764, 2014.
77. Liu, J.Y., Zheng, Q.Y., Goodman, M.D. *et al.*, Graphene sandwiched mesostructured Li-ion batteries. *Adv. Mater.*, 28, 7696–7702, 2016.

78. Huang, Y., Peng, J., Luo, J. *et al.*, Spherical G/Si/GO/C composite as high-performance anode material for Li-ion batteries. *Energy Fuels*, 34, 6, 7639–7647, 2020, https://doi.org/10.1021/acs.energyfuels.0c00982.
79. Yuan, F.W. and Tuan, H.Y., Scalable solution-grown high-germanium nanoparticle-loading graphene nanocomposites as high-performance lithium-ion battery electrodes: an example of a graphene-based platform toward practical full-cell applications. *Chem. Mater.*, 26, 2172–2179, 2014.
80. Medvedev, A.G., Mikhaylov, A.A., Grishanov, D.A. *et al.*, GeO_2 thin film deposition on graphene oxide by the hydrogen peroxide route: Evaluation for lithium ion battery anode. *ACS Appl. Mater.*, 9, 9152–9160, 2017.
81. Fang, S., Shen, L., Zheng, H. *et al.*, Ge-graphene-carbon nanotube composite anode for high-performance lithium-ion batteries. *J. Mater. Chem. A*, 3, 498–1503, 2015.
82. Qu, G.L., Geng, H.B., Ge, D.H. *et al.*, Graphene-coated mesoporous Co_3O_4 fibres as an efficient anode material for Li-ion batteries. *RSC Adv.*, 6, 71006–71011, 2016.
83. Yang, Q., Wu, J., Huang, K. *et al.*, Layer-by layer self-assembly of graphene-like Co_3O_4 nanosheet/graphene hybrids: Towards high-performance anode materials for lithium-ion batteries. *J. Alloys Compd.*, 667, 29–35, 2016.
84. Yang, Y., Huang, J., Zeng, J. *et al.*, Direct electrophoretic deposition of binder-free Co_3O_4/graphene sandwich-like hybrid electrode as remarkable lithium ion battery anode. *ACS Appl. Mater. Interfaces*, 9, 32801–32811, 2017.
85. Geng, H.B., Guo, Y.Y., Ding, X.G. *et al.*, Porous cubes constructed by cobalt oxide nanocrystals with graphene sheet coatings for enhanced lithium storage properties. *Nanoscale*, 8, 7688–7694, 2016.
86. Zhu, J., Tu, W., Pan, H. *et al.*, Self-templating synthesis of hollow Co_3O_4 nanoparticles embedded in N,S-dual reduced graphene oxide for lithium ion batteries. *ACS Nano*, 14, 5, 5780–5787, 2020, https://doi.org/10.1021/acsnano.0c00712.
87. Jiang, Y., Jiang, Z.J., Yang, L., A high-performance anode for lithium ion batteries: Fe_3O_4 microspheres encapsulated in hollow graphene shells. *J. Mater. Chem. A*, 3, 11847–11856, 2015.
88. Yang, Y., Li, J., Chen, D. *et al.*, A facile electrophoretic deposition route to the Fe_3O_4/CNTs/rGO composite electrode as a binder-free anode for lithium ion battery. *ACS Appl. Mater. Interfaces*, 8, 26730–26739, 2016.
89. Walsh, E.D., Han, X., Lacey, S.D. *et al.*, Dry-processed binder-free holey graphene electrodes for supercapacitors with ultrahigh areal loadings. *ACS Appl. Mater. Interfaces*, 8, 29478–29485, 2016.
90. Sun, H., Mei, L., Liang, J. *et al.*, Three-dimensional holey-graphene/niobia composite architectures for ultra-high rate energy storage. *Science*, 356, 599–604, 2017.
91. Yan, L., Chen, G., Zhou, M. *et al.*, Oxygen-deficient niobium oxide in carbon matrix as anode for lithium-ion battery. *ECS Trans.*, 66, 277–283, 2015.

92. Zhu, C., Hui, Z., Pan, H. *et al.*, Ultrafast Li-ion migration in holey-graphene-based composites constructed by a generalized ex situ methods towards high capacity energy storage. *J. Mater. Chem. A*, 7, 4788–4796, 2019.
93. Wang, J.X., Wang, G., Wang, H., Flexible free-standing Fe_2O_3/graphene/carbon nanotubes hybrid films as anode materials for high performance lithium-ion batteries. *Electrochim. Acta*, 182, 192–201, 2015.
94. Sheng, L., Liang, S., Wei, T. *et al.*, Space-confinement of MnO nanosheets in densely stacked graphene: Ultra-high volumetric capacity and rate performance for lithium-ion batteries. *Energy Storage Mater.*, 12, 94–102, 2018.
95. Xiao, Z., Ning, G., Ma, X. *et al.*, Hydrothermal assembly of MnO-graphene core–shell nanowires with superior anode performance. *Carbon*, 142, 461–467, 2019.
96. Zhang, Q., Gao, Q., Qian, W. *et al.*, Graphene-based carbon coated tin oxide as a lithium ion battery anode material with high performance. *J. Mater. Chem. A*, 5, 19136–19142, 2017.
97. Arnaiza, M., Botasa, C., Carriazoa, D. *et al.*, Reduced graphene oxide decorated with SnO_2 nanoparticles as negative electrode for lithium ion capacitors. *Electrochim. Acta*, 284, 542–550, 2018.
98. Wu, K., Shi, B., Qi, L. *et al.*, SnO_2 quantum dots@3D sulfur-doped reduced graphene oxides as active and durable anode for lithium ion batteries. *Electrochim. Acta*, 291, 24–30, 2018.
99. Xu, H., Shi, L., Wang, Z. *et al.*, Fluorine-doped tin oxide nanocrystal/reduced graphene oxide composites as lithium ion battery anode material with high capacity and cycling stability. *ACS Appl. Mater. Interfaces*, 7, 27486–27493, 2015.
100. Guo, J.X., Zhu, H.F., Sun, Y.F. *et al.*, Flexible foams of graphene entrapped SnO_2–Co_3O_4 nanocubes with remarkably large and fast lithium storage. *J. Mater. Chem. A*, 4, 16101–16107, 2016.
101. Jiang, X., Yang, X., Zhu, Y. *et al.*, 3D nitrogen-doped graphene foams embedded with ultrafine TiO_2 nanoparticles for high-performance lithium-ion batteries. *J. Mater. Chem. A*, 2, 11124–11133, 2014.
102. Wang, J.F., Zhang, J.J., He, D.N., Flower-like TiO_2-B particles wrapped by graphene with different contents as an anode material for lithium-ion batteries. *Nano-Struct. Nano-Objects*, 15, 216–223, 2018.
103. Zhang, F., Tang, Y.B., Liu, H. *et al.*, Uniform incorporation of flocculent molybdenum disulfide nanostructure into three-dimensional porous graphene as an anode for high-performance lithium ion batteries and hybrid supercapacitors. *ACS Appl. Mater. Interfaces*, 8, 4691–4699, 2016.
104. He, J.R., Li, P.J., Lv, W.Q. *et al.*, Three-dimensional hierarchically structured aerogels constructed with layered MoS_2/graphene nanoparticles as self-supported anodes for high-performance lithium ion batteries. *Electrochim. Acta*, 215, 12–18, 2016.
105. Chao, Y., Jalili, R., Ge, Y. *et al.*, Self-assembly of flexible free-standing 3D porous MoS_2-reduced graphene oxide structure for high-performance lithium-ion batteries. *Adv. Funct. Mater.*, 27, 1700234, 2017.

106. Mingjia, Z., Chengcheng, X., Jiaangtian, L. *et al.*, Nanostructured carbon-metal oxide composite electrodes for supercapacitors: A review. *Nanoscale*, 5, 72–88, 2013.
107. Zhi, M., Xiang, C., Li, J., Li, M., Wu, N., Nanostructured carbon–metal oxide composite electrodes for supercapacitors: A review. *Nanoscale*, 5, 72, 2013.
108. Fujisawa, K., Hasegawa, T., Shimamoto, D. *et al.*, Boron atoms as loop accelerator and surface stabilizer in platelet-type carbon nanofibers. *ChemPhysChem*, 11, 2345–2348, 2010.
109. Wu, G., Tan, P., Wu, X. *et al.*, High-performance wearable micro-supercapacitors based on microfluidic-directed Nitrogen-doped graphene fiber electrodes. *Adv. Funct. Mater.*, 27, 1702493, 2017.
110. Krishnan, D., Raidongia, K., Shao, J. *et al.*, Graphene oxide assisted hydrothermal carbonization of carbon hydrates. *ACS Nano*, 8, 449–457, 2013.
111. Lei, Z., Lu, L., Zhao, X.S., The electrocapacitive properties of graphene oxide reduced by urea. *Energy Environ. Sci.*, 5, 6391–6399, 2012.
112. Gao, K., Shao, Z., Li, J. *et al.*, Cellulose nanofiber-graphene all solid-state flexible supercapacitors. *J. Mater. Chem. A*, 1, 63–67, 2013.
113. Zeng, R., Deng, H., Xiao, Y. *et al.*, Cross-linked graphene/carbon nanotube networks with polydopamine "glue" for flexible supercapacitors. *Compos. Commun.*, 10, 73–80, 2018.
114. Li, B., Zhang, H., Zhang, C., Agricultural waste-derived activated carbon/graphene composites for high performance lithium-ion capacitors. *RSC Adv.*, 9, 29190–29194, 2019.
115. Lu, X., Li, L., Song, B. *et al.*, Mechanistic investigation of the graphene functionalization using p-phenylenediamine and its application for supercapacitors. *Nano Energy*, 17, 160–170, 2015.
116. Sk, M.M. and Yue, C.Y., Layer-by-layer (LBL) assembly of graphene with p-phenylenediamine (PPD) spacer for high performance supercapacitor applications. *RSC Adv.*, 4, 19908–19915, 2014.
117. Kang, H., Zang, C., Xu, Y. *et al.*, Diaminopyrene modified reduced graphene oxide as a novel electrode material for excellent performance supercapacitors. *RSC Adv.*, 10, 1507–1513, 2020.
118. Gao, L., Gan, S., Li, H. *et al.*, Self-assembling graphene-anthraquinone-2-sulphonate supramolecular nanostructures with enhanced energy density for supercapacitors. *Nanotechnology*, 28, 275602, 2017.
119. Boota, M., Chen, C., Bécuwe, M. *et al.*, Pseudocapacitance and excellent cyclability of 2,5-dimethoxy-1,4-benzoquinone on graphene. *Energy Environ. Sci.*, 9, 2586–2594, 2016.
120. An, N., Zhang, F., Hu, Z. *et al.*, Non-covalently functionalizing a graphene framework by anthraquinone for high-rate electrochemical energy storage. *RSC Adv.*, 5, 23942–23951, 2015.
121. Zhou, H., Uysal, A., Anjos, D.M. *et al.*, Understanding defect-stabilized non-covalent functionalization of graphene. *Adv. Mater. Interfaces*, 2, 1500277, 2015.

122. Aradilla, D., Estrany, F., Aleman, C., Symmetric supercapacitors based on multilayers of conducting polymers. *J. Phys. Chem. C*, 115, 8430–8438, 2011.
123. Solonaru, A.M. and Grigoras, M., Water-soluble polyaniline/graphene composites as materials for energy storage applications. *eXPRESS Polym. Lett.*, 11, 127–139, 2017.
124. Zhang, L., Huang, D., Hu, N. *et al.*, Three-dimensional structures of graphene/polyaniline hybrid films constructed by steamed water for high-performance supercapacitors. *J. Power Sources*, 342, 1–8, 2017.
125. Wu, J., Zhang, Q.E., Wang, J. *et al.*, A self-assembly route to porous polyaniline/reduced graphene oxide composite materials with molecular-level uniformity for high-performance supercapacitors. *Energy Environ. Sci.*, 11, 1280–1286, 2018.
126. Wang, C., Yang, Y., Li, R. *et al.*, Polyaniline functionalized reduced graphene oxide/carbon nanotube ternary nanocomposite as a supercapacitor electrode. *Chem. Commun.*, 56, 4003–4006, 2020.
127. Khani, H., Dowell, T.J., Wipf, D.O., Modifying current collectors to produce high volumetric energy density and power density storage devices. *ACS Appl. Mater. Interfaces*, 10, 21262–21280, 2018.
128. Chen, L., Li, D.P., Chen, L.N. *et al.*, Core–shell structured carbon nanofibers yarn@polypyrrole@graphene for high performance all-solid-state fiber supercapacitors. *Carbon*, 138, 264–270, 2018.
129. Zhu, J. and Xu, Y., Enhanced electrochemical performance of polypyrrole depending on morphology and structure optimization by reduced graphene oxide as support frameworks. *Electrochim. Acta*, 265, 47–55, 2018.
130. Qi, K., Hou, R., Zaman, S. *et al.*, A core/shell structured tubular graphene nanoflake-coated polypyrrole hybrid for all-solid-state flexible supercapacitors. *J. Mater. Chem. A*, 6, 3913–3918, 2018.
131. He, Y., Bai, Y., Yang, X. *et al.*, Holey graphene/polypyrrole nanoparticle hybrid aerogels with three-dimensional hierarchical porous structure for high performance supercapacitor. *J. Power Sources*, 317, 10–18, 2016.
132. Zhao, Q., Wang, G., Yan, K. *et al.*, Binder-free porous PEDOT electrodes for flexible supercapacitors. *J. Appl. Polym. Sci.*, 132, 42549, 2015.
133. Zhao, Z., Richardson, G.F., Meng, Q. *et al.*, PEDOT-based composites as electrode materials for supercapacitors. *Nanotechnology*, 27, 042001, 2016.
134. Zhou, H., Han, G., Fu, D. *et al.*, Petal-shaped poly(3,4-ethylenedioxythiophene)/sodium dodecyl sulfate-graphene oxide intercalation composites for high-performance electrochemical energy storage. *J. Power Sources*, 272, 203–210, 2014.
135. Zhou, H., Yao, W., Li, G. *et al.*, Graphene/poly(3,4-ethylenedioxythiophene) hydrogel with excellent mechanical performance and high conductivity. *Carbon*, 59, 495–502, 2013.
136. Xu, Z., Zhang, Z., Yin, H. *et al.*, Investiagation on the role of different conductive polymers in supercapacitors based on a zinc/sulphide reduced

graphene oxide/conductive polymer ternary composite electrode. *RSC Adv.*, 10, 3122–3129, 2020.
137. Majumdar, D., Maiyalagan, T., Jiang, Z., Recent progress in ruthenium oxide-based composites for supercapacitor applications. *ChemElectroChem*, 6, 4343–4372, 2019.
138. Hwang, J.Y., El Kady, M.F., Wang, Y. *et al.*, Direct preparation and processing of graphene/RuO_2 nanocomposite electrodes for high-performance capacitive energy storage. *Nano Energy*, 18, 57–70, 2015.
139. Patil, S.H., Gaikwad, A.P., Waghmode, B.J. *et al.*, A graphene-MnO_2 composite supercapacitor material accomplished tactically using liquid–liquid and solid–liquid interface reaction techniques. *New J. Chem.*, 44, 6853–6861, 2020.
140. Sadak, O., Wang, W., Guan, J. *et al.*, MnO_2 nanoflowers deposited on graphene paper as electrode materials for supercapacitors. *ACS Appl. Nanomater.*, 2, 4386–4394, 2019.
141. Cui, L., Cheng, C., Peng, F. *et al.*, A ternary-MnO_2 deposited RGO/lignin-based porous carbon composite electrode for flexible supercapacitor applications. *New J. Chem.*, 43, 14084–14092, 2019.
142. Trung, N.B., Tam, T.V., Dang, D.K. *et al.*, Facile synthesis of three-dimensional graphene/nickel oxide nanoparticles composites for high performance supercapacitor electrodes. *Chem. Eng. J.*, 264, 603–609, 2015.
143. Cao, P., Wang, L., Xu, Y. *et al.*, Facile hydrothermal synthesis of mesoporous nickel oxide/reduced graphene oxide composites for high performance electrochemical supercapacitor. *Electrochim. Acta*, 157, 359–368, 2015.
144. Liu, F., Wang, X., Hao, J. *et al.*, High density arrayed Ni/NiO core–shell nanospheres evenly distributed on graphene for ultrahigh performance supercapacitor. *Sci. Rep.*, 7, 17709, 2017.
145. Wu, Z., Huang, X.L., Wang, Z.L. *et al.*, Electrostatic induced stretch growth of homogeneous β-$Ni(OH)_2$ on graphene with enhanced high-rate cycling for supercapacitors. *Sci. Rep.*, 4, 3669, 2014.
146. Chen, S., Duan, J., Jaroniec, M. *et al.*, Hierarchically porous graphene-based hybrid electrodes with excellent electrochemical performance. *J. Mater. Chem. A*, 1, 9409–9413, 2013.
147. Liao, Q., Li, N., Jin, S. *et al.*, All-solid-state symmetric supercapacitor based on Co_3O_4 nanoparticles on vertically aligned graphene. *ACS Nano*, 5, 5310–5317, 2015.
148. Cheng, H., Dong, Z., Hu, C. *et al.*, Textile electrodes woven by carbon nanotube-graphene hybrid fibers for flexible electrochemical capacitors. *Nanoscale*, 5, 3428–3434, 2013.
149. Zhang, J., Zhang, G., Luo, W. *et al.*, Graphitic carbon coated CuO hollow nanospheres with penetrated mesochannels for high-performance asymmetric supercapacitors. *ACS Sustain. Chem. Eng.*, 5, 105–111, 2017.
150. Yang, S., Han, Z., Sun, J. *et al.*, Controllable $ZnFe_2O_4$/reduced graphene oxide hybrid for high-performance supercapacitor electrode. *Electrochim. Acta*, 268, 20–26, 2018.

151. Nagaraju, D.H., Wang, Q., Beaujuge, P. *et al.*, Two-dimensional heterostructures of V_2O_5 and reduced graphene oxide as electrodes for high energy density asymmetric supercapacitors. *J. Mater. Chem. A*, 2, 17146–17152, 2014.
152. Jafari, H., Ganjali, M.R., Dezfuli, A.S. *et al.*, A platform for electrochemical sensing of biomolecules based on europia/reduced graphene oxide nanocomposite. *J. Mater. Sci.: Mater. Electron.*, 29, 20639–20649, 2018.
153. Kong, L., Zhang, C., Wang, J. *et al.*, Free-standing TNb_2O_5/graphene composite papers with ultrahigh gravimetric/volumetric capacitance for Li-ion intercalation pseudocapacitor. *ACS Nano*, 9, 11200–11208, 2015.
154. Aryanrad, P., Naderi, H.R., Kohan, E. *et al.*, Europium oxide nanorod-reduced graphene oxide nanocomposites towards supercapacitors. *RSC Adv.*, 10, 17543–17551, 2020.
155. Chen, H.C., Jiang, J.J., Zhang, L. *et al.*, Facilely synthesized porous $NiCo_2O_4$ flowerlike nanostructure for high-rate supercapacitors. *J. Power Sources*, 248, 28–36, 2014.
156. Makkar, P. and Ghosh, N.N., Facile synthesis of MnFe2O4 hollow sphere-reduced graphene oxide nanocomposites as electrode materials for all-solid-state flexible high-performance asymmetric supercapacitors. *ACS Appl. Energy Mater.*, 3, 2653–2664, 2020.
157. Nonaka, L.H., Almeida, T.S.D., Aquino, C.B. *et al.*, Crumpled graphene decorated with manganese ferrite nanoparticles for hydrogen peroxide sensing and electrochemical supercapacitors. *ACS Appl. Nanomater.*, 3, 4859–4869, 2020.
158. Zhu, Y., Zhang, C., Tang, S. *et al.*, Graphene-anchored mesoporous Mn–Co oxide battery-like materials for ultrahigh performance hybrid supercapacitors. *ACS Appl. Energy Mater.*, 2, 7546–7553, 2019.
159. Zhu, Y., Huang, H., Li, G. *et al.*, Graphene-anchored $NiCoO_2$ nanoarrays as supercapacitor electrode for enhanced electrochemical performance. *Electrochim. Acta*, 248, 562–569, 2017.
160. Sun, T., Li, Z., Liu, X. *et al.*, Facile construction of 3D graphene/MoS_2 composites as advanced electrode materials for supercapacitors. *J. Power Sources*, 331, 180–188, 2016.
161. Wang, B., Wu, Q., Sun, H. *et al.*, An intercalated graphene/(molybdenum disulfide) hybrid fiber for capacitive energy storage. *J. Mater. Chem. A*, 5, 925–930, 2017.
162. Zhu, C., Yang, P., Chao, D. *et al.*, All metal nitrides solid-state asymmetric supercapacitors. *Adv. Mater.*, 27, 4566–4571, 2015.
163. Das, B., Behm, M., Lindbergh, G. *et al.*, High performance metal nitrides, MN (M=Cr, Co) nanoparticles for non-aqueous hybrid supercapacitors. *Adv. Powder Technol.*, 26, 783–788, 2015.
164. Chandel, M., Makkar, P., Ghosh, N.N., Ag–Ni nanoparticle anchored reduced graphene oxide nanocomposite as advanced electrode material for supercapacitor application. *ACS Appl. Electron. Mater.*, 1, 1215–1224, 2019.

165. Makkar, P., Chandel, M., Patra, M.K. *et al.*, A "one-pot" route for the synthesis of snowflake-like dendritic CoNi alloy-reduced graphene oxide-based multifunctional nanocomposites: An efficient magnetically separable versatile catalyst and electrode material for high-performance supercapacitors. *ACS Omega*, 4, 20672–20689, 2109.
166. Karimi, A., Kazeminezhad, I., Naderi, L. *et al.*, Construction of a ternary nanocomposite, polypyrrole/Fe-Co sulphide-reduced graphene oxide/nickel foam, as a novel binder-free electrode for high-performance asymmetric supercapacitors. *J. Phys. Chem. C*, 124, 4393–4407, 2020.
167. Krishnaveni, M., Suresh, C.M., Wu, J.J. *et al.*, Synthesis of 3D marigold flower-like rGO/BN/Ni(OH)2 ternary nanocomposites for supercapacitor applications. *Sustain. Energy Fuels*, 4, 3090–3101, 2020.

Index

Printed in the USA/Agawam, MA
September 16, 2026

910661.012